Advances in Deep-Learning-Based Sensing, Imaging, and Video Processing

Advances in Deep-Learning-Based Sensing, Imaging, and Video Processing

Editors

Yun Zhang
Sam Kwong
Xu Long
Tiesong Zhao

Basel • Beijing • Wuhan • Barcelona • Belgrade • Novi Sad • Cluj • Manchester

Editors

Yun Zhang
School of Electronics and
Communication Engineering
Sun Yat-Sen University
Shenzhen
China

Sam Kwong
Department of Computer
Science
City University of Hong Kong
Hong Kong
China

Xu Long
State Key Laboratory of Space
Weather
National Space Science Center
Chinese Academy of Sciences
Beijing
China

Tiesong Zhao
College of Physics and
Information Engineering
Fuzhou University
Fuzhou
China

Editorial Office
MDPI AG
Grosspeteranlage 5
4052 Basel, Switzerland

This is a reprint of articles from the Topical Collection published online in the open access journal *Sensors* (ISSN 1424-8220) (available at: https://www.mdpi.com/journal/sensors/topical_collections/DLBSIVP).

For citation purposes, cite each article independently as indicated on the article page online and as indicated below:

Lastname, A.A.; Lastname, B.B. Article Title. *Journal Name* **Year**, *Volume Number*, Page Range.

ISBN 978-3-7258-1781-8 (Hbk)
ISBN 978-3-7258-1782-5 (PDF)
doi.org/10.3390/books978-3-7258-1782-5

© 2024 by the authors. Articles in this book are Open Access and distributed under the Creative Commons Attribution (CC BY) license. The book as a whole is distributed by MDPI under the terms and conditions of the Creative Commons Attribution-NonCommercial-NoDerivs (CC BY-NC-ND) license.

Contents

About the Editors

Yun Zhang

Yun Zhang is currently Professor in the School of Electronics and Communication Engineering, Sun Yat Sen University, Shenzhen, China. Before that, he received his B.S. degrees in Automation, and M.S. degree in Circuits and System from Ningbo University, Ningbo, China, and the Ph.D. degree in Computer Science from Institute of Computing Technology, Chinese Academy of Sciences, Beijing, China, in 2010. From 2009 to 2014, he was a Postdoc Researcher with the Department of Computer Science, City University of Hong Kong, Kowloon, Hong Kong. From 2010 to 2022, he served as Assistant Professor, Associate Professor and Professor in Shenzhen Institutes of Advanced Technology (SIAT), Chinese Academy of Sciences, Shenzhen, China. He joined in the Sun Yat Sen University in 2022 as Professor. He is a senior member of IEEE and a member of Youth Innovation Promotion Association, Chinese Academy of Sciences. Due to the contributions on multimedia signal processing and communication, he was awarded with Second Prize of Natural and Science Award and Second Prize of Scientific & Technological Advancement from Ministry of Education (MOE), China, and First Prize of Science and Technology Award and Third Prize of Natural and Science Award from Department of Science and Technology of Zhejiang Province, China. His research interests are in the fields of video coding optimization, augmented/virtual reality, visual quality assessment and perceptual modeling, and machine learning based video processing. He has been the author of one book and over 150 publications in scientific journals and proceedings. He serves as an associate editor/editorial board member of IET Electronic Letters, IEEE Access, Sensors, and APSIPA Transactions on Signal and Information Processing.

Sam Kwong

Sam Kwong received his B.Sc. degree from the State University of New York at Buffalo, M.A.Sc. in electrical engineering from the University of Waterloo in Canada, and Ph.D. from Fernuniversität Hagen, Germany. He is currently Chair Professor at the Department of Computer Science, City University of Hong Kong, where he previously served as Department Head and Professor from 2012 to 2018. Prof. Kwong is currently the associate editor of leading IEEE transaction journals, including IEEE Transactions on Evolutionary Computation, IEEE Transactions on Industrial Informatics, and IEEE Transactions on Cybernetics. He has coauthored three research books, eight book chapters, and over 300 technical papers. His works have been cited over 23,000 times according to Google Scholar with an h-index of 67. In 2014, he was elevated to IEEE Fellow for his contributions to optimization techniques in cybernetics and video coding. He led the IEEE SMC Hong Kong Chapter to win the Best Chapter Award in 2011 and was awarded Outstanding Contribution Awards for his contributions to SMC 2015. He was the President-Elect of the IEEE SMC Society in 2021. Currently, he serves as the President of the IEEE SMC Society. His research interests are video and image coding and evolutionary algorithms.

Xu Long

Long Xu received his M.S. degree in applied mathematics from Xidian University, Xi'an, China, in 2002, and the Ph.D. degree from the Institute of Computing Technology, Chinese Academy of Sciences, Beijing, China. He was a Postdoc with the Department of Computer Science, City University of Hong Kong, the Department of Electronic Engineering, Chinese University of Hong Kong, from July Aug. 2009 to Dec. 2012. From Jan. 2013 to March 2014, he was a Postdoc with the School of Computer Engineering, Nanyang Technological University, Singapore. Currently, he is with the State Key Laboratory of Space Weather, National Space Science Center, Chinese Academy of Sciences. His research interests include image/video processing, wavelet, machine learning, and computer vision. He was selected into the 100-Talents Plan, Chinese Academy of Sciences, 2014.

Tiesong Zhao

Tiesong Zhao is currently a Minjiang Distinguished Professor in the College of Physics and Information Engineering, Fuzhou University, China. He also serves as the director of Fujian Key Lab for Intelligent Processing and Wireless Communication of Media Information. He received the B.S. degree in electrical engineering from the University of Science and Technology of China, Hefei, China, in 2006, and the Ph.D. degree in computer science from City University of Hong Kong, Hong Kong, in 2011. He has served as a Research Associate with the Department of Computer Science, City University of Hong Kong, from 2011 to 2012, a Postdoctoral Fellow with the Department of Electrical and Computer Engineering, University of Waterloo, from 2012 to 2013, and a Research Scientist with the Ubiquitous Multimedia Laboratory, The State University of New York at Buffalo, from 2014 to 2015. His research interests include multimedia communication and computer vision. Due to his contributions in multimedia coding and transmission, he received the Fujian Science and Technology Award for Young Scholars in 2017. He has been serving as an Associate Editor for IET Electronics Letters since 2019 and an Editor for CSIG Communciations since 2022.

Preface

Deep-learning-based sensing, imaging, and video processing have become essential tools in various applications, such as autonomous driving, robotics, medical diagnosis, and surveillance. The scope of this Topical Collection widely covers the fundamentals of deep learning and its applications, such as imaging, visual enhancement, image quality assessment, and encryption for low-level image processing category, object detection, semantic segmentation, image classification, and visual understanding for high-level pattern recognition category. This Topical Collection aims to provide a reference for researchers, students, and engineers who are interested in the latest developments and applications of deep-learning-based sensing, imaging, and video processing. This Topical Collection is also for professionals who work in industries that involve computer vision, machine learning, and image processing. The authors in this Topical Collection have extensive experience in the field of deep learning and provide the state-of-the-art research works, with particular emphasis on the latest advances in deep learning algorithms, architectures, and applications in the fields of sensing, imaging, and video processing.

We would like to express our gratitude to the authors and reviewers for their contributions to the Special Issue and the publisher, who has provided us with the opportunity to share our knowledge and insights with a wider audience. We also acknowledge the sponsorship of the National Natural Science Foundation of China under Grant 62172400, 62171134, and 11790305. We hope that this Topical Collection will be a valuable resource for researchers, students, and professionals who work in the field of deep-learning-based sensing, imaging, and video processing.

Yun Zhang, Sam Kwong, Xu Long, and Tiesong Zhao
Editors

Editorial

Advances in Deep-Learning-Based Sensing, Imaging, and Video Processing

Yun Zhang [1,*], Sam Kwong [2], Long Xu [3] and Tiesong Zhao [4]

1   Shenzhen Institutes of Advanced Technology, Chinese Academy of Sciences, Shenzhen 518055, China
2   Department of Computer Science, City University of Hong Kong, 83 Tatchee Ave., Kowloon, Hong Kong, China
3   State Key Laboratory of Space Weather, National Space Science Center, Chinese Academy of Sciences, Beijing 100190, China
4   College of Physics and Information Engineering, Fuzhou University, Fuzhou 350108, China
*   Correspondence: yun.zhang@siat.ac.cn

Citation: Zhang, Y.; Kwong, S.; Xu, L.; Zhao, T. Advances in Deep-Learning-Based Sensing, Imaging, and Video Processing. *Sensors* **2022**, *22*, 6192. https://doi.org/10.3390/s22166192

Received: 5 August 2022
Accepted: 16 August 2022
Published: 18 August 2022

Publisher's Note: MDPI stays neutral with regard to jurisdictional claims in published maps and institutional affiliations.

Copyright: © 2022 by the authors. Licensee MDPI, Basel, Switzerland. This article is an open access article distributed under the terms and conditions of the Creative Commons Attribution (CC BY) license (https://creativecommons.org/licenses/by/4.0/).

Deep learning techniques have shown their capabilities to discover knowledge from massive unstructured data, providing data-driven solutions for representation and decision making. They have demonstrated significant technical advancement potential for many research fields and applications, such as sensors and imaging, audio–visual signal processing, and pattern recognition. Today, with the rapid advancements of advanced deep learning models, such as conventional neural network (CNN), deep neural network (DNN), recurrent neural network (RNN), generative adversarial network (GAN), and transformer network, learning techniques, such as transfer learning, reinforcement learning, federal learning, multi-task learning, and meta-learning, and the increasing demands around effective visual signal processing, new opportunities are emerging in deep-learning-based sensing, imaging, and video processing.

After a careful peer-review process, this editorial presents the manuscripts accepted for publication in the Special Issue "Advances in Deep-Learning-Based Sensing, Imaging, and Video Processing" of *Sensors*, which includes fourteen articles. These articles are original research papers describing current challenges, innovative methodologies, technical solutions, and real-world applications related to advances in deep-learning-based sensing, imaging, and video processing. They can generally be divided into two categories.

The first category is the deep-learning-based image and video processing by exploiting low-level visual features, including five articles [1–5]. Inspired by biological structure of avian retinas, Zhao et al. [1] developed a chromatic LED array with a geometric arrangement of multi-hyper uniformity to suppress frequency aliasing and color misregistration. The proposed concept provides insights for designing and manufacturing future bionic imaging sensors. To enhance image quality of imaging systems, Wang et al. [2] developed a novel color-dense illumination adjustment network (CIANet) for removing haze and smoke from fire scenario images. Schiopu et al. [3] explored a novel filtering method based on deep attention networks for the quality enhancement of light field (LF) images captured by plenoptic cameras and compressed by the high efficiency video coding (HEVC) standard. Tian et al. [4] proposed a dynamic neighborhood network (DNet) to dynamically select the neighborhood for local region feature learning in point clouds which improved the performances of point cloud classification and segmentation tasks. To access visual quality of videos, Lin et al. [5] proposed a no-reference objective video quality metric called saliency-aware artifact measurement (SAAM), which consists of an attentive CNN-LSTM network for video saliency detection, Densenet for distortion type classification, and support vector regression for quality prediction. These works reveal that deep learning models can exploit low-level visual features and promote imaging, image/video enhancement, segmentation, and quality assessment.

The second category relates to deep-learning-based visual object detection and analysis by exploiting higher-level visual and cognitive features. It contains nine articles [6–14]. Li et al. [6] developed a wheat ear recognition method based on RetinaNet and transfer learning by detecting the number of wheat ears as an essential indicator. This method can be used for automatic wheat ear recognition and yield estimation. To detect surface defects with variable scales, Xu et al. [7] proposed a multi-scale feature learning network (MSF-Net) based on a dual module feature (DMF) extractor, which classified the surface defects with multifarious sizes. In addition, Yu et al. [8] developed a deep-learning-based automatic pipe damage detection system for pipe maintenance. This detection system was composed of a laser-scanned pipe's ultrasonic wave propagation imaging (UWPI) and CNN-based object detection algorithms. To inspect condition of hull surfaces by using underwater images acquired from a remotely controlled underwater vehicle (ROUV), Kim et al. [9] proposed a binary classification method by resembling multiple CNN classifiers which were transfer-learned from larger natural image datasets. Kim et al. [10] proposed a neg-region attention network (NRA-Net) to suppress negative areas and emphasize the texture information of objects in positive areas, which was then applied in an auto-encoder architecture based salient objects detection. He et al. [11] developed a small object detection algorithm named YOLO-MXANet for traffic scenes, which reduced the computational complexity of the object detection and meanwhile improved the detection accuracy. Alia et al. [12] proposed a hybrid deep learning and visualization framework of pushing behavior detection for pedestrian videos, which comprised a recurrent all-pairs field transforms (RAFT)-based motion extraction and an EfficientNet-B0-based pushing patches annotation. Deepfakes may cause information abuse by creating fake visual information. To verify video integrity, Lee et al. [13] presented a deep learning-based deepfake detection method by measuring changing rate of a number of visual features among adjacent frames. Then, a learned DNN was used to identify whether a video was manipulated. Xu et al. [14] proposed a timestamp-independent synchronization method for haptic–visual signals by exploiting a sequential cross-modality correlation between haptic and visual signals, where the deep learning network YOLO V3 was employed in visual object detection. In these works, deep learning technologies were applied to promote the performances of defect detection, object detection, anomaly detection, and recognition tasks in practical sensing, imaging, and video processing applications.

We would like to thank all the authors and reviewers for their contributions to the Special Issue. We hope this Special Issue can provide some research insights, useful solutions, and exciting applications to scholars in academics and researchers in the industry interested in Deep-Learning-Based Sensing, Imaging, and Video Processing.

Author Contributions: All the authors contributed equally to this editorial. All authors have read and agreed to the published version of the manuscript.

Funding: This work was supported in part by the National Natural Science Foundation of China under Grant 62172400, 62171134, and 11790305.

Institutional Review Board Statement: Not applicable.

Informed Consent Statement: Not applicable.

Data Availability Statement: Not applicable.

Conflicts of Interest: The authors declare no conflict of interest.

References

1. Zhao, X.-Y.; Li, L.-J.; Cao, L.; Sun, M.-J. Bionic Birdlike Imaging Using a Multi-Hyperuniform LED Array. *Sensors* **2021**, *21*, 4084. [CrossRef]
2. Wang, C.; Hu, J.; Luo, X.; Kwan, M.-P.; Chen, W.; Wang, H. Color-Dense Illumination Adjustment Network for Removing Haze and Smoke from Fire Scenario Images. *Sensors* **2022**, *22*, 911. [CrossRef]
3. Schiopu, I.; Munteanu, A. Attention Networks for the Quality Enhancement of Light Field Images. *Sensors* **2021**, *21*, 3246. [CrossRef]

4. Tian, F.; Jiang, Z.; Jiang, G.D. Net: Dynamic Neighborhood Feature Learning in Point Cloud. *Sensors* **2021**, *21*, 2327. [CrossRef] [PubMed]
5. Lin, L.; Yang, J.; Wang, Z.; Zhou, L.; Chen, W.; Xu, Y. Compressed Video Quality Index Based on Saliency-Aware Artifact Detection. *Sensors* **2021**, *21*, 6429. [CrossRef] [PubMed]
6. Li, J.; Li, C.; Fei, S.; Ma, C.; Chen, W.; Ding, F.; Wang, Y.; Li, Y.; Shi, J.; Xiao, Z. Wheat Ear Recognition Based on RetinaNet and Transfer Learning. *Sensors* **2021**, *21*, 4845. [CrossRef]
7. Xu, P.; Guo, Z.; Liang, L.; Xu, X. MSF-Net: Multi-Scale Feature Learning Network for Classification of Surface Defects of Multifarious Sizes. *Sensors* **2021**, *21*, 5125. [CrossRef]
8. Yu, B.; Tola, K.D.; Lee, C.; Park, S. Improving the Ability of a Laser Ultrasonic Wave-Based Detection of Damage on the Curved Surface of a Pipe Using a Deep Learning Technique. *Sensors* **2021**, *21*, 7105. [CrossRef]
9. Kim, B.C.; Kim, H.C.; Han, S.; Park, D.K. Inspection of Underwater Hull Surface Condition Using the Soft Voting Ensemble of the Transfer-Learned Models. *Sensors* **2022**, *22*, 4392. [CrossRef]
10. Kim, H.; Kwon, S.; Lee, S. NRA-Net—Neg-Region Attention Network for Salient Object Detection with Gaze Tracking. *Sensors* **2021**, *21*, 1753. [CrossRef] [PubMed]
11. He, X.; Cheng, R.; Zheng, Z.; Wang, Z. Small Object Detection in Traffic Scenes Based on YOLO-MXANet. *Sensors* **2021**, *21*, 7422. [CrossRef]
12. Alia, A.; Maree, M.; Chraibi, M. A Hybrid Deep Learning and Visualization Framework for Pushing Behavior Detection in Pedestrian Dynamics. *Sensors* **2022**, *22*, 4040. [CrossRef] [PubMed]
13. Lee, G.; Kim, M. Deepfake Detection Using the Rate of Change between Frames Based on Computer Vision. *Sensors* **2021**, *21*, 7367. [CrossRef] [PubMed]
14. Xu, Y.; Huang, L.; Zhao, T.; Fang, Y.; Lin, L.A. Timestamp-Independent Haptic–Visual Synchronization Method for Haptic-Based Interaction System. *Sensors* **2022**, *22*, 5502. [CrossRef] [PubMed]

sensors

Communication

Bionic Birdlike Imaging Using a Multi-Hyperuniform LED Array

Xin-Yu Zhao, Li-Jing Li, Lei Cao and Ming-Jie Sun *

School of Instrumentation and Optoelectronic Engineering, Beihang University, Beijing 100191, China; zhaoxinyu@buaa.edu.cn (X.-Y.Z.); lilijing@buaa.edu.cn (L.-J.L.); caolei_bh17@163.com (L.C.)
* Correspondence: mingjie.sun@buaa.edu.cn; Tel.: +86-10-8231-6547 (ext. 812)

Abstract: Digital cameras obtain color information of the scene using a chromatic filter, usually a Bayer filter, overlaid on a pixelated detector. However, the periodic arrangement of both the filter array and the detector array introduces frequency aliasing in sampling and color misregistration during demosaicking process which causes degradation of image quality. Inspired by the biological structure of the avian retinas, we developed a chromatic LED array which has a geometric arrangement of multi-hyperuniformity, which exhibits an irregularity on small-length scales but a quasi-uniformity on large scales, to suppress frequency aliasing and color misregistration in full color image retrieval. Experiments were performed with a single-pixel imaging system using the multi-hyperuniform chromatic LED array to provide structured illumination, and 208 fps frame rate was achieved at 32×32 pixel resolution. By comparing the experimental results with the images captured with a conventional digital camera, it has been demonstrated that the proposed imaging system forms images with less chromatic moiré patterns and color misregistration artifacts. The concept proposed verified here could provide insights for the design and the manufacturing of future bionic imaging sensors.

Keywords: multi-hyperuniform; single-pixel imaging; frequency aliasing; color misregistration

Citation: Zhao, X.-Y.; Li, L.-J.; Cao, L.; Sun, M.-J. Bionic Birdlike Imaging Using a Multi-Hyperuniform LED Array. *Sensors* **2021**, *21*, 4084. https://doi.org/10.3390/s21124084

Academic Editors: KWONG Tak Wu Sam, Xu Long, Tiesong Zhao and Yun Zhang

Received: 31 March 2021
Accepted: 11 June 2021
Published: 14 June 2021

Publisher's Note: MDPI stays neutral with regard to jurisdictional claims in published maps and institutional affiliations.

Copyright: © 2021 by the authors. Licensee MDPI, Basel, Switzerland. This article is an open access article distributed under the terms and conditions of the Creative Commons Attribution (CC BY) license (https://creativecommons.org/licenses/by/4.0/).

1. Introduction

Both spatial and spectral information provides us crucial knowledge of the world. Chromatic digital cameras obtain spatial and spectral information simultaneously by placing an absorbing color-filter array (a.k.a. the Bayer filter) on top of a detector array [1], because photoelectronic sensors are only sensitive to light intensity, regardless of its wavelength. Due to technical limitations and commercial considerations, the elements of the pixelated detectors, as well as those of the filter arrays, are usually arranged in a Cartesian geometry. Such periodic arrangement of both the detector array and the Bayer filter array introduces artificial effects decreasing the fidelity of the captured images. Frequency aliasing, which converts frequencies above the Nyquist limit into moiré fringes during the optical sampling, is one such effect. Color misregistration, which causes inauthentic color shifts during the demosaicking process, is another. Many post-processing algorithms have been proposed [2–9] to suppress these artificial effects in the captured images, which increase the computational burden on imaging system. However, these effects could be avoided or suppressed if raw image data could be sampled in a different manner with low cost.

The evolution of species is governed by neither technical limitation nor commercial consideration, but environmental requirements, and environment requires diurnal animals to evolve eyes which can obtain images with high fidelity and dynamic performance. Some researches about human retinas showed that arrangement of the photoreceptors is random and uniform, which is able to yield images with better reconstruction quality by suppressing frequency aliasing [10]. The similar structures exist in the avian vision system as well. Recent biological investigations have found that birds, being the vertebrate with the most sophisticated vision, have retinas consisting of five types of cones, each of which independently exhibits a disorder on small-length scales but a quasi-uniformity on large scales [11,12]. The fantastic structure of correlated disorder is known as hyperuniformity [13], and the fact that such arrangement can obtain high fidelity images is explained by the theory that a slight irregularity in the optical sampling arrangement can avoid frequency aliasing [14]. It has been shown that the evolution of the avian vision system may be most sophisticated among all animals. Another biological research [15] shows that the birds can achieve continuous imaging with up to 145 fps frame rate. A recent hyperuniform sampling experiment [16] further verified the feasibility of the theory. However, the experiment was performed with a single-pixel imaging [17–21] system using a digital-micromirror-device which is commonly used in many applications [22,23]. While the sampling patterns were designed to be hyperuniform, the micromirrors forming the patterns were arranged in Cartesian coordinates and displayed at an up-to-22 KHz modulational rate. Consequently, the frequency aliasing in the sampled image was not suppressed completely and the system cannot perform high dynamic tasks.

In this paper, we addressed that the periodic sampling caused frequency aliasing and color misregistration by utilizing a customed chromatic LED array in which red-green-blue luminous points formed a multi-hyperuniform arrangement [11], that is, luminous points of one color form a hyperuniform point pattern, and all points together, regardless of their colors, exhibit hyperuniformity as well. Such arrangement of the chromatic LED array was designed to mimic the multi-hyperuniform structure of the chicken retina system [12]. Optical sampling using multi-hyperuniformity was performed experimentally via a single-pixel imaging system. The high-speed hyperuniform LED array developed in this work, which have a maximum illumination rate of 2.5 MHz, can effectively improve the dynamic performance of the imaging system. Both numerical and experimental results indicated that the images retrieved by multi-hyperuniform sampling contained less chromatic moiré patterns at high frequencies and less color misregistration artifacts at the edge of color transition, where the proposed imaging system achieved 208 fps frame rate in experiment. The work is different from the methods of optimizing interpolation algorithm, which can solve these artificial effects in hardware through a simple imaging system. The proof-of-principle system demonstrated here might push us one step closer to the biomimetic digital camera which the imaging community aimed to invent for so long.

2. Theory

Hyperuniform structure exists in not only avian retinas but also physical systems such as crystal [24], or even the large-scale structure of the universe [13]. The property of a hyperuniform system can be quantified as the density fluctuation in its corresponding point patterns. For a 2D point pattern with hyperuniform distribution, the variance of the number of points $\sigma^2(R)$ within a circular domain S is approximately proportional to the R [13] i.e.,

$$\sigma^2(R) = \left\langle N_S^2 \right\rangle - \left\langle N_S \right\rangle^2 \propto R \tag{1}$$

where N_s is the number of points contained in S, and angular brackets represent an ensemble average. R is the radius of a circular observation window. Equation (1) indicates that the variance of the hyperuniform point patterns grows more slowly than the area of the domain, while for any statistically homogeneous and isotropic point pattern, the variance cannot grow more slowly than the area of circle S or other strictly convex domains [25,26].

A special hyperuniformity, known as multi-hyperuniformity [11], contains more than one type of points. For example, five types of cone photoreceptors exist in chicken retina: violet, blue, green, red species to sense color and double species to detect luminance. The point patterns of these five types of cones are arranged individually and never occurred in the near vicinity of other cones of the same type, which ensure each cone pattern achieves a much more uniform arrangement [11]. All types of photoreceptors grow simultaneously with the constraints of cell size, and such competing interactions ensure that all cone patterns are arranged in a hyperuniformity. Multi-hyperuniform structure contains multiple point species where both the total population and the individual point types are simultaneously hyperuniform. It is worth noting that the overall point arrangement in multi-hyperuniform obey Equation (1) as well, which means that the point patterns are independent with each other and display hyperuniform in total whether the individual species is removed or not. Such multi-hyperuniform structure is believed to be the main reason that birds have the most sophisticated vision of any vertebrate. It would be interesting to exploit the concept in optical sampling by using such geometry, and here it was performed in the manner of structured illumination in a single-pixel imaging scheme.

A chromatic LED array was developed, in which its red-green-blue luminous points were arranged in multi-hyperuniformity. Specifically, a hyperuniform point pattern was generated by a 'cell-growing' random procedure based on a regular hexagonal geometry [16]; the green luminous points, being the center of the LED chips, were then arranged following the hyperuniform point pattern, as shown in Figure 1a. As the red-green-blue luminous points in an LED chip have a fixed geometry (Figure 1b), a periodicity would exist in the LED array if each chip were arranged in the same manner. Therefore, an extra irregularity was introduced by rotating the LED chip of the ith column and the jth row with a randomly generated angle θ_{ij} (Figure 1c). The LED array, as shown in Figure 1d, was fabricated by placing 32×32 LED chips on corresponding positions and integrating them on the printed circuit board, where the red-green-blue luminous points on each chip have center wavelengths of 632 nm, 518 nm, and 468 nm, respectively. It is worth mentioning that, for the purpose of variance $\sigma^2(R)$ estimation and further numerical simulation, the multi-hyperuniform arrangement were generated on an underlying Cartesian grid of 544 $\times$ 544 pixels with one pixel corresponding to 0.1 mm $\times$ 0.1 mm. The actual LED array, however, was not limited by Cartesian coordinates. Recent research has found that the quantities of five types of cones are different in avian retina, where the double cones were the most abundant cone type (40.7%) followed by green (21.1%), red (17.1%), blue (12.6%) and violet (8.5%) single cones. Due to the fixed arrangement of R-G-B channels of each LED chip, the three channels have the same spatial density on the multi-hyperuniform LED array with 32×32 chips.

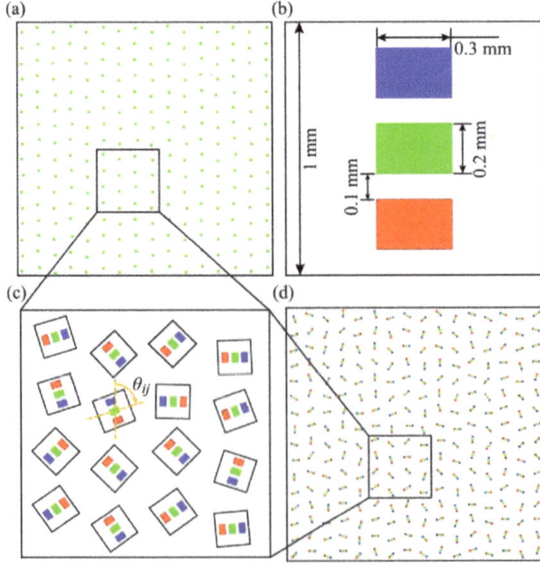

Figure 1. Schematics of multi-hyperuniform sampling arrangement: (**a**) Hyperuniform point pattern generated by the randomization procedure for green luminous point; (**b**) The geometry of a single LED chip; (**c**) Randomized rotations were introduced to each LED chip; (**d**) Multi-hyperuniform point pattern of the LED array.

To evaluate the hyperuniformity of the chromatic LED array, as in the calculating process in [11], the variances $\sigma^2(R)$ were computed directly for each monochromatic point patterns separately, and for overall point pattern, as shown in Figure 2. Specifically, for each R value, 2500 circular domains S were randomly placed in the pattern without overlapping the system boundary. The maximum radius was chosen to be $R_{max} = L/2$, limited by the pattern size L. The variances of each pattern were fitted (dashed lines in Figure 2) using the fitting function:

$$\sigma^2\left(\frac{R}{D}\right) = P\left(\frac{R}{D}\right)\left(1 + Q\cos\left(\frac{\pi}{2}\frac{R}{D} + \frac{\pi}{3}\right)\right) \tag{2}$$

where a cosine term represented that the patterns were originated from the regular hexagonal arrangement [16], the window size R is normalized by the averaged points' distance $D = 17$ for monochromatic point patterns and $D = 5.67$ for overall point pattern. According to the previous research [11], the structural properties of individual and overall point patterns could be obtained in the same manner, so the Equation (2) is suitable for hyperuniform evaluation of chromatic LED point array.

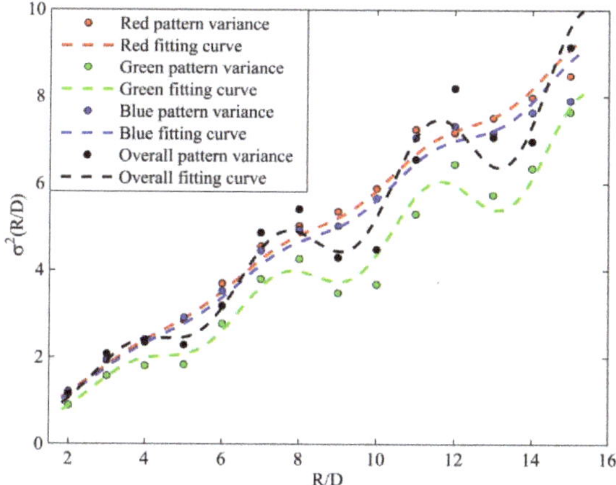

Figure 2. Number variance $\sigma^2(R/D)$ of all monochromatic point patterns and the overall point pattern, and their corresponding fitting curves.

The parameters of the fitting curves in Figure 2 are listed in Table 1. The fitting curves and their fitting parameters, listed in Table 1, indicated that the variances of all four patterns grew proportionally to R rather than R^2, which met the criterion of hyperuniformity described by Equation (1). Each monochromatic luminous point pattern exhibits hyperuniformity individually, and combined as an overall point pattern, the LED array remains hyperuniform, therefore, the arrangement of the LED array was multi-hyperuniform.

Table 1. Fitting coefficients for different types of point patterns.

	Green	Red	Blue	Overall
P	0.4675	0.5932	0.5725	0.5663
Q	0.1274	0.0271	0.0335	0.1505

It is worth mentioning that Q is the coefficient for the cosine term in Equation (2), representing the hexagonal geometry. In Table 1, coefficient Q of the green point pattern is larger than the red and blue ones because the latter ones had an extra random rotation introduced during the generation of the pattern, meaning the red and blue point patterns had a larger deviation from the regular hexagonal geometry than the green point pattern. The parameter P is a coefficient for fitting curves.

3. Numerical Simulations

Numerical simulations were performed using a multi-hyperuniform sampling point pattern, which was generated based on the multi-hyperuniform LED array. The point pattern, having a 544 × 544 underlying pixel grid, contained 3072 sampling points, 1024 for red, green, and blue each. Figure 3a showed a partial area of the arrangement with 4 red, 4 green, and 4 blue points. Each LED luminous point, not strictly a point, had the size of 6 pixel grids, representing its 0.3 mm × 0.2 mm physical size.

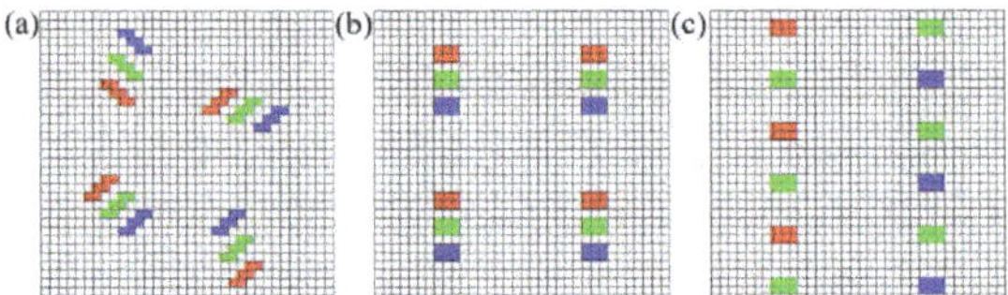

Figure 3. Partial areas of sampling patterns containing 12 sampling points: (**a**) Multi-hyperuniform sampling pattern; (**b**) Regular sampling pattern; (**c**) Bayer sampling pattern.

For comparison, two other types of sampling patterns were also used in the numerical simulation. The regular LED pattern, a part of which was shown in Figure 3b, represented that the LED chips were arranged in a regular geometry. The regular LED pattern also contained 3072 sampling points, 1024 for red, green, and blue each. The Bayer pattern, a part of which was shown in Figure 3c, is a common arrangement used in conventional chromatic digital cameras. The Bayer pattern, having a 1:2:1 ratio of red, green, and blue points, contained 3072 sampling points which consisted of 768 red, 1536 green, and 768 blue points.

To ensure that the comparison is fair, the three sampling patterns had the same number of sampling points and the same size of underlying pixel grid, therefore, the same sampling frequency and the same field-of-view for the optical sampling.

In numerical simulation, a group of 35 chromatic images, whose pixel resolution is 544 × 544, were used as the objects. Each chromatic image I was under-sampled by the three sampling patterns, and a demosaicking algorithm [4,5] was applied to the under-sampled monochromatic data of red, green, and blue to reconstruct the chromatic image I'. There are many sophisticated demosaicking methods for various applications [6–8] and the gradient-based interpolation algorithm [9] was applied in this work. The gradients of different directions are calculated according to the sampling structure, which can ensure for selecting the proper direction to estimate the missing pixel values of the images. This algorithm is efficient to reduce the pseudo color of color-transition area in reconstructed images for three sampling patterns without much time cost. It is worth mentioning that the demosaicking algorithm used here is not best for three sampling structures but valid and simple, which can ensure the fair comparison of image reconstruction with three sampling patterns. There are some complicated algorithms to improve the quality of images as well, which would cause more time costs of imaging. It is verified that the multi-hyperuniform sampling structure could be used to improve the image quality in a hardware way without more computational burden.

The qualities of reconstruction images were evaluated using the root mean square error (RMSE) between each original image I and its reconstruction I' as:

$$RMSE_{channel} = \sqrt{\frac{\sum_{i,j=1}^{m,n} \left(I'_{channel}(i,j) - I_{channel}(i,j) \right)^2}{m \times n}} \tag{3}$$

where $m = n = 544$ were the pixel resolution of the images, and channel, being red, green, or blue, represented the monochromatic data of different channels. The final RMSE is the averaging value of the RMSEs for three monochromatic channels, i.e.,

$$RMSE = \sum_{channel}^{R,G,B} RMSE_{channel}/3 \tag{4}$$

Both original images and the reconstructions were normalized to the same scale.

The RMSEs of all resulting images, sorted in descending order of multi-hyperuniform pattern RMSEs, are shown in Figure 4. In most cases, 33 out of 35, the proposed multi-hyperuniform pattern yielded the best resulting images among the three sampling patterns. The average RMSE for all 35 images reconstructed from multi-hyperuniform sampling, being 0.1214, is the smallest of the RMSEs yielded from the three sampling patterns. Regular and Bayer patterns each yielded the lowest RMSE in two cases, where the original images contain large blocky areas with no color transitions and high-frequency details. The slight degradation of image quality by using multi-hyperuniform sampling in such cases, was predicted by the fact that the irregularity in optical sampling degrades image quality [27]. The RMSEs of the reconstructed images by multi-hyperuniform structure demonstrate the improvement of the image quality on a pixel-wise level.

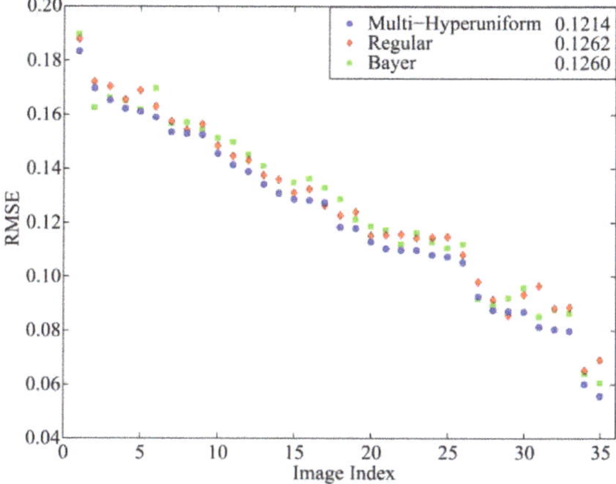

Figure 4. RMSEs of the reconstructed chromatic images using multi-hyperuniform (**blue dot**), regular (**red diamond**), and Bayer (**green square**) sampling patterns. The averaged values of 35 image RMSEs are listed on the label.

For considering larger-scale features of reconstructed images, the structural similarity (SSIM) between each original image I and its reconstruction I' is calculated as:

$$SSIM(I'_{channel}, I_{channel}) = \frac{\left(2\mu_{I_{channel}}\mu_{I'_{channel}} + c_1\right)\left(2\sigma_{I_{channel},I'_{channel}} + c_2\right)}{\left(\mu_{I_{channel}}^2 + \mu_{I'_{channel}} + c_1\right)\left(\sigma_{I_{channel}}^2 + \sigma_{I'_{channel}}^2 + c_2\right)} \tag{5}$$

where, μ is the average value of images, σ^2 is the variance of images, c_1 and c_2 are constants. $\sigma_{I,I'}$ is covariance of I and I', $c_1 = 0.01$ and $c_2 = 0.03$ are constants for ensuring the validation of this equation. The meaning of the channel is the same as the previous calculation process. The final SSIM is the averaging value of the SSIMs for three channels, i.e.,

$$SSIM = \sum_{channel}^{R,G,B} SSIM(I'_{channel}, I_{channel})/3 \tag{6}$$

The SSIMs of all images reconstructed are shown in Figure 5, where image indexes are consistent with the results in Figure 4. The averaging SSIM for all 35 images from multi-hyperuniform sampling is the largest compared with other results yielded from regular and Bayer sampling. The two images with the large block can be reconstructed better by regular sampling pattern and their indexes are the same with the two cases where we find lower RMSEs by using regular and Bayer patterns in Figure 4. The SSIMs of the reconstructed images demonstrate that image quality can be improved on larger-scale features by multi-hyperuniform sampling as well.

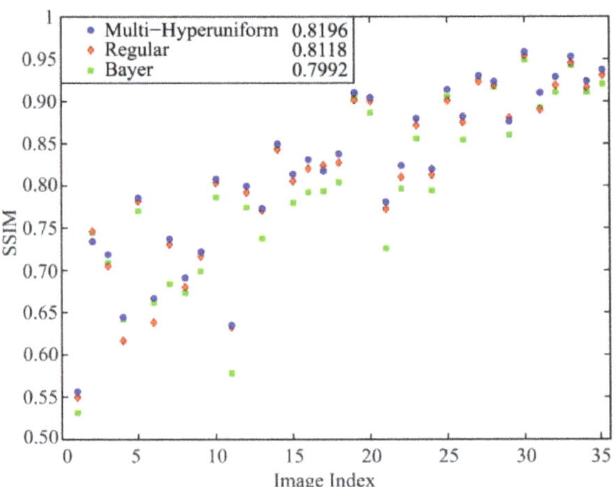

Figure 5. The SSIM values of the reconstructed chromatic images using multi-hyperuniform (**blue dot**), regular (**red diamond**), and Bayer (**green square**) sampling patterns. The averaged values of 35 image SSIMs are listed on the label.

Since aliasing errors are usually caused by insufficient sampling, it is necessary to compare the proposed sampling with other random sampling strategies. The blue-noise sampling strategy is another random strategy, which is derived by the human vision system. This method is used here and the reconstruction results are listed in Figures 6 and 7. The points in blue-noise sampling pattern are arranged randomly but the distances of any two adjacent points are uniform, which causes the reconstructed images to be close to the results using multi-hyperuniform sampling structure.

Due to the fact that different images have different power distribution of image spatial frequency [28,29], more various scenes are used for comparisons, where the reconstructed images of artificial scenes and natural scenes are obtained by different patterns in Figures 6 and 7. In Figure 6, it is obvious that less color misregistration artifacts and moiré fringes were observed in reconstructed images by multi-hyperuniform sampling pattern compared with the other structures. It is worth noting that the color misregistration could be considered as one of the small-scale features and the chromatic moiré fringes could be treated as one of the large-scale features. The RMSEs and SSIMs are sequentially listed under the resulting images, which indicate that multi-hyperuniform sampling structure can suppress the color misregistration and chromatic moiré fringes in different scale features. The reconstructed results in Figure 7 demonstrated that the multi-hyperuniform sampling structure is valid to improve the qualities of natural scenes with sparse periodic patterns as well.

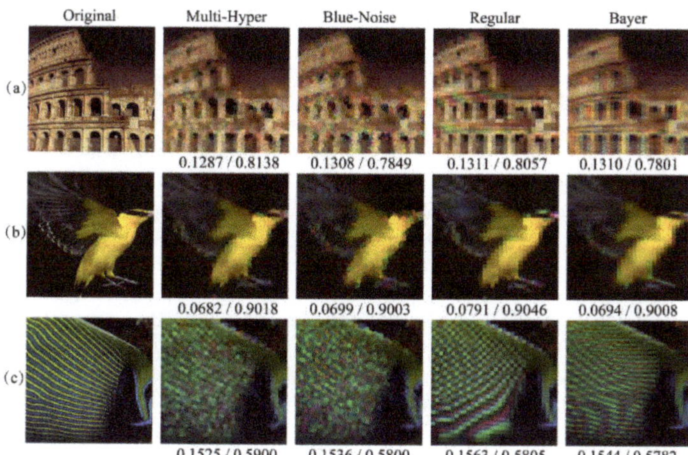

Figure 6. Comparison of reconstructed images of artificial scenes using different sampling patterns. From the left to right are the original images, reconstructions from multi-hyperuniform, regular and Bayer sampling patterns. The corresponding RMSEs and SSIMs are listed below the images, respectively. (**a**) The scene of Colosseum with color transition area and corresponding reconstructed images in numerical simulations; (**b**) The scene of bird with periodic patterns on the wings and corresponding reconstructed images; (**c**) The scene of fish with periodic stripes and corresponding reconstructed results.

Figure 7. Comparison of reconstructed images of natural scenes using different sampling patterns. From the left to right are the original images, reconstructions from multi-hyperuniform, blue-noise sampling patterns, regular and Bayer sampling patterns. The corresponding RMSEs and SSIMs are listed below the images, respectively. (**a**) The scene of forest and corresponding reconstructed image; (**b**) The scene of sky and corresponding reconstructed images; (**c**) Natural scene with the different contents and corresponding reconstructed results.

4. Experimental Results

It is difficult to perform the multi-hyperuniform optical sampling with an actual detector array, because such a chromatic detector array would be impossible to manufacture. However, it is possible to perform a multi-hyperuniform optical sampling experiment utilizing the computational imaging methods which have emerged during the last decade [30–33]. Single-pixel imaging, being a typical computational imaging method, forms an image by sampling the scene with varying structured illuminations, and associating the illumination patterns with the corresponding light intensities recorded with a single-pixel detector. This imaging strategy provides advantages for imaging in situations that are challenging with a detector array, such as special spectrum imaging [34,35], adaptive imaging [36–38], optical phased array imaging [39] and 3D profiling [40–43].

In this work, a single-pixel imaging experimental system was set up as shown in Figure 8. The multi-hyperuniform chromatic LED structured illumination module was used in the system to sample the scene, which consists of a field-programmable gate array (Xilinx Spartan XC6SLX9-2FTG256C), a drive circuit and an LED array developed in Section 2. During the experiment, for each monochromatic channel, a projection lens (f = 150 mm) projected N masks P_i (I = 1, . . . , N) displayed on the LED array. The mask P_i is orthonormal and derived from the Hadamard matrix, a square matrix with elements ± 1 whose rows (or columns) are orthogonal to one another [44]. The LED array displays monochromatic illumination masks at the rate of 1.25 MHz. The high illumination rate was achieved by using the line control modulational strategy proposed in our previous work [45]. A single-pixel bucket detector (Thorlabs PMT2102) and a digitizer (PicoScope 6404D) were used to record the corresponding reflecting total light intensities S_i (I = 1, . . . , N). The image $I'_{channel}$ for the monochromatic channel can be reconstructed as:

$$I'_{channel} = \sum_{i=1}^{N} S_i \cdot P_i \tag{7}$$

where the N is the quantity of the Hadamard basis masks. The problem of reconstructing the single channel image of the scene becomes a problem of solving N independent unknowns using a set of linear equations. Due to the orthonormal property of the mask, the Equation (7) can be solved perfectly if the number of Hadamard basis masks is equal to pixels of the image [16,21,44,45]. Provided the scene is sparse, compressive sensing [18,21,44] can be used to reconstruct image with $N < 32 \times 32$ measurements by sub-sampling the scene and solving the optimization problem. To yield one 32×32 monochromatic image, 2048 masks (1024 Hadamard masks and their inverses) were used to perform a fully Hadamard sampling. After images $I'_{channel}$ for red, green, and blue were reconstructed separately, the gradient-based interpolation algorithm was applied to yield the chromatic image I' of the object. A chromatic image required 6144 illumination masks, or 4.8 ms acquisition time, resulting in a 208 fps frame rate for the multi-hyperuniform chromatic LED array-based single-pixel imaging system. Due to the lab manufacturing limitation, the imaging experiment only can achieve 32×32 pixels resolution.

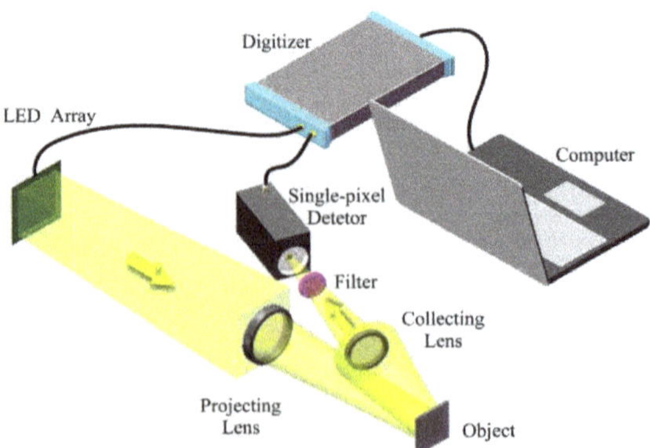

Figure 8. Experimental set-up. A multi-hyperuniform chromatic $32 \times 32 \times 3$ LED array provided structured illumination to sample the object through a projection lens (f = 150 mm). A single-pixel bucket detector (Thorlabs PMT2102) and a digitizer (PicoScope 6404D) collected the reflected light intensity and transferred it to a computer for reconstruction.

For comparison, another LED array with regular sampling arrangement, as shown in Figure 3b, was used in the experimental system to obtain images with regular sampling. An ordinary smartphone camera was also used to capture images with Bayer sampling, where the method of sub-sampling is chosen to ensure the same sampling structure compared with the simulation. All images were obtained with the same number of spatial sampling points to make sure the comparison was fair.

Figure 9 illustrated the resulting images yielded from multi-hyperuniform, regular and Bayer sampling patterns. Like numerical simulation, color misregistration and chromatic moiré fringes were observed in images yielded by regular and Bayer sampling, while multi-hyperuniform sampling suppressed both artifacts. The calculated RMSEs and SSIMs listed below the images are in a good agreement with the numerical simulation. It is showed below that the images reconstructed by multi-hyperuniform have the lowest RMSE and highest SSIM values, demonstrating the effectiveness of suppression of color misregistration and frequency aliasing.

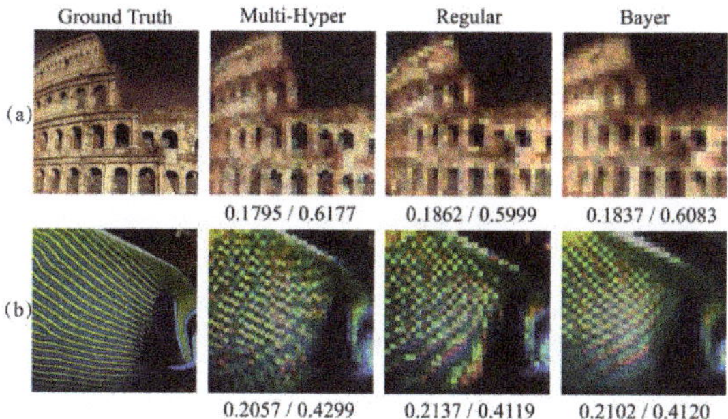

Figure 9. Experimental results of two group scenes. Chromatic images are reconstructed experimentally by multi-hyperuniform, regular and Bayer sampling patterns with their RMSEs and SSIMs listed below. (**a**) The scene of Colosseum with color transition area and corresponding reconstructed results in experiments; (**b**) The scene of fish with periodic stripes and corresponding recon-structed results.

5. Discussions and Conclusions

In summary, we developed a chromatic LED array with multi-hyperuniform structure, that is, luminous points of each monochromatic chip exhibited hyperuniformity independently, and red-green-blue luminous points combined to show hyperuniformity as well. The chromatic LED array was developed to mimic the virtues of avian retina optical sampling, specifically, suppressing color misregistration and chromatic moiré fringes caused by periodic optical sampling. The placement and orientation of LED array has an impact on whether the arrangement of LED array is multi-hyperuniform, and two random procedures are used here to ensure LED array is of multi-hyperuniformity. Comparisons were performed numerically and experimentally by reconstructing images by different sampling methods in a single-pixel imaging system. Both numerical and experimental results indicated that the multi-hyperuniform sampling method yielded images with better image quality compared to the other two methods by using the same and basic interpolation algorithm. Besides the improvement of the images' quality, the proposed imaging system achieved 208 fps frame rate experimentally, which has a potential in high dynamic applications.

This work is a proof-of-principle to demonstrate the feasibility of multi-hyperuniformity in high dynamic chromatic optical sampling. The chromatic LED array developed in this work contains only 1024 chromatic luminous LED chips due to the lab manufacturing limitation. The capability of improvement of image quality is almost the same by 32×32 multi-hyperuniform LED array with different orientations and placements, which can be enhanced by integrating more LED chips on the sampling array due to the property of the hyperuniform structure. Although the low-resolution color images are reconstructed, the method in this paper can offer a new solution to suppress the artificial effects in high resolution imaging which would not increase imaging time. Cutting-edge LED manufacturing techniques such as micro-LED or OLED could be used to develop multi-hyperuniform LED array with a much larger chip number and a higher density to take full advantages of such a sampling structure in high-resolution imaging. In this work, the multi-hyperuniform pattern has been verified to be able to improve image quality with a high frame rate. In future, LEDs could be used not only for illumination and display, but also for the development of new bionic imaging sensors.

Author Contributions: Conceptualization, M.-J.S. and L.-J.L.; methodology, M.-J.S.; software, X.-Y.Z.; validation, X.-Y.Z. and L.C.; data curation, X.-Y.Z.; writing—review and editing, X.-Y.Z. and M.-J.S. All authors have read and agreed to the published version of the manuscript.

Funding: This research was funded by National Natural Science Foundation of China, grant number 61922011, and Open Research Projects of Zhejiang Lab, grant number 2021MC0AB03.

Conflicts of Interest: The authors declare no conflict of interest.

References

1. Bayer, B.E. Color Imaging Array. U.S. Patent 3971065, 20 July 1976.
2. Takamatsu, M.; Ito, M. Color Image Forming Apparatus Providing Registration Control for Individual Color Images. U.S. Patent 5550625, 27 August 1996.
3. Costenza, D.W. Image Registration for a Raster Output Scanner Color Printer. U.S. Patent 5412409, 2 May 1995.
4. Rajeev, R.; Snyder, W.E.; Griff, B.L. Demosaicking methods for Bayer color arrays. *J. Electron. Imaging* **2002**, *11*, 306–315.
5. Chung, K.H.; Chan, Y.H. Color Demosaicing Using Variance of Color Differences. *IEEE Trans. Image Process.* **2006**, *15*, 2944–2955. [CrossRef]
6. Kimmel, R. Demosaicing: Image reconstruction from color CCD samples. *IEEE Trans Image Process.* **1999**, *8*, 1221–1228. [CrossRef] [PubMed]
7. Malvar, H.S.; He, L.W.; Cutler, R. High-quality linear interpolation for demosaicing of Bayer-patterned color images. In Proceedings of the IEEE International Conference on Acoustics, Speech, and Signal Processing, Montreal, QC, Canada, 17–21 May 2004.
8. Alleysson, D.; Süsstrunk, S. Linear demosaicing inspired by the human visual system. *IEEE Trans Image Process.* **2005**, *14*, 439–449. [PubMed]
9. Laroche, C.A.; Prescott, M.A. Apparatus and method for adaptively interpolating a full color image utilizing chrominance gradients. U.S. Patent 5373322, 13 December 1994.
10. Yellott, J. Spectral analysis of spital sampling photoreceptors: Topological disorder prevents aliasing. *Vis. Res.* **1982**, *22*, 1205–1210. [CrossRef]
11. Jiao, Y.; Lau, T.; Hatzikirou, H.; Meyer-Hermann, M.; Corbo, J.C.; Torquato, S. Avian photoreceptor patterns represent a disordered hyperuniform solution to a multiscale packing problem. *Phys. Rev. E* **2014**, *89*, 022721. [CrossRef]
12. Hart, N.S. The visual ecology of avian photoreceptors. *Prog. Retin. Eye Res.* **2001**, *20*, 675–703. [CrossRef]
13. Torquato, S.; Stillinger, F.H. Local density fluctuations, hyperuniformity, and order metrics. *Phys. Rev. E* **2003**, *68*, 041113. [CrossRef]
14. Yellott, J. Spectral consequences of photoreceptor sampling in the rhesus retina. *Science* **1983**, *221*, 382–385. [CrossRef]
15. Boström, J.E.; Dimitrova, M.; Canton, C.; Håstad, O.; Qvarnström, A.; Ödeen, A. Ultra-Rapid Vision in Birds. *PLoS ONE* **2016**, *11*, e0151099. [CrossRef] [PubMed]
16. Sun, M.J.; Zhao, X.Y.; Li, L.J. Imaging using hyperuniform sampling with a single-pixel camera. *Opt. Lett.* **2018**, *43*, 4049–4052. [CrossRef] [PubMed]
17. Shapiro, J.H. Computational ghost imaging. *Phys. Rev. A* **2008**, *78*, 061802. [CrossRef]
18. Duarte, M.F.; Davenport, M.A.; Takhar, D.; Laska, J.N.; Sun, T.; Kelly, K.F.; Baraniuk, R.G. Single-pixel imaging via compressive sampling. *IEEE Signal Process. Mag.* **2008**, *25*, 83–91. [CrossRef]
19. Bromberg, Y.; Katz, O.; Silberberg, Y. Ghost imaging with a single detector. *Phys. Rev. A* **2009**, *79*, 053840. [CrossRef]
20. Zhang, Z.B.; Ma, X.; Zhong, J.G. Single-pixel imaging by means of Fourier spectrum acquisition. *Nat. Commun.* **2015**, *6*, 6225–6230. [CrossRef] [PubMed]
21. Sun, M.J.; Zhang, J.M. Single-pixel imaging and its application in three-dimensional reconstruction: A brief review. *Sensors* **2019**, *19*, 732. [CrossRef]
22. Xu, C.; Xu, T.F.; Yan, G.; Ma, X.; Zhang, Y.H.; Wang, X.; Zhao, F.; Arce, G.R. Super-resolution compressive spectral imaging via two-tone adaptive coding. *Photonics Res.* **2020**, *3*, 395–411. [CrossRef]
23. Li, J.J.; Matlock, A.; Li, Y.Z.; Chen, Q.; Zuo, C.; Tian, L. High-speed in vitro intensity diffraction tomography. *Adv. Photonics* **2019**, *1*, 066004. [CrossRef]

24. DeckLéger, Z.; Chamanara, N.; Skorobogatiy, M.; Silveirinha, M.G.; Caloz, C. Uniform-velocity spacetime crystals. *Adv. Photonics* **2019**, *1*, 056002.
25. Beck, J. Irregularities of distribution. I. *Acta Math.* **1987**, *159*, 1–49. [CrossRef]
26. Beck, J. Randomness in lattice point problems. *Discrete Math.* **2001**, *229*, 29–55. [CrossRef]
27. French, A.S.; Snyder, A.W.; Stavenga, D.G. Image degradation by an irregular retinal mosaic. *Biol. Cybern.* **1977**, *27*, 229–233. [CrossRef]
28. Farinella, G.M.; Battiato, S.; Gallo, G.; Cipolla, R. Natural versus artificial scene classification by ordering discrete fourier power spectra. *SSPR SPR* **2008**, *5342*, 137–146.
29. Li, Y.Q.; Majumder, A.; Zhang, H.; Gopi, M. Optimized multi-spectral filter array based imaging of natural scenes. *Sensors* **2018**, *18*, 1172. [CrossRef]
30. Barbastathis, G.; Ozcan, A.; Situ, G.H. On the use of deep learning for computational imaging. *Optica* **2019**, *6*, 921–943. [CrossRef]
31. Feng, S.J.; Chen, Q.; Gu, G.H.; Tao, T.Y.; Zhang, L.; Hu, Y.; Hu, W.; Zuo, C. Fringe pattern analysis using deep learning. *Adv. Photonics* **2019**, *1*, 025001. [CrossRef]
32. Gibson, G.M.; Johnson, S.D.; Padgett, M.J. Single-pixel imaging 12 years on: A review. *Opt. Express* **2020**, *28*, 28190–28208. [CrossRef]
33. Jiao, S.M.; Sun, M.J.; Gao, Y.; Lei, T.; Xie, Z.W.; Yuan, X.C. Motion estimation and quality enhancement for a single image in dynamic single-pixel imaging. *Opt. Express* **2019**, *27*, 12841–12854. [CrossRef] [PubMed]
34. Watts, C.M.; Shrekenhamer, D.; Montoya, J.; Lipworth, G.; Hunt, J.; Sleasman, T.; Krishna, S.; Smith, D.R.; Padilla, W.J. Terahertz compressive imaging with metamaterial spatial light modulators. *Nat. Photonics* **2014**, *8*, 605–609. [CrossRef]
35. Zhang, A.X.; He, Y.H.; Wu, L.A.; Chen, L.M.; Wang, B.B. Tabletop x-ray ghost imaging with ultra-low radiation. *Optica* **2018**, *5*, 374–377. [CrossRef]
36. Phillips, D.B.; Sun, M.J.; Taylor, J.M.; Edgar, M.P.; Barnett, S.M.; Gibson, G.M.; Padgett, M.J. Adaptive foveated single-pixel imaging with dynamic supersampling. *Sci. Adv.* **2017**, *3*, e1601782. [CrossRef]
37. Sun, M.J.; Edgar, M.P.; Phillips, D.B.; Gibson, G.M.; Padgett, M.J. Improving the signal-to-noise ratio of single-pixel imaging using digital microscanning. *Opt. Express* **2016**, *24*, 10476–10485. [CrossRef]
38. Sun, M.J.; Wang, H.Y.; Huang, J.Y. Improving the performance of computational ghost imaging by using a quadrant detector and digital micro-scanning. *Sci. Rep.* **2019**, *9*, 4105. [CrossRef] [PubMed]
39. Li, L.J.; Chen, W.; Zhao, X.Y.; Sun, M.J. Fast Optical Phased Array Calibration Technique for Random Phase Modulation LiDAR. *IEEE Photonics J.* **2018**, *11*, 1–10. [CrossRef]
40. Howland, G.A.; Lum, D.J.; Ware, M.R.; Howell, J.C. Photon counting compressive depth mapping. *Opt. Express* **2013**, *21*, 23822–23837. [CrossRef] [PubMed]
41. Sun, B.; Edgar, M.P.; Bowman, R.; Vittert, L.E.; Welsh, S.; Bowman, A.; Padgett, M.J. 3D computational imaging with single-pixel detectors. *Science* **2013**, *340*, 844–847. [CrossRef]
42. Sun, M.J.; Edgar, M.P.; Gibson, G.M.; Sun, B.Q.; Radwell, N.; Lamb, R.; Padgett, M.J. Single-pixel three-dimensional imaging with a time-based depth resolution. *Nat. Commun.* **2016**, *7*, 12010. [CrossRef] [PubMed]
43. Wang, M.; Sun, M.J.; Huang, C. Single-pixel 3D reconstruction via a high-speed LED array. *J. Phys. Photonics* **2020**, *2*, 025006. [CrossRef]
44. Sun, M.J.; Meng, L.T.; Edgar, M.P.; Padgett, M.J.; Radwell, N. A Russian Dolls ordering of the Hadamard basis for compressive single-pixel imaging. *Sci. Rep.* **2017**, *7*, 3464. [CrossRef]
45. Xu, Z.H.; Chen, W.; Penuelas, J.; Padgett, M.; Sun, M.J. 1000 fps computational ghost imaging using LED-based structured illumination. *Opt. Express* **2018**, *26*, 2427–2434. [CrossRef]

Article

Color-Dense Illumination Adjustment Network for Removing Haze and Smoke from Fire Scenario Images

Chuansheng Wang [1,2], Jinxing Hu [2,*], Xiaowei Luo [3], Mei-Po Kwan [4], Weihua Chen [1] and Hao Wang [1]

1   State Key Laboratory of Nuclear Power Safety Monitoring Technology and Equipment, China Nuclear Power Engineering Co., Ltd., Shenzhen 518172, China; wangcs95@163.com (C.W.); chenweihua@cgnpc.com.cn (W.C.); wang_h@cgnpc.com.cn (H.W.)
2   Shenzhen Institute of Advanced Technology, Chinese Academy of Sciences, Shenzhen 518055, China
3   Department of Architecture and Civil Engineering, City University of Hong Kong, Kowloon, Hong Kong 999077, China; xiaowluo@cityu.edu.hk
4   Institute of Space and Earth Information Science, The Chinese University of Hong Kong, Shatin, Hong Kong 999077, China; mpk654@gmail.com
*   Correspondence: jinxing.hu@siat.ac.cn

Abstract: The atmospheric particles and aerosols from burning usually cause visual artifacts in single images captured from fire scenarios. Most existing haze removal methods exploit the atmospheric scattering model (ASM) for visual enhancement, which inevitably leads to inaccurate estimation of the atmosphere light and transmission matrix of the smoky and hazy inputs. To solve these problems, we present a novel color-dense illumination adjustment network (CIANet) for joint recovery of transmission matrix, illumination intensity, and the dominant color of aerosols from a single image. Meanwhile, to improve the visual effects of the recovered images, the proposed CIANet jointly optimizes the transmission map, atmospheric optical value, the color of aerosol, and a preliminary recovered scene. Furthermore, we designed a reformulated ASM, called the aerosol scattering model (ESM), to smooth out the enhancement results while keeping the visual effects and the semantic information of different objects. Experimental results on both the proposed RFSIE and NTIRE'20 demonstrate our superior performance favorably against state-of-the-art dehazing methods regarding PSNR, SSIM and subjective visual quality. Furthermore, when concatenating CIANet with Faster R-CNN, we witness an improvement of the objection performance with a large margin.

Keywords: haze removal; visual enhancement; aerosol scattering model

Citation: Wang, C.; Hu, J.; Luo, X.; Kwan, M.-P.; Chen, W.; Wang, H. Color-Dense Illumination Adjustment Network for Removing Haze and Smoke from Fire Scenario Images. *Sensors* **2022**, *22*, 911. https://doi.org/10.3390/s22030911

Academic Editor: Christophoros Nikou

Received: 23 December 2021
Accepted: 21 January 2022
Published: 25 January 2022

Publisher's Note: MDPI stays neutral with regard to jurisdictional claims in published maps and institutional affiliations.

Copyright: © 2022 by the authors. Licensee MDPI, Basel, Switzerland. This article is an open access article distributed under the terms and conditions of the Creative Commons Attribution (CC BY) license (https://creativecommons.org/licenses/by/4.0/).

1. Introduction

The phenomenon of images degradation from fire scenarios is usually caused by the large number of suspended particles generated during combustion. When executing the robot rescue in such scenes, the quality of the images collected from the fire scenarios will be seriously affected [1]. For example, most of the current research in the computer vision community is based on the assumption that the input datasets are clear images or videos. However, burning is usually accompanied by uneven light and smoke, reducing the scene's visibility and failing many high-level vision algorithms [2,3]. Therefore, removing haze and smoke from fire scenario scenes is very important to improve the detection performance for rescue robots and monitoring equipment.

Generally, the brightness distribution in the fire scenarios is uneven, and different kinds of materials will produce different colors of smoke when burning [4]. Therefore, the degradation of the images in fire scenarios is more variable than common hazy scenes. Optically, poor visibility in fire scenarios is due to the substantial presence of solid and aerosol particles of significant size and distribution in the participating medium [5,6]. Light from the illuminant reflected tends to be absorbed and scattered by those particles, causing degraded visibility of a scene. The brightness is not evenly distributed in the background.

The uncertain light source is the two main factors that cause the hazy fire scenarios to be far different from the common hazy scenarios.

Recently, many dehazing algorithms for single images have been proposed [7–10], aimed at improving the quality of the images captured from hazy or foggy weather. Image dehazing algorithms can be used as a preprocessing step for many high-level computer vision tasks, such as video coding [11–13], image compression [14,15] and object detection [16], etc. The dehazing algorithms can be roughly divided into two categories: the traditional prior-based methods and the modern learning-based methods [17]. The conventional techniques get plausible dehazing results by designing some hand-crafted priors, which lead to color distortion due to lack of consideration and comprehensive understanding of the imaging mechanism of hazy scenarios [18–20]. Therefore, traditional prior-based dehazing methods are difficult to achieve desirable dehazing effects.

Learning-based dehazing methods adopt convolution neural networks (CNNs) to simulate the mapping relationships between the hazy images and the clear images [21]. However, since the parameters and weights of the model are fixed after training, the datasets will seriously affect the performance of learning-based dehazing methods. Therefore, the learning-based dehazing methods lack sufficient flexibility to deal with the changeable fire environment. In addition, many synthesized training datasets for the dehazing algorithms are based on the atmospheric scattering model (ASM) [22,23], in which only white haze can be synthesized, and the other potential colors of smoke cannot be synthesized. These limitations will affect the application of current leaning-based dehazing models in fire monitoring systems [24]. Therefore, both prior-based methods and the learning-based methods are limited in the fire scenario dehazing.

This paper modifies ASM and proposes a new imaging model named aerosol scattering model (ESM) for enhancing the quality of images or videos captured from fire scenarios. In addition, this paper also presents a novel deep learning model for fire scenarios image dehazing. Specifically, instead of directly learning an image-to-image mapping function, we design a three-branch network to handle the transmission, suspend the particle color, and obtain a preliminary dehazing result separately.

This strategy is based on two observations. Firstly, the illumination intensity of most fire scenarios is uneven, the area near the fire source is significantly brighter than other areas. Second, different types of combustion usually produce various forms of smoke. For example, solid combustion usually produces white smoke, while combustible liquids generally generate black smoke. Therefore, the degraded images of fire scenarios present generally different styles. The reliability of most current dehazing methods is higher only under even illumination conditions [25]. To address the above-mentioned problems, this paper proposed a novel CIANet that can effectively improve the haze images captured from fire scenarios.

The proposed method can be seen as a compensation process that can enhance the quality of the images affected by combustion. The network learns the features of the images from the training data. The three branches of the structure generate an intermediate result, a transmission map and a color value, respectively. To a certain extent, our method integrates all the conditions for compensating or repairing the loss caused by scattering. The improved ESM is post-processing to transform the intermediate results to higher-quality images. After ESM processes the intermediate results, the color of the image is brighter, and the contrast is more elevated. ESM can also be employed in conventional image dehazing tasks, especially natural conditions.

The contributions of this work are summarized as follows:

- This paper proposes a novel learning-based dehazing model to improve the quality of images captured from fire scenarios, built with CNN and a physical imaging model. Combining the modern learning-based strategy with a traditional ASM makes the proposed model handle various hazy images in the fire scenarios without incurring additional parameters and computational burden.

- To improve the effect of image dehazing, we improve the existing ASM and propose a new ASM called the aerosol scattering model (ESM). The ESM uses brightness, color, and the transmission information of the images and can generate a more realistic images without causing over enhancement.
- We conducted extensive experiments on multiple datasets, and experiments show that the proposed CIANet achieves better performance quantitatively and qualitatively. The detailed analysis and experiments show the limitation of the classical dehazing algorithms in fire scenarios. Moreover, the insights from the experimental results confirm what is useful in more complex scenarios and suggest new research directions in image enhancement and image dehazing.

The remaining part of this paper is organized as follows. In Section 1, we review the ASM and some state-of-the-art image dehazing algorithms. In Section 2, we present the proposed CIANet in detail. Experiments are presented in Section 3, before conclusion is drawn in Section 4.

2. Related Works

Generally, the existing image haze removal methods can be roughly divided into two categories: the prior-based and learning-based methods. The prior-based strategy use hand-crafted prior inspired by statistical assumptions, while the learning-based methods automatically obtain the nonlinear mapping between images pairs from the training data [26]. We will discuss the differences between the two paradigms in this chapter.

2.1. Atmospheric Scattering Model

The prior-based dehazing algorithms can be regarded as an ill-posed problem. In this line of methods, the physical imaging model and various prior statistics are used to estimate the expected results from degraded inputs. In the dehazing community, the most authoritative model is the ASM proposed by McCartney [27], which can be formulated as:

$$I(x) = J(x)t(x) + a(1 - t(x)) \tag{1}$$

where, $J(x)$ is the clear images to be recovered, $I(x)$ is the captured hazy images, α is the global atmospheric light, and $t(x)$ is the transmission map. Equation (4) suggests that the clear images $J(x)$ can be recovered after $t(x)$ and α are estimated.

The transmission map $t(x)$ describes that the light reaches the camera instead of being scattered and absorbed, $t(x)$ is defined as:

$$t(x) = e^{-\beta d(x)} \tag{2}$$

where $d(x)$ is the distance between the scene point and the imaging devices, and β is the scattering coefficient of the atmosphere. Equation (2) shows that $t(x)$ approaches 0 as $d(x)$ approaches infinity.

2.2. Prior-Based Methods

The unknown items α and $t(x)$ are the main factors that cause the image dehazing problem to be ill-posed. Therefore, various traditional prior-based methods [28] have been proposed to obtain an approximate dehazed result. In [29], the authors adopted haze-lines prior for estimating the transmission. In [30], the transmission map is calculated by proposing a color attenuation prior, which exploited the attenuation difference among the RGB channels. The paradigm of these kind of methods are illustrated in Figure 1a.

However, these prior-based methods often achieve sub-optimal restoration performance. For example, He et al. [31] utilize the dark channel prior to estimate the transmission map $t(x)$ and employ a simple method to estimate the atmospheric light value a for restoring the clear images according to Equation (1). However, the sky region in the hazy images suffers from a negative visual effect when a dark channel prior is used. Zhu et al. [30] propose the color attenuation prior (CAP) for estimating the depth information of the hazy images to estimate the transmission maps $t(x)$. Berman et al. [29] propose the non-local prior for estimating the transmission maps $t(x)$ of hazy images in RGB space by varying the color distances. Even though these prior-based methods can restore clear images from the hazy images, the process can easily lead to incorrect estimation of the atmospheric lights a and transmission map $t(x)$, which cause the color distortion in the restored images.

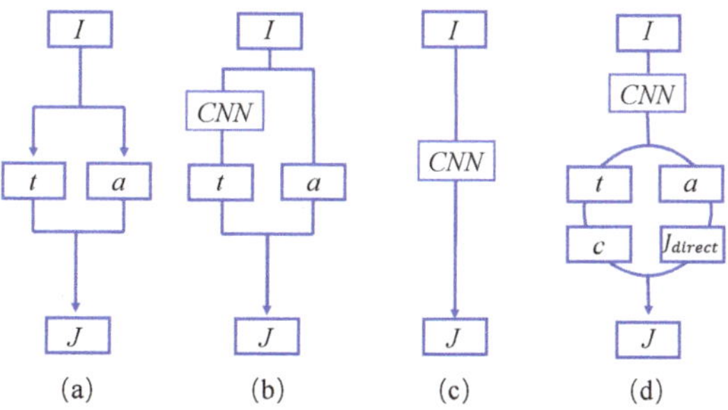

Figure 1. Different diagrams of dehazing schemes: (**a**) traditional two-step dehazing strategy; (**b**) estimate the transmission matrix through CNN; (**c**) end-to-end diagrams; (**d**) the proposed diagrams that estimates the transmission t, atmospheric light a, aerosol color c, and preliminary enhancement results J_{direct}.

2.3. Learning-Based Methods

Some learning-based methods are proposed to estimate the transmission maps $t(x)$. For example, Cai et al. [17] and Ren et al. [32] first employ the CNNs for estimating the transmission map $t(x)$ and use simple methods to calculate the atmospheric light a from the single hazy images. The paradigm of such models is shown in Figure 1b.

Although such CNN-based methods can remove haze by separately estimating the transmission map and the atmospheric light, it will introduce errors that affecting the image restoration. To avoid this problem, Zhang et al. [33]. adopted two CNN branches to estimate the atmospheric light and transmission map, respectively, and restoring clear images from the hazy images according to Equation (1). Compared with a separate estimation of atmospheric lights and transmission maps, the strategy proposed in [33] can significantly improve the dehazed results.

As shown in Figure 1c, several CNN-based algorithms regard image dehazing as enhancement tasks and directly recover clear images from hazy inputs. The GCANet [34] was proposed by Chen et al. for image dehazing with a new smoothed dilated convolution. The experimental results show that this method can achieve the better performances than previous representative dehazing methods. The training dataset mainly determines the performances of the algorithms. For example, when the image rain removal dataset replaces the training data, the algorithm can still achieve a good image rain removal performance as long as there is sufficient training. However, the pixel value of aerosols default to $(255, 255, 255)$ in the traditional ASM and assume that the intensity of light in the scene is uniform.

The existing image dehazing methods and the traditional ASM give us the following inspirations:

- The image dehazing task can be viewed as a case of the decomposing images into clear layer and haze layer [35]. In the traditional ASM [27], the haze layer color is white by default, so many classical prior-based methods, such as [31], fail on white objects [2]. Therefore, it is necessary to improve the atmospheric model for adapting the different haze scenarios.
- The haze-free images obtained by Equation (4) has obvious defects when the value of atmospheric light received by the prior-based methods and the transmission maps obtained by the learning-based methods are used, due to they fail to cooperate with each other when two independent systems calculate two separate projects.

- The learning-based algorithms can directly output a clear images without using ASM. Such a strategy can achieve good dehazing performance on some datasets [36–38]. As CNN can have multiple outputs, one of the branches can directly output haze-free images.

As shown in Figure 1d, we propose another image dehazing paradigm for fire scenarios. In this paradigm, the deep learning model outputs four variables at the same time, and these four variables will act on the final dehazing result.

3. Proposed Method

To solve the problems encountered by the traditional image dehazing algorithms in the fire scenario, a novel network build with CNN and a new physical model are proposed in this paper. Unlike the general visual enhancement model, the proposed method is committed to adapting to the image degradation caused by different colors of the haze. Firstly, the proposed method adopts CNN similar to extract the low-dimensional features of the inputs and then outputs the scene transmittance map $t(x)$, the atmospheric light value a, the color value of haze $c(x)$, and the preliminary image recovered results J_{Direct}. After obtaining these crucial factors, CIANet adopts ESM to complete the fire scenario image dehazing task. Different from the traditional image dehazing methods, the proposed method can deal with the different scenes with different colors and degrees of haze and adapt to the overall atmospheric light value of the environment. This section will introduce the proposed CIANet in detail and explain how to use the ESM to restore the haze images in the fire scenario.

3.1. Color-Dense Illumination Adjustment Network

As described in [39], the hazy-to-transmission paradigms can achieve better performance than hazy-to-clear paradigms in uneven haze and changing illumination regions. Therefore, the proposed network utilizes these parameters to directly estimate clear images and ASM-related maps, i.e., illumination intensity, haze color, and transmission map. The proposed network is mainly composed of the following building blocks: (1) one shared encoder, which is constructed based on feature pyramid networks [40]; (2) three bottleneck block branches used to bifurcate features from the encoder to specific flows for decoders; (3) three separate decoders with different outputs. The complete network structure is shown in Figure 2.

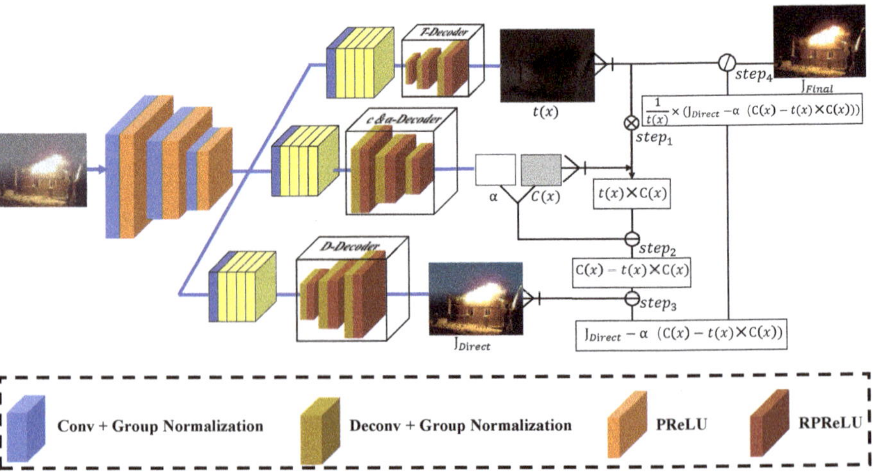

Figure 2. The structure of the proposed CIANet. All decoders are identical except for the *c&α decoder*, which outputs two floating point numbers The *T-decoder* and *D-decoder* outputs images.

Encoder: The structure of the shared encoder is shown in Table 1. DRHNet was initially being proposed for image dehazing and deraining, which proved that such an encoder could extract detailed features effectively by achieving a good performance in dehazing and deraining tasks. Therefore, the proposed CIANet utilizes the same model structure as the encoder part of DRHNET as the encoder.

Bottleneck: The bottleneck structure is used to connect the encoder and decoders. Ren et al. [32] prove that representing features at multiple scales is of great importance for image dehazing tasks. The Res2Net [41] represents multi-scale features that can expand the range of receptive fields for each layer. Gao et al. prove that the Res2Net can be plugged into the state-of-the-art CNN models, e.g., ResNet [42], ResNetXt [43] and DLA [44]. Due to the performance of the algorithms can be improved by increasing the receptive field of the convolution layer, Res2Net can significantly improve the receptive field of the CNN layer without incurring a significant increase in parameters. Different bottleneck structures connect to different decoders according to the function of the decoders. We use a shared bottleneck to estimate the global atmospheric light α and color c, which reducing the number of parameters in the network.

Decoders: The network includes three different decoders: the t-decoder, $c\&\alpha$-decoder and J-decoder, for predicting the color value c, global atmospheric light α, transmission map t and intermediate result J, respectively. The decoders share similar structures as the encoder but have different intermediate structures. In the $c\&\alpha$-decoder, we add a specially designed dilation inception module for the J-decoder, which we will describe in detail in the next section. Table 2 shows the details of the decoders.

Table 1. Encoder structure.

	Enc. 1	Enc. 2	Enc. 3	Enc. 4	Enc. 5	Enc. 6	Enc. 7
Input	Input	Input	Input	Input	Input	Input	Input
Structure	3×3 Conv. Stride $= 2$, Pool $= 0$	1×1 Conv 3×3 Conv	3×3 Conv Stride $= 2$, Pool $= 0$	1×1 Conv 3×3 Conv	3×3 Conv Stride $= 2$, Pool $= 1$	1×1 Conv 3×3 Conv	3×3 Conv Stride $= 2$, Pool $= 1$
Output	$310 \times 230 \times 32$	$310 \times 230 \times 32$	$155 \times 115 \times 64$	$155 \times 115 \times 64$	$78 \times 58 \times 128$	$78 \times 58 \times 128$	$39 \times 29 \times 256$

Table 2. Decoder structure.

	Dec. 1	Dec. 2	Dec. 3	Dec. 4	Dec. 5	Dec. 6
T-Decoder	[Res. 1] $\begin{bmatrix} 1 \times 1 \text{ Conv, } 256 \\ 3 \times 3 \text{ Conv, } 256 \\ 1 \times 1 \text{ Conv, } 512 \end{bmatrix}$ $39 \times 29 \times 512$	[Res. 2] $\begin{bmatrix} 1 \times 1 \text{ Conv, } 512 \\ \text{upsample } 2 \end{bmatrix}$ $78 \times 58 \times 512$	[Res. 3] $\begin{bmatrix} 1 \times 1 \text{ Conv, } 256 \\ 3 \times 3 \text{ Conv, } 256 \\ 1 \times 1 \text{ Conv, } 512 \end{bmatrix}$ $78 \times 58 \times 512$	[Res. 4] $\begin{bmatrix} 1 \times 1 \text{ Conv, } 512 \\ \text{upsample } 2 \end{bmatrix}$ $156 \times 116 \times 512$	[Res. 5] $\begin{bmatrix} 1 \times 1 \text{ Conv, } 256 \\ 3 \times 3 \text{ Conv, } 256 \\ 1 \times 1 \text{ Conv, } 512 \end{bmatrix}$ $155 \times 115 \times 512$	[Res. 6] $\begin{bmatrix} 1 \times 1 \text{ Conv, } 512 \\ \text{upsample } 2 \end{bmatrix}$ $310 \times 230 \times 512$
	Dec. 1	Dec. 2	Dec. 3	Dec. 4	Dec. 5	Dec. 6
c&a-Decoder	[Res, 4, Trans. 2] $\begin{bmatrix} 1 \times 1 \text{ Conv, } 256 \\ 3 \times 3 \text{ Conv, } 256 \\ 1 \times 1 \text{ Conv, } 512 \end{bmatrix}$ $39 \times 29 \times 512$	[Res. 4, Trans. 2] $\begin{bmatrix} 1 \times 1 \text{ Conv, } 256 \\ \text{downsample } 2 \end{bmatrix}$ $20 \times 14 \times 512$	[Res. 4, Trans. 2] $\begin{bmatrix} 1 \times 1 \text{ Conv, } 256 \\ 3 \times 3 \text{ Conv, } 256 \\ 1 \times 1 \text{ Conv, } 512 \end{bmatrix}$ $20 \times 14 \times 512$	[Res. 4, Trans. 2] $\begin{bmatrix} 1 \times 1 \text{ Conv, } 256 \\ \text{downsample } 2 \end{bmatrix}$ $10 \times 7 \times 512$	[Res. 4, Trans. 2] $\begin{bmatrix} 1 \times 1 \text{ Conv, } 256 \\ 3 \times 3 \text{ Conv, } 256 \\ 1 \times 1 \text{ Conv, } 512 \end{bmatrix}$ $10 \times 7 \times 512$	[Res. 4, Trans. 2] $\begin{bmatrix} 1 \times 1 \text{ Conv, } 256 \\ 3 \times 3 \text{ Conv, } 256 \\ 1 \times 1 \text{ Conv, } 512 \end{bmatrix}$ $10 \times 7 \times 512$
	Dec. 1	Dec. 2	Dec. 3	Dec. 4	Dec. 5	Dec. 6
J-Decoder	[Res, 4, Trans. 2] $\begin{bmatrix} 1 \times 1 \text{ Conv, } 256 \\ 3 \times 3 \text{ Conv, } 256 \\ 1 \times 1 \text{ Conv, } 512 \end{bmatrix}$ $39 \times 29 \times 512$	[Res. 4, Trans. 2] $\begin{bmatrix} 1 \times 1 \text{ Conv, } 256 \\ \text{downsample } 2 \end{bmatrix}$ $78 \times 58 \times 512$	[Res. 4, Trans. 2] $\begin{bmatrix} 1 \times 1 \text{ Conv, } 256 \\ 3 \times 3 \text{ Conv, } 256 \\ 1 \times 1 \text{ Conv, } 512 \end{bmatrix}$ $78 \times 58 \times 512$	[Res. 4, Trans. 2] $\begin{bmatrix} 1 \times 1 \text{ Conv, } 256 \\ \text{downsample } 2 \end{bmatrix}$ $156 \times 116 \times 512$	[Res. 4, Trans. 2] $\begin{bmatrix} 1 \times 1 \text{ Conv, } 256 \\ 3 \times 3 \text{ Conv, } 256 \\ 1 \times 1 \text{ Conv, } 512 \end{bmatrix}$ $155 \times 115 \times 512$	[Res. 4, Trans. 2] $\begin{bmatrix} 1 \times 1 \text{ Conv, } 512 \\ \text{upsample } 2 \end{bmatrix}$ $310 \times 230 \times 512$

3.2. Aerosol Scattering Model

The traditional ASM has been widely used in the image dehazing community, which can reasonably describe the imaging process in a hazy environment. However, many ASM-based dehazing algorithms suffer from the same limitation that may be invalid when the scene is inherently similar to the airlight [31]. The ineffectiveness is due to the assumption that the color of haze is white in the traditional ASM, which does not apply to all hazy environments due to the aerosols in the air may be mixed with some colored suspended particles. Therefore, the haze in different scenes has different color characteristics.

Due to the aerosol suspended in the air has a greater impact on the imaging results, we modified the traditional ASM and propose a new ESM. The default pixel value of haze in the air of traditional ASM is (255,255,255), which obviously does not conform to the appearance characteristics of degraded images in fire scenarios. In order to solve this problem, the ESM proposed in this paper combines color information $c(x)$ to the haze and smoke generated in the fire scenarios. The schematic diagram of ESM is shown in Figure 3, and the formula expression is as follows:

$$I(x) = J(x)t(x) + \alpha(1 - t(x))c(x) \tag{3}$$

where, $I(x)$ is the images captured by the devices, and $J(x)$ is the clear images. Consistent with the traditional ASM, $t(x)$ represents the transmission map, and α) represents the airlight value. The difference is that ESM introduces color information $c(x)$, which is a $1 * 3$ array, including RGB values of haze color. In the model, Equation (3) is rewritten as follows:

$$J_{Final} = \frac{J_{Direct} - \alpha(c(x) - c(x)t(x))}{t(x)} \tag{4}$$

where, J_{Final} is the final output result, $t(x)$ is the estimated transmission map, J_{Direct} is the intermediate result produced by the proposed network directly, α is the global atmospheric light, and $c(x)$ is the color.

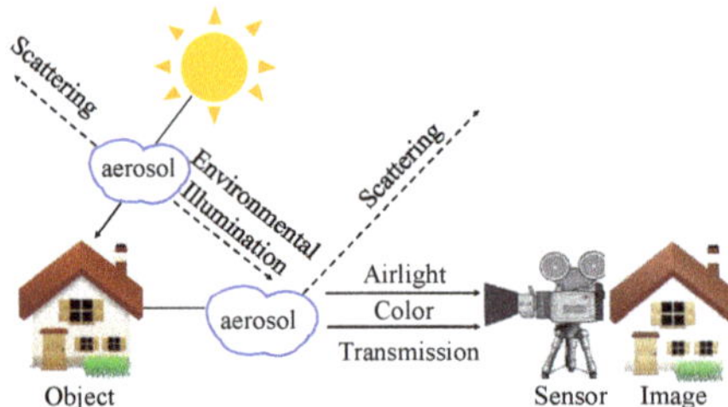

Figure 3. The imaging process in hazy fire scenarios. The transmission attenuation $J(x)t(x)$ is caused by reducing reflection energy and making the color distortion and low brightness. The color value of aerosol in traditional ASM [27] is $(255, 255, 255)$ by default, but the proposed ESM presents different visual characteristics of aerosols in different scenes.

3.3. Loss Function

3.3.1. Mean Square Error

Recently, many data-driven image enhancement algorithms have used the mean square error (MSE) as the loss function to guide the direction of optimization [17,33]. To clearly describe the image pairs needed in calculating the loss function, let $J_n = (J_n, n = 1, ..., N)$ represent the dehazing result of the proposed model, where $G_n = (G_n, n = 1, ..., N)$ is the corresponding ground truth for the corresponding images. In the sequel, we omit the subscript n due to the inputs are independent. The mathematical expression of MSE is as follows:

$$L = \| G - J \| \tag{5}$$

where G is the ground truth images, J is the dehazed images.

3.3.2. Feature Reconstruction Loss

We use both feature reconstruction loss [45] and MSE as the loss function. Li et al. prove that similar images are close to each other in their underlying and high-level features extracted from the deep learning model. This model is called "loss network" [45]. In this paper, we chose the VGG-16 model [46] as the loss network and used the reciprocal first, second, and third layers of the network as measurements to determine the loss function. The formula is as follows:

$$L_p = \sum_{i=1}^{3} \frac{1}{C_i H_i W_i} \| VGG_i(R) - VGG_i(DRHNet(I)) \|_2^2 \tag{6}$$

where $VGG()$ is the VGG-16 model, and R is the residual between the ground truth and the hazy images. H, W, and C represent the length, width, and the number of the feature map channels, respectively.

The final loss function can be described as follows:

$$L_{total} = L + \gamma L_p \tag{7}$$

where, γ is set to 0.5 in this paper, it should be noted that the design of loss function is not important in this paper, but the CIANet still can achieve good results with such simple loss function.

4. Experiment Result

We first introduce the experimental details in this section including the experimental datasets and the comparative algorithms, and then analyze and validate the effectiveness of different modules in the proposed CIANet. Finally, we compare with state-of-the-art dehazing methods by conducting extensive experiments both in synthetic and real-world datasets.

4.1. Experimental Settings

Network training setting: We adopt the same initialization scheme as DehazeNet [17] due to it is an effective dehazing algorithm based on the ASM and CNN. The weights of each layer are initialized by drawing randomly from a standard normal distribution, and the biases are set to 0. The initial learning rate is 0.1 and decreases by 0.01 every 40 epochs. The "Adam" optimization method [47] is adopted to optimize two networks.

The proposed network was trained end-to-end and implemented in the PyTorch platform, and all experiments were performed on a laptop with Intel(R) Core(TM) i7-8750H CPU @ 2.20GHz 2.20 GHz, 16GB RAM, and NVIDIA GeForce GTX 1070.

Dataset: Regarding the training data, 500 fire scenarios images with low image degradation were used as the training data. We uniformly sampled ten random gray values $color \in [140, 230]$ to generate the hazy images for each images. Therefore, a total of 5000 hazy images were generated for training. We named this dataset the realistic fire single image enhancement (RFSIE). Besides, the haze in the fire scenarios is usually non-homogeneous. Therefore, the training set provided in NTIRE'20 competition [48] can also be used as the training set for fire scenarios image dehazing algorithms. The images provided by NTIRE'20 were collected by a professional camera and haze generators to ensure the captured image pairs are the same except for the haze information. Moreover, due to the non-homogeneous haze was captured by [49], it has some similarities with the images of fire scenarios, so it is suitable for image enhancement for fire scenarios.

Compared methods: We compare our model with several state-of-the-art methods both on RFSIE and NTIRE'20, including He [31], Zhu [30], Ren [32], Cai [17], Li [2], Meng [49], Ma [50], Berman [29], Chen [34], Zhang [33] and Zheng [5].

4.2. Ablation Study

This section discusses different modules in CIANet and evaluates its impacts on the enhancement results.

Effects of ESM: Three groups of experiments were designed to verify the effectiveness of ESM. Table 3 presents the quantitative evaluation results of the proposed CIANet with different physical scattering models. In Table 3, J_{Direct} represents the output of the *J-decoder*, and J_{ASM} is the dehazing result obtained by using the traditional ASM with the default color of aerosol being white, J_{ESM} represents the output of CIANet using the proposed ESM.

Table 3. Average PSNR (dB) and SSIM results of different outputs from CIANet on RFSIE and NTIRE'20. The first and second best results are highlighted in red and blue.

Metric		NTIRE'20		RFSIE		Time/Epoch
		PSNR	SSIM	PSNR	SSIM	
CIANet	J_{direct}	13.11	0.56	24.81	0.82	63 min
	J_{ASM}	14.23	0.58	25.34	0.81	
	J_{ESM}	18.34	0.62	31.22	0.91	
J_{direct}-only		12.11	0.51	24.96	0.78	21 min
J_{ASM}-only		14.21	0.59	25.91	0.80	21 min

According to Table 3, J_{ESM} achieves the best PSNR and SSIM values. The reasons are: (1) The image-to-image strategy usually disable to estimate the depth information of the image accurately, which leads to the inconspicuous dehazing performances in the area with dense haze. Therefore, the PSNR and SSIM values obtained by $J_{Direct} - only$ are slightly lower. (2) ESM can propose appropriate image restoration strategies for images with different styles and degrees of damage, while the traditional ASM assumes the color of aerosol is white by default, so it is easy to estimate the degree of damage falsely. Therefore, $J_{ASM} - only$ cannot achieve good dehazing performances. When the *T-decoder, c&a-decoder*, and *D-decoder* are trained together, the common backpropagation will promote each effect of the decoder and ensure that the encoder can extract the most effective haze features. Therefore, PSNR and SSIM values of J_{Direct} and J_{ASM} are slightly higher than those of $J_{Direct} - only$ and $J_{ASM} - only$.

Figure 4 shows the effectiveness of ESM intuitively. As can be seen from the first line of images in Figure 4, when the color of aerosol becomes milky white, the dehazing results of J_{ASM} and J_{ESM} are very similar. However, due to the obvious highlighted area in the fire scenarios, the traditional algorithm used to estimate the value of atmospheric light tends to overestimate the brightness, resulting in lower brightness of the dehazed result. When the color of aerosols in the air is dark, the dehazing effect of J_{ESM} is better than that of J_{ASM}. The second row of images is from the position circled by the red rectangle. It can be seen that J_{ASM} cannot restore the color of grass very well due to the default color of aerosol in ASM is lighter than the actual color. Hence, the performance of haze removal using ASM is lower than the real pixel value, and the overall color is dark.

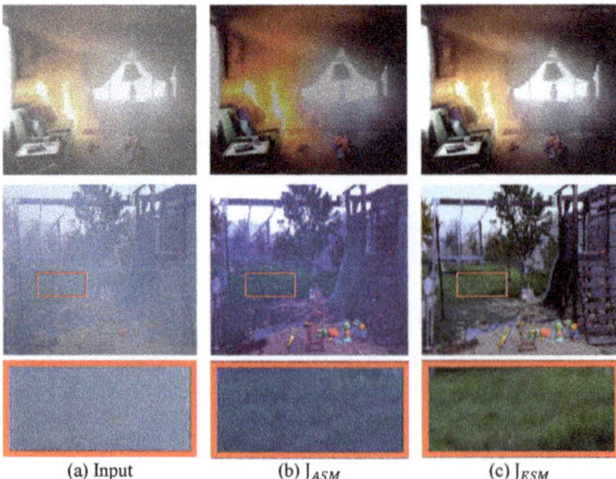

Figure 4. The effectiveness of the ESM on real-world images. The dehazed results of J_{ESM} are much clearer than J_{ASM}.

4.3. Evaluation on Synthetic Images

We compare the proposed CIANet with some of the most advanced single-image dehazing methods on RFSIE, and adopt the indices of the peak signal-to-noise ratio (PSNR) and structural similarity index (SSIM) [51] to evaluate the quality of restored images.

Table 4 shows the quantitative evaluation results on our synthetic test dataset. Compared with other state-of-the-art baselines, the proposed CIANet can obtain higher PSNR and SSIM values. The average PSNR and SSIM of our dehazing algorithm are 18.37 db and 0.13 higher than the input hazy images, which indicates that the proposed algorithm can effectively remove haze and generate high-quality images.

Table 4. Average PSNR and SSIM results on RFSIE. The first, second, and third best results are highlighted in red, blue, and bold, respectively.

Methods	Hazy	He	Zhu	Ren	Cai	Li	Meng	Ma	Berman	Chen	Zhang	Zheng	Ours
PSNR	12.85	17.42	19.67	23.68	21.95	23.92	24.94	26.19	26.95	**27.25**	27.36	26.11	31.22
SSIM	0.78	0.80	0.82	**0.85**	0.87	0.82	0.82	0.82	**0.85**	**0.85**	**0.85**	0.82	0.91

As shown in Figure 5, the proposed method can generate clearer enhancement results than the most advanced image dehazing algorithms on hazy indoor images. The first and second rows of Figure 5 are the synthetic smoke and haze images of the indoor fire scenarios, and the third and fourth rows are the indoor haze images taken from the landmark dataset NTIRE2020. The dark channel algorithm proposed by He et al. produces some color distortion or brightness reduction (such as the first row and the third row of walls). The results show that the color attenuation prior algorithm proposed by Zhu et al. is not very effective, some images have a large amount of haze residues (such as the first row of images). The BCCR image dehazing algorithm proposed by Meng et al. can also cause image distortion or brightness reduction (such as the third and fourth rows of images). The Dehazenet algorithm proposed by Cai et al. can achieve a good dehazing effect in most cases, but there is an obvious haze residual in the first row of scene. The NLD algorithm proposed by Berman et al. can better complete the image dehazing task for indoor scenes, but there is less color distortion in the third row of images compared with the ground truth.

Figure 5. The performances of different image dehazing algorithms on synthetic indoor images.

Figure 6 shows that the proposed method can also generate clearer images for outdoor scenes. Figure 6 can be divided into two parts. The first row and the second row of Figure 6 are taken from the landmark image dehazing database NTIRE2020, and the third and fourth rows of images are taken from the fire-related videos of traffic scenes captured by monitoring equipment. The image dehazing algorithm based on dark channel prior proposed by He et al. tends to estimate the transmission rate of the images and the atmospheric light value, resulting in large distortion in some parts of the images (such as the third and fourth rows of images). The color attenuation prior algorithm proposed by Zhu et al. can complete the image dehazing task to a certain extent, but there are still haze residues in the images (such as the second, third and fourth rows of images). The BCCR algorithm proposed by Meng et al. usually overestimates the brightness of the result. Although this estimation method can improve the detailed information of images, the result obtained is not similar to the ground truth. Dehazenet proposed by Cai et al. is based on the combination of deep learning and the traditional image dehazing algorithm. The results obtained by Dehazenet are very similar to that of CAP, and there are many haze residues. For outdoor images, the NLD algorithm can achieve good image dehazing effects, but some dehazing results are not as good as the results obtained with the proposed algorithm (such as the first row of images). Furthermore, the proposed algorithm can achieve better dehazing effects for outdoor scenes.

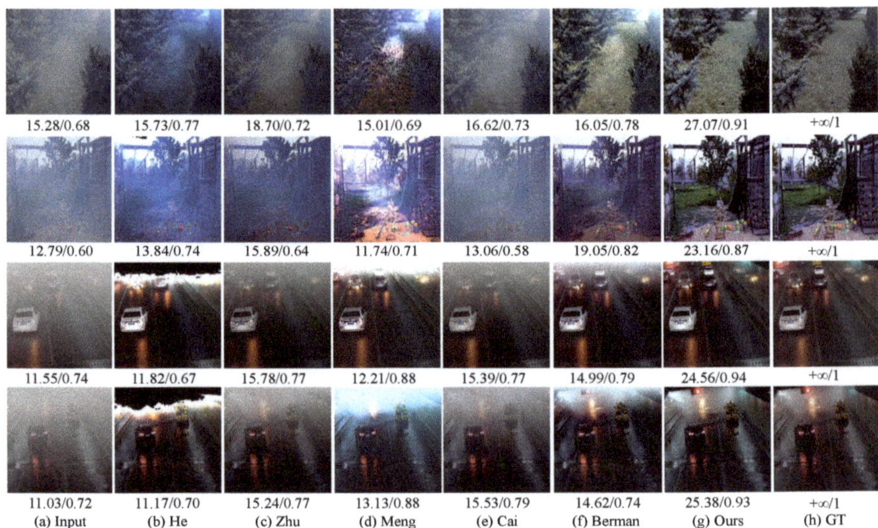

Figure 6. The performances of different image dehazing algorithms on synthetic outdoor images.

4.4. Evaluation on Real-World Images

Figure 7 shows the dehazing effect of the proposed algorithm compared with other state-of-the-art algorithms from real-world images. The first three rows of images are taken from NTIRE'20, and the last three rows are taken from the real images of fire scenarios. The most evident characteristic of the first three rows of images is that the thickness of the haze in image is uneven. For example, in the image of the second row in Figure 7, the haze thickness in the upper left corner of the image is obviously higher than that in the image area. The following three rows of images have similar characteristics with the first three rows of images, that is, different areas have varied degrees of damage. In addition, there is another remarkable feature on the last three rows of images, that is, each image has obvious highlighted area. These two characteristics basically cover all the features of the hazy images with the scenes of fire and smoke.

From the first row of images in Figure 7, we can see that the clarity of images obtained the proposed algorithm is significantly higher than that obtained with other algorithms, and there is no obvious artifact area. From the annotated area, the definition of this area is significantly higher than that of images obtained with other algorithms for comparison. On the second row of images in Figure 7, we can see that most algorithms cannot achieve the image dehazing well due to the left half of the images is seriously affected by haze. Compared with other algorithms, the proposed algorithm can still generate good performances. As shown in the marked area, the color saturation and clarity in the area can be reflected. The algorithm proposed in this paper can obtain better image dehazing effects. From the third row of images, we can still see that the algorithm can achieve better image dehazing effects. Both the brightness and the saturation of color are significantly higher than that of other algorithms.

In Figure 7, the images in the last three rows are obviously more complex than those in the first three rows. First of all, the images in the last three rows have obvious highlighted areas. Secondly, the haze color of the images in the last three rows is darker, which is more challenging than the images in the first three rows. It can be seen from the fifth row of Figure 7 that the algorithm proposed in this paper can remove most of the haze in the images, and basically maintain the structural information of the images, while other algorithms in comparison, such as AODNet, hardly remove any haze from the images. The image on the sixth row reflects that the proposed algorithm can almost remove all the haze in the images, and ensure that the result of the image will not change. In contrast, the dehazing effects of other algorithms are not obvious in this images, and it can be considered that the image dehazing task is completed to a large extent. Therefore, the CIANet proposed in this paper can be used to achieve dehazing of real hazy images to a certain extent.

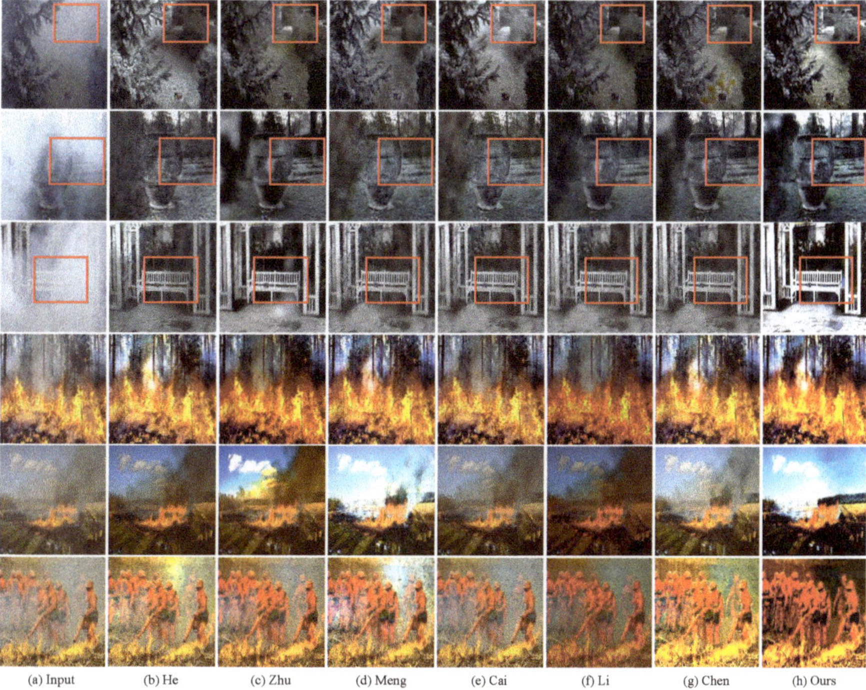

Figure 7. Visual comparisons on real-world images. The proposed method can effectively enhance the quality of different real-world hazy images with naturalness preservation.

4.5. Qualitative Visual Results on Challenging Images

Haze-free images: In order to prove the robustness of the proposed algorithm for all scenarios, we input the fire images which are not affected by the air particles into the network model. It can be seen from Figure 8 that the algorithm proposed in this paper has little effect on the fire scenarios image without fog, and it only slightly changes the color of the images, increasing the saturation of the image without damaging the structural information of the images. Therefore, this experiment proves the robustness of the algorithm. Hence, when the algorithm is embedded in the intelligent edge computing devices, it is not necessary to choose whether to run CIANet according to the change of situation.

(a) Input clear images   (b) Our results

Figure 8. Examples for haze-free scenarios enhancement. (**a**): haze-free real photos with fire and. (**b**):enhancement results by CIANet.

4.6. Potential Applications

The CIANet proposed in this paper can effectively improve the visibility and clarity of the scene to promote the performance of other high-level visual tasks, which is the application significance of the algorithm proposed in this paper. To verify the proposed CIANet could benefit other vision tasks, we perform two applications: fire scenarios object detection and local keypoints matching. As can be seen from Figures 9 and 10, the algorithm proposed in this paper can not only improve the visual quality and the quality of the input image, but also significantly improve the performance of subsequent important high-dimensional vision. The following two sections will discuss in detail the improvement of CIANet on object detection and local keypoint matching tasks.

4.6.1. Object Detection

Most existing deep models for high-level vision tasks are trained using clear images. Such learned models will have low robustness when applied to degraded hazy fire scenarios images. In this case, the enhanced results can be useful for these high-level vision applications. To prove the proposed model can improve the detection precision, we analyze the performances of the object detection on our dehazed results. Figure 9 shows that using the proposed model as pre-processing can improve the detection performance.

 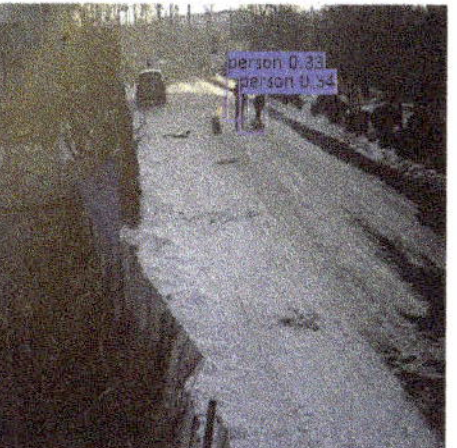

(a) original hazy image   (b) enhancement result

Figure 9. Pre-processing for object detection (Faster R-CNN [52], threshold = 0.3). (**a**): detection on hazy fire scenarios; (**b**): detection on the enhancement result.

4.6.2. Local Keypoint Matching

We also adopt local keypoints matching, which aims to find correspondences between two similar scenarios, to test the effectiveness of the proposed CIANet. We utilize the SIFT operator for a pair or hazy fire scenarios images and as well as for the corresponding dehazed images. The matching result are shown in Figure 10. It is clear that the number of matched keypoints is significantly increased in the dehaze fire scenarios image pairs. This verifies that the proposed CIANet can recover the important features of the hazy images.

(a) Hazy: 138 matches   (b) Our result: 440 matches

Figure 10. Local keypoints matching by applying the SIFT operator. Compared with the hazy images, the matching results shown that the proposed method can improved the quality of inputs significantly.

4.7. Runtime Analysis

The light-weight structure of CIANet leads to faster image enhancement. We select only one image from real-world and then repeat runing 100 times by different dehazing algorithms, on the same machine (Intel(R) Core(TM) i7-8750H CPU @2.20GHz and 16GB memory), without GPU acceleration. The per-image average running time of all models are shown in Table 5. Despite other slower MATLAB implementations, it is fair to compare DehazeNet (Pytorch version) and ours methods. The results illustrate the promising efficiency of CIANet.

Table 5. Comparison of average model running time (in seconds).

Image Size	480 × 640	Platform
He	26.03	Matlab
Berman	8.43	Matlab
Meng	2.19	Matlab
Ren	2.01	Matlab
Zhu	1.02	Matlab
Cai (Matlab)	2.09	Matlab
Cai (Pytorch)	6.31	Pytorch
CIANet	4.77	Pytorch

5. Conclusions

This paper proposes CIANet, a color-dense illumination adjustment network that reconstructs clear fire scenario images via a novel ESM. We compare CIANet with the state-of-the-art dehazing methods, both on synthetic and real-world images both quantitatively (PSNR, SSIM) qualitatively (subjective measurements). The experimental results have shown that the superiority of the CIANet. Moreover, the experiments show that the proposed ESM is more reasonable than the traditional ASM in the fire scenarios imaging process. In the future, we will study the image enhancement algorithm of fire scenario on lidar image, so as to solve the problem that the traditional computer vision algorithm can not deal with the scene where there is a large amount of smoke and all visual information is lost.

Author Contributions: C.W., Conceptualization, Data Curation, Methodology, Writing—Original Draft; J.H., Funding Acquisition, Supervision, Methodology.; X.L., Writing—Review and Editing.; M.-P.K., Writing—Review and Editing.; W.C., Visualization, Investigation, Project Administration; H.W., Visualization, Formal Analysis, Software. All authors have read and agreed to the published version of the manuscript.

Funding: This research was funded by Key Area Research and Development Program of Guangdong Province: 2019B111102002; National Key Research and Development Program of China: 2019YFC0810704; Shenzhen Science and Technology Program: KCXFZ202002011007040.

Conflicts of Interest: The authors declare there is no conflict of interest regarding the publication of this paper.

References

1. Liang, J.X.; Zhao, J.F.; Sun, N.; Shi, B.J. Random Forest Feature Selection and Back Propagation Neural Network to Detect Fire Using Video. *J. Sens.* **2022**, *2022*, 5160050. [CrossRef]
2. Li, B.; Peng, X.; Wang, Z.; Xu, J.; Feng, D. Aod-net: All-in-one dehazing network. In Proceedings of the IEEE International Conference on Computer Vision, Venice, Italy, 22 October 2017; pp. 4770–4778.
3. Xu, G.; Zhang, Y.; Zhang, Q.; Lin, G.; Wang, J. Deep domain adaptation based video smoke detection using synthetic smoke images. *Fire Saf. J.* **2017**, *93*, 53–59. [CrossRef]
4. Chen, T.H.; Yin, Y.H.; Huang, S.F.; Ye, Y.T. The smoke detection for early fire-alarming system base on video processing. In Proceedings of the 2006 International Conference on Intelligent Information Hiding and Multimedia, Pasadena, CA, USA, 18–20 December 2006; pp. 427–430.
5. Zheng, Z.; Ren, W.; Cao, X.; Hu, X.; Wang, T.; Song, F.; Jia, X. Ultra-High-Definition Image Dehazing via Multi-Guided Bilateral Learning. In Proceedings of the 2021 IEEE/CVF Conference on Computer Vision and Pattern Recognition (CVPR), Nashville, TN, USA, 20–25 June 2021; pp. 16180–16189.
6. Yoon, I.; Jeong, S.; Jeong, J.; Seo, D.; Paik, J. Wavelength-adaptive dehazing using histogram merging-based classification for UAV images. *Sensors* **2015**, *15*, 6633–6651. [CrossRef] [PubMed]
7. Dong, T.; Zhao, G.; Wu, J.; Ye, Y.; Shen, Y. Efficient traffic video dehazing using adaptive dark channel prior and spatial–temporal correlations. *Sensors* **2019**, *19*, 1593. [CrossRef]
8. Liu, K.; He, L.; Ma, S.; Gao, S.; Bi, D. A sensor image dehazing algorithm based on feature learning. *Sensors* **2018**, *18*, 2606. [CrossRef]
9. Qu, C.; Bi, D.Y.; Sui, P.; Chao, A.N.; Wang, Y.F. Robust dehaze algorithm for degraded image of CMOS image sensors. *Sensors* **2017**, *17*, 2175. [CrossRef]
10. Hsieh, P.W.; Shao, P.C. Variational contrast-saturation enhancement model for effective single image dehazing. *Signal Process.* **2022**, *192*, 108396. [CrossRef]
11. Zhang, Y.; Kwong, S.; Wang, S. Machine learning based video coding optimizations: A survey. *Inf. Sci.* **2020**, *506*, 395–423. [CrossRef]
12. Zhang, Y.; Zhang, H.; Yu, M.; Kwong, S.; Ho, Y.S. Sparse representation-based video quality assessment for synthesized 3D videos. *IEEE Trans. Image Process.* **2019**, *29*, 509–524. [CrossRef]
13. Zhang, Y.; Pan, Z.; Zhou, Y.; Zhu, L. Allowable depth distortion based fast mode decision and reference frame selection for 3D depth coding. *Multimed. Tools Appl.* **2017**, *76*, 1101–1120. [CrossRef]
14. Zhu, L.; Kwong, S.; Zhang, Y.; Wang, S.; Wang, X. Generative adversarial network-based intra prediction for video coding. *IEEE Trans. Multimed.* **2019**, *22*, 45–58. [CrossRef]
15. Liu, H.; Zhang, Y.; Zhang, H.; Fan, C.; Kwong, S.; Kuo, C.C.J.; Fan, X. Deep learning-based picture-wise just noticeable distortion prediction model for image compression. *IEEE Trans. Image Process.* **2019**, *29*, 641–656. [CrossRef] [PubMed]
16. Astua, C.; Barber, R.; Crespo, J.; Jardon, A. Object Detection Techniques Applied on Mobile Robot Semantic Navigation. *Sensors* **2014**, *14*, 6734–6757. [CrossRef] [PubMed]
17. Cai, B.; Xu, X.; Jia, K.; Qing, C.; Tao, D. Dehazenet: An end-to-end system for single image haze removal. *IEEE Trans. Image Process.* **2016**, *25*, 5187–5198. [CrossRef] [PubMed]

18. Thanh, L.T.; Thanh, D.N.H.; Hue, N.M.; Prasath, V.B.S. Single Image Dehazing Based on Adaptive Histogram Equalization and Linearization of Gamma Correction. In Proceedings of the 2019 25th Asia-Pacific Conference on Communications, Ho Chi Minh City, Vietnam, 6–8 November 2019; pp. 36–40.

19. Golts, A.; Freedman, D.; Elad, M. Unsupervised Single Image Dehazing Using Dark Channel Prior Loss. *IEEE Trans. Image Process.* **2020**, *29*, 2692–2701. [CrossRef]

20. Parihar, A.S.; Gupta, Y.K.; Singodia, Y.; Singh, V.; Singh, K. A Comparative Study of Image Dehazing Algorithms. In Proceedings of the 2020 5th International Conference on Communication and Electronics Systems , Hammamet, Tunisia, 10–12 March 2020; pp. 766–771. [CrossRef]

21. Hou, G.; Li, J.; Wang, G.; Pan, Z.; Zhao, X. Underwater image dehazing and denoising via curvature variation regularization. *Multimed. Tools Appl.* **2020**, *79*, 20199–20219. [CrossRef]

22. Nayar, S.K.; Narasimhan, S.G. Vision in bad weather. In Proceedings of the IEEE Seventh IEEE International Conference on Computer Vision, Kerkyra, Greece, 20–27 September 1999; Volume 2, pp. 820–827.

23. Narasimhan, S.G.; Nayar, S.K. Contrast restoration of weather degraded images. *IEEE Trans. Pattern Anal. Mach. Intell.* **2003**, *25*, 713–724. [CrossRef]

24. Lin, G.; Zhang, Y.; Xu, G.; Zhang, Q. Smoke detection on video sequences using 3D convolutional neural networks. *Fire Technol.* **2019**, *55*, 1827–1847. [CrossRef]

25. Shu, Q.; Wu, C.; Xiao, Z.; Liu, R.W. Variational Regularized Transmission Refinement for Image Dehazing. In Proceedings of the 2019 IEEE International Conference on Image Processing (ICIP), Taipei, Taiwan, 22–29 September 2019; pp. 2781–2785. [CrossRef]

26. Fu, X.; Cao, X. Underwater image enhancement with global-local networks and compressed-histogram equalization. *Signal Process. Image Commun.* **2020**, *86*, 115892. [CrossRef]

27. McCartney, E.J. Optics of the Atmosphere: Scattering by Molecules and Particles. *Phys. Bull.* **1977**, *28*, 521–531. [CrossRef]

28. Tang, K.; Yang, J.; Wang, J. Investigating haze-relevant features in a learning framework for image dehazing. In Proceedings of the IEEE Conference on Computer Vision and Pattern Recognition, Columbus, OH, USA, 24–27 June 2014; pp. 2995–3000.

29. Berman, D.; Avidan, S. Non-local image dehazing. In Proceedings of the IEEE Conference on Computer Vision and Pattern Recognition, Las Vegas, NV, USA, 26 June–1 July 2016; pp. 1674–1682.

30. Zhu, Q.; Mai, J.; Shao, L. A fast single image haze removal algorithm using color attenuation prior. *IEEE Trans. Image Process.* **2015**, *24*, 3522–3533. [PubMed]

31. He, K.; Sun, J.; Tang, X. Single image haze removal using dark channel prior. *IEEE Trans. Pattern Anal. Mach. Intell.* **2011**, *33*, 2341–2353. [PubMed]

32. Ren, W.; Pan, J.; Zhang, H.; Cao, X.; Yang, M.h. Single Image Dehazing via Multi-Scale Convolutional Neural Networks with Holistic Edges. *Int. J. Comput. Vis.* **2020**, *128*, 240–259. [CrossRef]

33. Zhang, H.; Patel, V.M. Densely Connected Pyramid Dehazing Network. In Proceedings of the 2018 IEEE/CVF Conference on Computer Vision and Pattern Recognition, Salt Lake City, Utah, USA, 18–22 June 2018; pp. 3194–3203. [CrossRef]

34. Chen, D.; He, M.; Fan, Q.; Liao, J.; Zhang, L.; Hou, D.; Yuan, L.; Hua, G. Gated Context Aggregation Network for Image Dehazing and Deraining. In Proceedings of the IEEE Winter Conference on Applications of Computer Vision, Waikola, HI, USA, 7–11 January 2019; pp. 1375–1383.

35. Gandelsman, Y.; Shocher, A.; Irani, M. "Double-DIP": Unsupervised Image Decomposition via Coupled Deep-Image-Priors. In Proceedings of the 2019 IEEE/CVF Conference on Computer Vision and Pattern Recognition (CVPR), Long Beach, CA, USA, 16–20 June 2019; pp. 11018–11027. [CrossRef]

36. Wang, C.; Li, Z.; Wu, J.; Fan, H.; Xiao, G.; Zhang, H. Deep residual haze network for image dehazing and deraining. *IEEE Access* **2020**, *8*, 9488–9500. [CrossRef]

37. Li, B.; Ren, W.; Fu, D.; Tao, D.; Feng, D.; Zeng, W.; Wang, Z. Benchmarking single-image dehazing and beyond. *IEEE Trans. Image Process.* **2018**, *28*, 492–505. [CrossRef] [PubMed]

38. Zhang, H.; Patel, V.M. Density-aware single image de-raining using a multi-stream dense network. In Proceedings of the IEEE Conference on Computer Vision and Pattern Recognition, Salt Lake City, UH, USA, 18–22 June 2018; pp. 695–704.

39. Sakaridis, C.; Dai, D.; Van Gool, L. Semantic foggy scene understanding with synthetic data. *Int. J. Comput. Vis.* **2018**, *126*, 973–992. [CrossRef]

40. Lin, T.Y.; Dollár, P.; Girshick, R.; He, K.; Hariharan, B.; Belongie, S. Feature pyramid networks for object detection. In Proceedings of the IEEE Conference on Computer Vision and Pattern Recognition, Hawaii Convention Center, Honolulu, Hawaii, 21–26 July 2017; pp. 2117–2125.

41. Gao, S.; Cheng, M.M.; Zhao, K.; Zhang, X.Y.; Yang, M.H.; Torr, P.H. Res2net: A new multi-scale backbone architecture. *IEEE Trans. Pattern Anal. Mach. Intell.* **2019**, *43*, 652–662. [CrossRef]

42. He, K.; Zhang, X.; Ren, S.; Sun, J. Deep residual learning for image recognition. In Proceedings of the IEEE Conference on Computer Vision and Pattern Recognition, Las Vegas, NV, USA, 26 June–1 July 2016; pp. 770–778.

43. Xie, S.; Girshick, R.; Dollár, P.; Tu, Z.; He, K. Aggregated residual transformations for deep neural networks. In Proceedings of the IEEE Conference on Computer Vision and Pattern Recognition, 2017; pp. 1492–1500.

44. Fisher, Y.; Dequan, W.; Evan, S.; Trevor, D. Deep layer aggregation. In Proceedings of the IEEE Conference on Computer Vision and Pattern Recognition, Salt Lake City, UT, USA, 18–22 June 2018; pp. 2403–2412.

45. Johnson, J.; Alahi, A.; Fei-Fei, L. Perceptual Losses for Real-Time Style Transfer and Super-Resolution. In *European Conference on Computer Vision*; Springer: Cham, Switzerland, 2016.
46. Simonyan, K.; Zisserman, A. Very deep convolutional networks for large-scale image recognition. *arXiv* **2014**, arXiv:1409.1556.
47. Kingma, D.P.; Ba, J. Adam: A method for stochastic optimization. *arXiv* **2014**, arXiv:1412.6980.
48. Ancuti, C.O.; Ancuti, C.; Timofte, R. NH-HAZE: An Image Dehazing Benchmark with Non-Homogeneous Hazy and Haze-Free Images. In Proceedings of the 2020 IEEE/CVF Conference on Computer Vision and Pattern Recognition Workshops (CVPRW), Online, 14–19 June 2020; pp. 1798–1805. [CrossRef]
49. Meng, G.; Wang, Y.; Duan, J.; Xiang, S.; Pan, C. Efficient image dehazing with boundary constraint and contextual regularization. Proceedings of the IEEE International Conference on Computer Vision, Sydney Conference Centre in Darling Harbour, Sydney, 3–6 December 2013; pp. 617–624.
50. Ren, W.; Ma, L.; Zhang, J.; Pan, J.; Cao, X.; Liu, W.; Yang, M.H. Gated fusion network for single image dehazing. In Proceedings of the IEEE Conference on Computer Vision and Pattern Recognition, Salt Lake City, UT, USA, 18–22 June 2018; pp. 3253–3261.
51. Wang, Z.; Bovik, A.C.; Sheikh, H.R.; Simoncelli, E.P. Image quality assessment: from error visibility to structural similarity. *IEEE Trans. Image Process.* **2004**, *13*, 600–612. [CrossRef]
52. Ren, S.; He, K.; Girshick, R.; Sun, J. Faster r-cnn: Towards real-time object detection with region proposal networks. *Adv. Neural Inf. Process. Syst.* **2015**, *28*, 91–99. [CrossRef] [PubMed]

 *sensors*

MDPI

Article

Attention Networks for the Quality Enhancement of Light Field Images

Ionut Schiopu * and Adrian Munteanu

Department of Electronics and Informatics (ETRO), Vrije Universiteit Brussel (VUB), Pleinlaan 2, 1050 Brussels, Belgium; acmuntea@etrovub.be
* Correspondence: ischiopu@etrovub.be

Abstract: In this paper, we propose a novel filtering method based on deep attention networks for the quality enhancement of light field (LF) images captured by plenoptic cameras and compressed using the High Efficiency Video Coding (HEVC) standard. The proposed architecture was built using efficient complex processing blocks and novel attention-based residual blocks. The network takes advantage of the macro-pixel (MP) structure, specific to LF images, and processes each reconstructed MP in the luminance (Y) channel. The input patch is represented as a tensor that collects, from an MP neighbourhood, four Epipolar Plane Images (EPIs) at four different angles. The experimental results on a common LF image database showed high improvements over HEVC in terms of the structural similarity index (SSIM), with an average Y-Bjøntegaard Delta (BD)-rate savings of 36.57%, and an average Y-BD-PSNR improvement of 2.301 dB. Increased performance was achieved when the HEVC built-in filtering methods were skipped. The visual results illustrate that the enhanced image contains sharper edges and more texture details. The ablation study provides two robust solutions to reduce the inference time by 44.6% and the network complexity by 74.7%. The results demonstrate the potential of attention networks for the quality enhancement of LF images encoded by HEVC.

Keywords: attention network; quality enhancement; light field images; video coding

Citation: Schiopu, I.; Munteanu, A. Attention Networks for the Quality Enhancement of Light Field Images. *Sensors* **2021**, *21*, 3246. https://doi.org/10.3390/s21093246

Academic Editor: Yun Zhang

Received: 6 April 2021
Accepted: 1 May 2021
Published: 7 May 2021

Publisher's Note: MDPI stays neutral with regard to jurisdictional claims in published maps and institutional affiliations.

Copyright: © 2021 by the authors. Licensee MDPI, Basel, Switzerland. This article is an open access article distributed under the terms and conditions of the Creative Commons Attribution (CC BY) license (https://creativecommons.org/licenses/by/4.0/).

1. Introduction

In recent years, the technological breakthroughs in the sensor domain have made possible the development of new camera systems with steadily increasing resolutions and affordable prices for users. In contrast to conventional Red-Green-Blue (RGB) cameras, which only capture light intensity, plenoptic cameras provide the unique ability of distinguishing between the light rays that hit the camera sensor from different directions using microlens technology. To this end, the main lens of plenoptic cameras focus light rays onto a microlens plane, and each microlens captures the incoming light rays from different angles and directs them onto the camera sensor.

For each microlens, a camera sensor produces a so-called Macro-Pixel (MP). The raw LF image contains the entire information captured by the camera sensor, where the array of microlenses generates a corresponding array of MPs, a structure also known as lenslet images. Since each pixel in the MP corresponds to a specific direction of the incoming light, the lenslet image is typically arranged as an array of SubAperture Images (SAIs), where each SAI collects, from all MPs, one pixel at a specific position corresponding to a specific direction of the incoming light. The captured LF image can, thus, be represented as an array of SAIs corresponding to a camera array with a narrow baseline.

LF cameras have proven to be efficient passive devices for depth estimation. A broad variety of depth estimation techniques based on LF cameras have been proposed in the literature, including multi-stereo techniques [1,2], artificial intelligence-based methods [3] as well as combinations of multi-stereo and artificial intelligence-based techniques [4]. Accurately estimating depth is of paramount importance in view synthesis [5] and 3D reconstruction [6,7].

The LF domain was intensively studied during recent decades, and many solutions were proposed for each module in the LF processing pipeline, such as LF acquisition, representation, rendering, display, and LF coding. The LF coding approaches are usually divided into two major classes, including transform-based approaches and predictive-based approaches, depending on which module in the image or video codec is responsible for exploiting the LF correlations.

The transform-based approaches are designed to apply a specific type of transform, such as Discrete Cosine Transform [8,9], Discrete Wavelet Transform [10,11], Karhunen Loéve Transform [12,13], or Graph Fourier Transform [14,15], to exploit the LF correlations.

However, the predictive-based approaches received more attention as they propose a more straightforward solution where different prediction methods are proposed to take advantage of the LF structure. These approaches propose to exploit the correlations between the SAIs using the coding tools in the High Efficiency Video Coding (HEVC) standard [16].

The pseudo-video-sequence-based approach proposes to select a set of evenly distributed SAIs as intra-coded frames and the remaining SAIs as inter-coded frames, e.g., [17,18]. In [19,20], the non-local spatial correlation is exploited when using the lenslet representation. The view-synthesis-based approach proposes to encode only a sparse set of reference SAIs and additional geometry information and then to synthesize the remaining SAIs at the decoder side [21,22]. In this work, we first employ HEVC [16] to encode the SAI video sequence and then to enhance the reconstructed lenslet image. The proposed Convolutional Neural Network (CNN)-based filtering method can be used to post-process any HEVC-based solution.

The attention mechanism was first proposed in the machine translation domain [23]. The main idea is that instead of building a single context vector, it is better to create weighted shortcuts between the context vector and the entire source input. This revolutionary concept now provides outstanding improvements in different domains, such as hyperspectral image classification [24], deblurring [25], image super-resolution [26], traffic sign recognition [27], and small object detection [28], to name a few. Many different network architectures have leveraged the attention mechanism to significantly improve over the state-of-the-art. In this work, an attention-based residual block is introduced to help the architecture learn and focus more on the most important information in the current MP context.

In our prior work, research efforts were invested to provide innovative solutions for LF coding based on efficient Deep-Learning (DL)-based prediction methods [20,29–32] and CNN-based filtering methods for quality enhancement [33,34]. In [29], we introduced a lossless codec for LF images based on context modeling of SAI images. In [30], we proposed an MP prediction method based on neural networks for the lossless compression of LF images.

In [31], we proposed to employ a DL-based method to synthesize an entire LF image based on different configurations of reference SAIs and then to employ an MP-wise prediction method to losslessly encode the remaining views. In [32], we proposed a residual-error prediction method based on deep learning and a context-tree based bit-plane codec, where the experimental evaluation was carried out on photographic images, LF images, and video sequences. In [20], the MP was used as an elementary coding unit instead of HEVC's traditional block-based coding structure for lossy compression of LF images. In recent work, we focused on researching novel CNN-based filtering methods.

In [33], we proposed a frame-wise CNN-based filtering method for enhancing the quality of HEVC-decoded videos. In [34], we proposed an MP-wise CNN-based filtering method for the quality enhancement of LF images. The goal of this paper is to further advance our findings in [34] by introducing a novel filtering method based on attention networks, where the proposed architecture is built based on efficient processing blocks and attention-based residual blocks and operates on Epipolar Plane Images (EPI)-based input patches.

In summary, the novel contributions of this paper are as follows:

(1) A novel CNN-based filtering method is proposed for enhancing the quality of LF images encoded using HEVC [16].
(2) A novel neural network architecture design for the quality enhancement of LF images is proposed using an efficient complex Processing Block (PB) and a novel Attention-based Residual Block (ARB).
(3) The proposed CNN-based filtering method follows an MP-wise filtering approach to take advantage of the specific LF structure.
(4) The input patch is designed as a tensor of four MP volumes corresponding to four EPIs at four different angles ($0°, 45°, 90°$, and $135°$).
(5) The elaborated experimental validation carried out on the EPFL LF dataset [35] demonstrates the potential of attention networks for the quality enhancement of LF images.

The remainder of this paper is organized as follows. Section 2 presents an overview of the state-of-the-art methods for quality enhancement. In Section 3, we describe the proposed CNN-based filtering method. Section 4 presents the experimental validation on LF images. Finally, in Section 5, we draw our conclusions from this work.

2. Related Work

In recent years, many coding solutions based on machine learning techniques have rapidly gained popularity by proposing to simply replace specific task-oriented coding tools in the HEVC coding framework [16] with powerful DL-based equivalents. The filtering task was widely studied, and many DL-based filtering tools for quality enhancement were introduced to reduce the effects of coding artifacts in the reconstructed video.

The first DL-based quality enhancement tools were proposed for image post-filtering. In [36], the Artifact Reduction CNN (AR-CNN) architecture was proposed to reduce the effect of the coding artifacts in JPEG compressed images. In [37], a more complex architecture with hierarchical skip connections was proposed. A dual (pixel and transform) domain-based filtering method was proposed in [38]. A discriminator loss, as in Generative Adversarial Networks (GANs), was proposed in [39]. An iterative post-filtering method based on a recurrent neural network was proposed in [40].

Inspired by AR-CNN [36], the Variable-filter-size Residue-learning CNN (VRCNN) architecture was proposed in [41]. The inter-picture correlation is used by processing multiple neighboring frames to enhance one frame using a CNN [42]. In [43], the authors proposed to make use of mean- and boundary-based masks generated by HEVC partitioning. In [44], a CNN processes the intra prediction signal and the decoded residual signal. In [45], a CNN processes the QP value and the decoded frame. In [46], the CNN operates on input patches designed based on additional information extracted from the HEVC decoder, which specifies the current QP value and the CU partitioning maps.

In another approach, the authors proposed to replace the HEVC built-in in-loop filtering, the Deblocking Filter (DBF) [47], and the Sample Adaptive Offset (SAO) [48]. This is a more demanding task as, in this case, the filtered frame enters the coding loop and serves as a reference to other frames. In [49], a CNN was used to replace the SAO filter. Similarly, in [50], a deep CNN was applied after SAO and was controlled by the frame- and coding tree unit (CTU)-level flags.

In [51], the authors used a deep residual network to estimate the lost details. In [52], the Multistage Attention CNN (MACNN) architecture was introduced to replace the HEVC in-loop filters. Other coding solutions focus on inserting new filtering blocks in the HEVC framework. In [53], an adaptive, in-loop filtering algorithm was proposed using an image nonlocal prior, which collaborates with the existing DBF and SAO in HEVC. In [54], a residual highway CNN (RHCNN) was applied after the SAO filter. In [55], a content-aware CNN-based in-loop filtering method was integrated in HEVC after the SAO built-in filter.

In this work, we propose to employ the attention mechanism for the quality enhancement of LF images (represented as lenslet images) by following an MP-wise filtering approach. Our experiments show that an increased coding performance was achieved when the SAI video sequence was encoded by running HEVC without its built-in filtering methods, DBF [47] and SAO [48].

3. Proposed Method

In the literature, the LF image is usually represented as a 5D structure denoted by $LF(p, q, x, y, c)$, where the (p, q) pair denotes the pixel location in an MP matrix, usually of $N \times N$ resolution; the (x, y) pair denotes the pixel location in an SAI matrix of size $W \times H$; and c denotes the primary color channel, $c = 1, 2, 3$. Let us denote $MP_{x,y} = LF(:, :, x, y, c)$ as the MP captured by the microlens at position (x, y) in the microlens array; $SAI_{p,q} = LF(p, q, :, :, c)$ as the SAI corresponding to view (p, q) in the SAI stack; and LL as the lenslet image of size $NH \times NW$, which is defined as follows:

$$LL((x-1)N + 1 : xN, (y-1)N + 1 : yN, c) = MP_{x,y}, \forall x = 1 : W, \forall y = 1 : H. \quad (1)$$

The experiments were conducted using the EPFL LF dataset [35] where $N = 15$ and $W \times H = 625 \times 434$. The LF images were first color-transformed from the RGB color-space to the YUV color-space, and only the Y (luminance) channel was enhanced. Therefore, $c = 1$ and $MP_{x,y}$ were of size 15×15.

In this paper, a novel CNN-based filtering method is proposed to enhance the quality of LF images encoded using the HEVC video coding standard [16]. Figure 1 depicts the proposed CNN-based filtering scheme. The LF image, represented as an array of SAIs, is first arranged as an SAI video sequence and then encoded by the reference software implementation of HEVC called HM (HEVC Test Model) [56] under the All Intra (AI) profile [57]. Any profile can be used to encode the SAI video sequence as the proposed CNN-based filtering scheme is applied to the entire SAI video sequence. Therefore, in this work, a raster scan order is used to generate the SAI video sequence, while in the literature, a spiral order starting from the center view and looping in a clockwise manner towards the edge views is used to generate the SAI video sequence. Next, the reconstructed SAI sequence is arranged as a lenslet image using Equation (1), and EPI-based input patches were extracted from the reconstructed lenslet image, see Section 3.1.

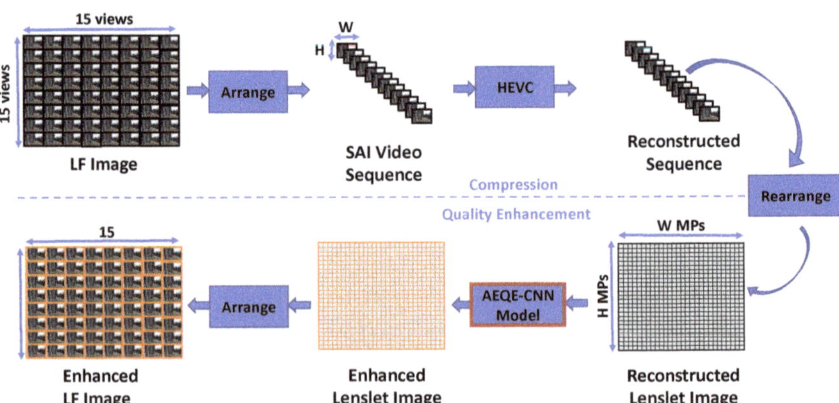

Figure 1. The proposed CNN-based filtering scheme. (**Top**) Compression: The LF Image (represented as an array of SAI) is arranged as a SAI video sequence and then encoded by HEVC. (**Bottom**) Quality Enhancement: The reconstructed sequence is arranged as a lenslet image (represented as an array of MPs) and each MP is enhanced by the proposed CNN-based filtering method using an AEQE-CNN model.

A CNN model with the proposed novel deep neural architecture called Attention-aware EPI-based Quality Enhancement Convolutional Neural Network (AEQE-CNN), see Section 3.2, processed the input patches to enhance the MPs and obtain the enhanced lenslet image. Finally, the enhanced lenslet image is arranged as a LF image to be easily consumed by users.

Section 3.1 presents the proposed algorithm used to extract the EPI-based input patches. Section 3.2 describes in detail the network design of the proposed AEQE-CNN architecture. Section 3.3 presents the training details.

3.1. Input Patch

In this paper, input patches of size $15 \times 15 \times 9 \times 4$ were extracted from the reconstructed lenslet image. More exactly, the input patch concatenated four EPIs corresponding to 0° (horizontal EPI), 45° (first diagonal EPI), 90° (vertical EPI), and 135° (second diagonal EPI) from the MP neighbourhood of $b = 4$ MPs around the current MP, as depicted in Figure 2. Let us denote $\mathcal{N}_{x,y}$ as the MP neighbourhood around the current MP, $MP_{x,y}$, where

$$\mathcal{N}_{x,y} = \begin{bmatrix} MP_{x-b,y-b} & \cdots & MP_{x-b,y} & \cdots & MP_{x-b,y+b} \\ \vdots & & \vdots & & \vdots \\ MP_{x,y-b} & \cdots & MP_{x,y} & \cdots & MP_{x,y+b} \\ \vdots & & \vdots & & \vdots \\ MP_{x+b,y-b} & \cdots & MP_{x+b,y} & \cdots & MP_{x+b,y+b} \end{bmatrix}. \tag{2}$$

Four EPIs of size $N \times N \times (2b + 1) = 15 \times 15 \times 9$ were extracted from $\mathcal{N}_{x,y}$ as follows:

(1) The 0° EPI of MP volume: $[MP_{x,y-b} \, MP_{x,y-b+1} \, \cdots \, MP_{x,y+b}]$;
(2) The 45° EPI of MP volume: $[MP_{x-b,y-b} \, MP_{x-b+1,y-b+1} \, \cdots \, MP_{x+b,y+b}]$;
(3) The 90° EPI of MP volume: $[MP_{x-b,y} \, MP_{x-b+1,y} \, \cdots \, MP_{x+b,y}]$; and
(4) The 135° EPI of MP volume: $[MP_{x+b,y-b} \, MP_{x+(b-1),y-(b-1)} \, \cdots \, MP_{x-b,y+b}]$.

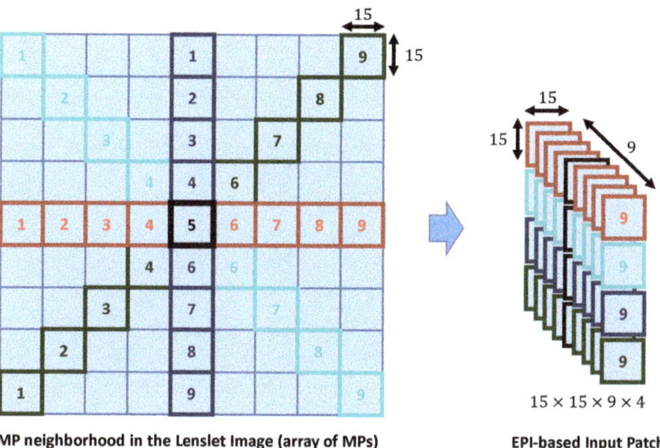

Figure 2. Extraction of the EPI-based input patch from the lenslet image represented as an array of MPs. Four EPIs are selected: 0° (horizontal) EPI marked with red, 45° (first diagonal) EPI marked with cyan, 90° (vertical) EPI marked with blue, and 135° (second diagonal) EPI marked with green. The current MP is marked with black.

The four EPIs were processed separately by the AEQE-CNN architecture as described in the following section.

3.2. Network Design

Figure 3 depicts the proposed deep neural network architecture. AEQE-CNN is designed to process the EPI-based input patches using efficient processing blocks and attention-based residual blocks. 3D Convolutional layers (Conv3D) equipped with $3 \times 3 \times 3$ kernels are used throughout the network architecture.

AEQE-CNN was built using the following types of blocks depicted in Figure 4: *(i)* the Convolutional Block (CB) contains a sequence of one Conv3D, one batch normalization (BN) layer [58], and one Rectified Linear Unit (ReLU) activation function; *(ii)* the proposed Processing Block (PB) contains a two branch design with one and two CB blocks where the feature maps of the two branches are concatenated to obtain the output feature maps; *(iii)* the proposed Attention-based Residual Block (APB) contains a sequence of two PB blocks and one Convolutional Block Attention Module (CBAM), see Figure 5, and a skip connection to process the current patch.

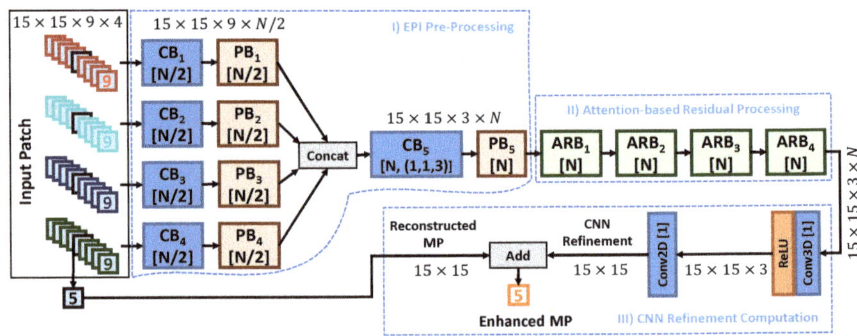

Figure 3. The proposed network architecture called Attention-aware EPI-based Quality Enhancement Convolutional Neural Network (AEQE-CNN).

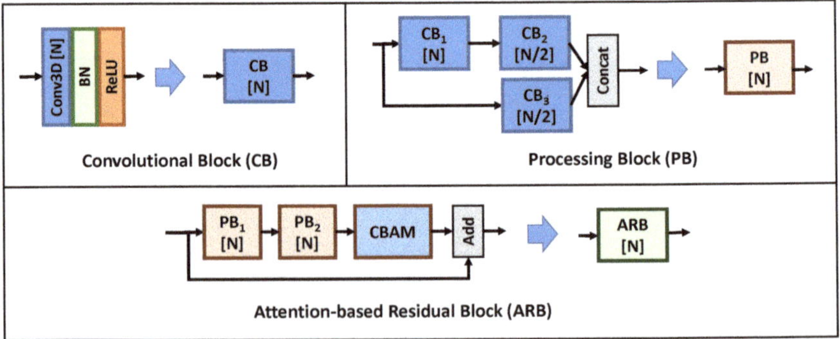

Figure 4. The layer structure of the three blocks used to build the proposed architecture: (**top-left**) Convolutional Block (CB); (**top-right**) Processing Block (PB); and (**bottom**) Attention-based Residual Block (APB), where the Convolutional Block Attention Module (CBAM) was proposed [59] and modified here as depicted in Figure 5.

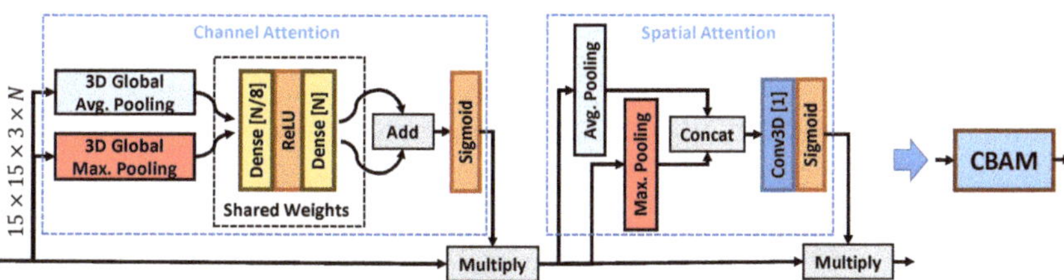

Figure 5. The layer structure of Convolutional Block Attention Module (CBAM), which uses both channel and spatial attention. The module was proposed in [59] and was modified here to compute the attention map for an MP volume.

Figure 3 shows that the AEQE-CNN architecture processes the EPI-based input patch using three stages. In the first stage, called EPI Pre-Processing, the MP volume corresponding to an EPI is processed using one CB block and one PB block, each equipped with $N/2$ filters, to extract the EPI feature maps, which are then concatenated and further processed by CB_5 and PB_5, which are both equipped with N filters. CB_5 uses the stride $s = (1, 1, 3)$ to reduce the current patch resolution from $15 \times 15 \times 9$ to $15 \times 15 \times 3$ to decrease the inference time and to reduce the MP neighbourhood from 9 MPs to 3 MPs.

In the second stage, called Attention-based Residual Processing, a sequence of four APB blocks with N filters are used to further process the patch and extract the final feature maps of size $15 \times 15 \times N$. The final stage, called CNN Refinement Computation, is used to extract the final CNN-refinement using one Conv3D layer with ReLU activation and one Conv2D layer (equipped with a 3×3 kernel) with one filter. The CNN-refinement is then added to the currently reconstructed MP to obtain the enhanced MP.

In this paper, we propose to employ an attention-based module designed based on the CBAM module introduced in [59]. Figure 5 depicts the layer structure of CBAM. CBAM proposes the use of both channel attention and spatial attention. The channel attention uses the shared weights of two dense layers to process the two feature vectors extracted using global average pooling and global maximum pooling, respectively. The spatial attention uses a Conv3D layer to process the feature maps extracted using average pooling and maximum pooling. The two types of attention maps are obtained using a sigmoid activation layer and then applied in turn using a multiplication layer. The CBAM block was proposed in [59] for the processing of two-dimensional patches, while, here, the CBAM design was modified to be applied to MP volumes (three-dimensional patches).

3.3. Training Details

The AEQE-CNN models were trained using the Mean Squared Error (MSE) loss function equipped with an ℓ_2 regularization procedure to prevent model over-fitting. Let us denote: $\Theta_{\text{AEQE-CNN}}$ as the set of all learned parameters of the AEQE-CNN model; $\mathbf{X}^{(i)}$ as the i-th EPI-based input patch in the training set of size $15 \times 15 \times 9 \times 4$; and $\mathbf{Y}^{(i)}$ as the corresponding MP in the original LF image of size 15×15. Let $F(\cdot)$ be the function that processes $\mathbf{X}^{(i)}$ using $\Theta_{\text{AEQE-CNN}}$ to compute the enhanced MP as $\hat{\mathbf{Y}}^{(i)} = F(\mathbf{X}^{(i)}, \Theta_{\text{AEQE-CNN}})$. The loss function is formulated as follows:

$$\mathcal{L}(\Theta_{\text{AEQE-CNN}}) = \frac{1}{L} \sum_{i=1}^{L} \|vec(\mathbf{Y}^{(i)}) - vec(\hat{\mathbf{Y}}^{(i)})\|_2^2 + \lambda \|\Theta_{\text{AEQE-CNN}}\|_2^2, \qquad (3)$$

where L is the number of input patches, λ is the regularization term that is set empirically as $\lambda = 0.001$, and vec is the vectorization operator. Here, the Adam optimization algorithm [60] is employed.

By setting $N = 32$, the AEQE-CNN models contain 782,661 parameters that must be trained. Experiments using a more lightweight AEQE-CNN architecture were also performed, see Section 4.4. Version *HM 16.18* of the reference software implementation is used for the HEVC codec [16]. Note that other software implementations of HEVC, such as FFmpeg [61], Kvazaar [62], and OpenHEVC [63,64] are available; however, in this work, the reference software implementation of HEVC was used due to its high popularity within the research community. The proposed CNN-based filtering method trained four AEQE-CNN models, one for each of the four standard QP values, $QP = \{22, 27, 32, 37\}$.

The proposed neural network was implemented in the Python programming language using the Keras open-source deep-learning library, and was run on a machine equipped with Titan Xp Graphical Processing Units (GPUs).

In our previous work [33,34], the experimental results showed that an improved performance was obtained when HEVC was modified to skip its built-in in-loop filters, DBF [47] and SAO [48]. Therefore, here, four models were trained using EPI-based input patches extracted from reconstructed LF images obtained by running HEVC with its built-in in-loop filters, called AEQE-CNN + DBF&SAO, and four models were trained using EPI-based input patches extracted from reconstructed LF images obtained by running HEVC without its built-in in-loop filters, called AEQE-CNN. This training strategy demonstrates that the proposed CNN-based filtering method can be integrated into video coding systems where no modifications to the HEVC anchor are allowed.

The proposed AEQE-CNN architecture differs from our previous architecture design named MP-wise quality enhancement CNN (MPQE-CNN) [34] as follows. MPQE-CNN operates on MP volumes extracted from the closest 3×3 MP neighbourhood, while AEQE-CNN operates on EPI-based input patches extracted from an 9×9 MP neighbourhood. MPQE-CNN follows a multi-resolution design with simple CB blocks, while AEQE-CNN follows a design of multi-EPI branch processing and sequential residual block processing built based on more efficient PB blocks and novel attention-aware ARB blocks.

4. Experimental Validation

Section 4.1 describes the experimental setup used to compare the proposed CNN-based filtering method with the state-of-the-art methods. Section 4.2 illustrates the experimental results obtained over the test. Section 4.3 presents the visual results of the proposed CNN-based filtering method in comparison with the HEVC anchor. Finally, Section 4.4 presents an ablation study that analyses the possibility to reduce the network complexity and runtime using different approaches.

4.1. Experimental Setup

LF image Dataset. The experimental validation was carried out on the EPFL LF dataset [35], which contained 118 LF images in the RGB format, divided into 10 categories. Similar to [34], here, only the first 8 bits of the RGB color channels were encoded, and, similar to [29], 32 corner SAIs (8 from each corner) were dropped from the array of SAIs as they contained sparse information due to the shape of the microlens used by the plenoptic camera. Since the SAIs were color-transformed to the YUV format and only the Y channel was enhanced, the SAI video sequence contained 193 Y-frames. The closest frame resolution that HEVC [16] accepted as input was $W \times H = 632 \times 440$.

For a fair comparison with MPQE-CNN [34], the experiments were carried out on the same Training set (10 LF images) and Test set (108 LF images) as defined in [34], i.e., the Training set contained the following LF images: *Black_Fence, Chain_link_fence_1, ISO_chart_1, Houses_&_lake, Backlight_1, Broken_mirror, Bush, Fountain_&_Vincent_1, Ankylosaurus_&_Diplodocus_1,* and *Bench_in_Paris.* A total number of $625 \times 434 \times 10 = 2{,}712{,}500$ EPI-based input patches were collected from the 10 training images, and a 90%–10% ratio was used for splitting the training set into training–validation data. A batch size of 350 EPI-based input patches was used.

Comparison with the state-of-the-art methods. The two proposed methods, AEQE-CNN + DBF&SAO and AEQE-CNN, were compared with *(i)* the HEVC [16] anchor, denoted by HEVC + DBF&SAO; *(ii)* the FQE-CNN architecture from [33] where each SAI in the LF image was enhanced in turn; and *(iii)* the MPQE-CNN architecture from [34] based on a similar MP-wise filtering approach. The distortion was measured using the Peak Signal-to-Noise Ratio (PSNR) and the Structural Similarity Index Measure (SSIM) [65]. The standard Bjøntegaard delta bitrate (BD-rate) savings and Bjøntegaard delta PSNR (BD-PSNR) improvement [66] were computed using the four standard QP values: $QP = \{22, 27, 32, 37\}$.

4.2. Experimental Results

Figure 6 shows the compression results over the test set (108 LF images) for the rate-distortion curves computed as Y-PSNR-vs.-bitrate and SSIM-vs.-bitrate. Figure 7 shows the Y-BD-PSNR and Y-BD-rate values computed for each LF image in the test set. The proposed methods provide an improved performance compared with HEVC [16] + DBF&SAO, FQE-CNN [33], and MPQE-CNN [34] at both low and high bitrates. The results show that AEQE-CNN provided a small improvement over AEQE-CNN + DBF&SAO. The proposed CNN-based filtering method was able to provide a large improvement even when no modification was applied to the HEVC video codec.

Table 1 shows the average results obtained over the test set. AEQE-CNN provided Y-BD-rate savings of 36.57% and Y-BD-PSNR improvements of 2.301 dB over HEVC [16], i.e., a more than 40% improvement was achieved compared with MPQE-CNN [33].

Table 1. Average results obtained over the test set.

Method	Bjøntegaard Metric	
	Y-BD-PSNR (dB)	**Y-BD-Rate (%)**
FQE-CNN [33]	0.4515	−9.1921
MPQE-CNN [34]	1.5478	−25.5285
AEQE-CNN + DBF&SAO	2.2044	−35.3142
AEQE-CNN	**2.3006**	**−36.5713**

Figure 8 shows the Rate-Distortion (RD) results for three randomly selected LF images in the test set, *Chain_link_fence_2*, *Flowers*, and *Palais_du_Luxembourg*. AEQE-CNN provided an Y-BD-PSNR improvement of around 2 dB at both low and high bitrates. The SSIM-vs.-bitrate results show that the visual quality at low bitrates was highly improved of around 0.08.

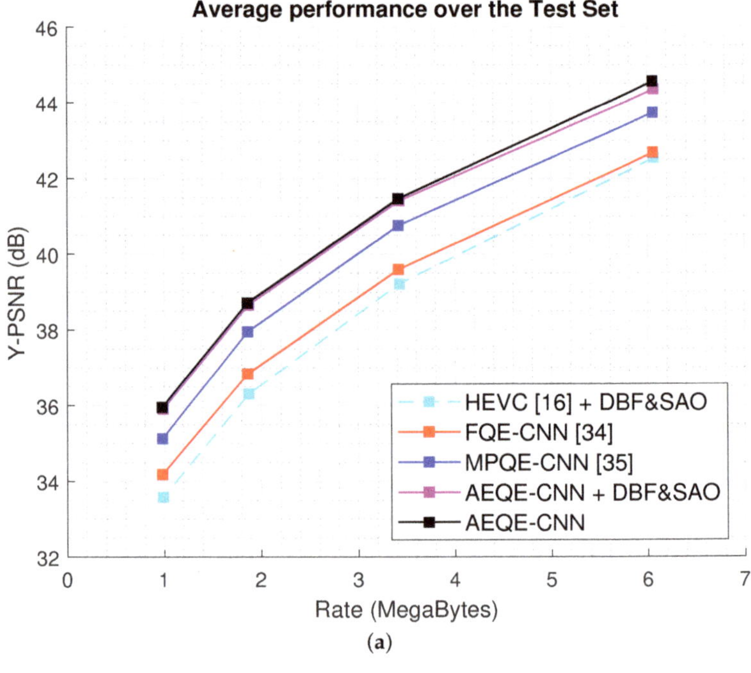

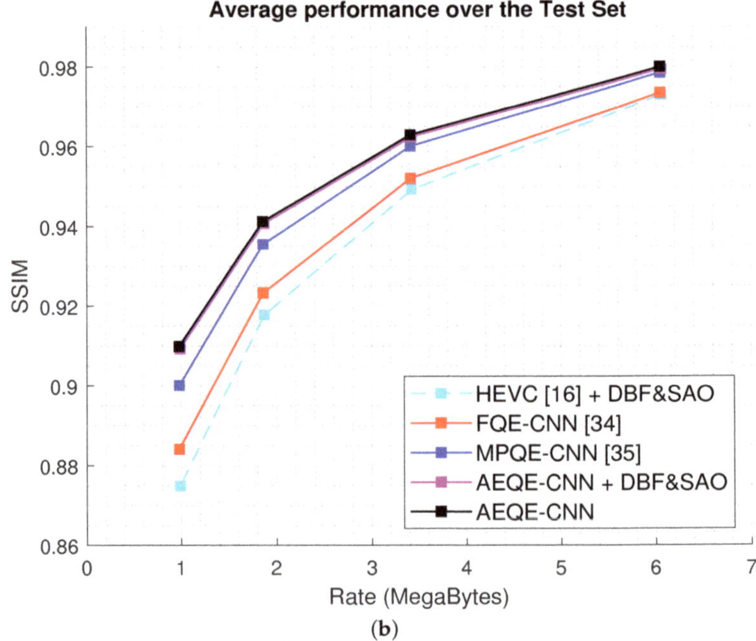

Figure 6. The Rate-Distortion results over the test set. (**a**) Y-PSNR-vs.-bitrate. (**b**) SSIM-vs.-bitrate.

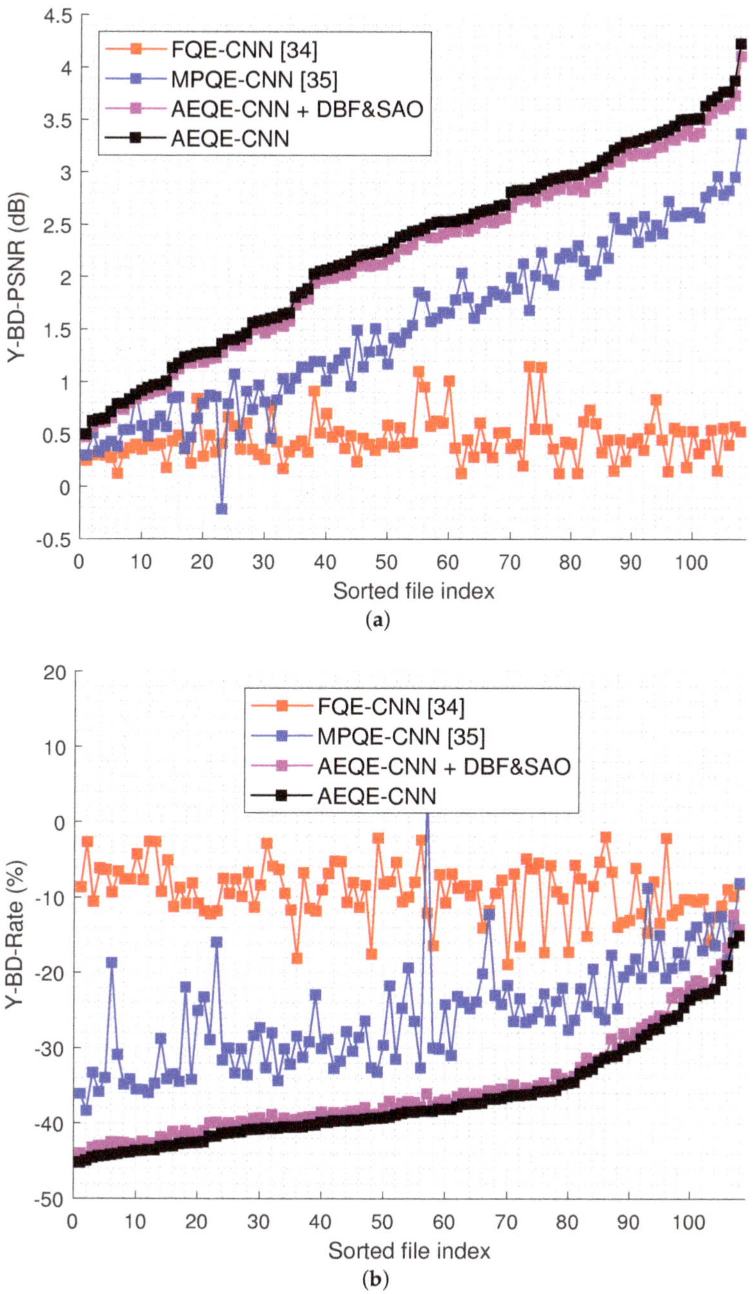

Figure 7. The Bjøntegaard metric results for every LF image in the test set: (**a**) Y-BD-PSNR gains (dB); (**b**) Y-BD-rate savings (%).

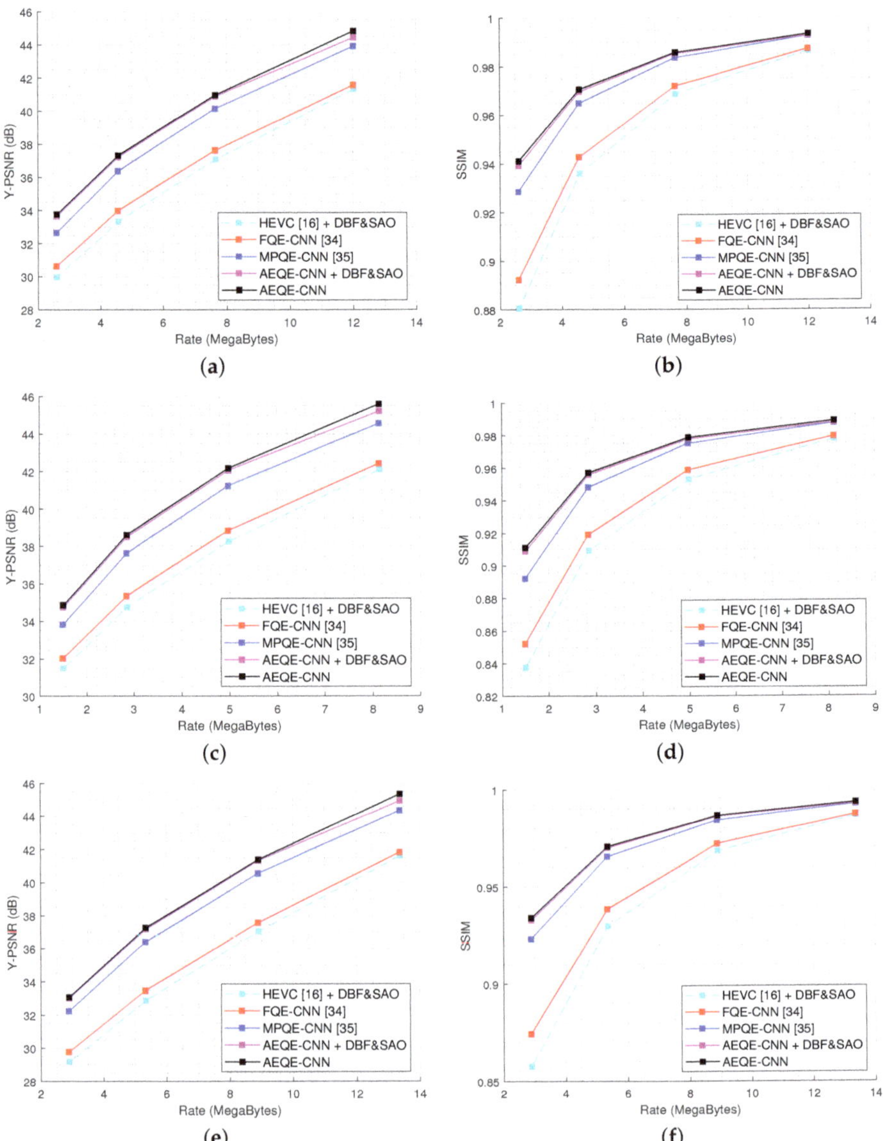

Figure 8. The Rate-Distortion results for three LF images in the test set. (**a**) Y-PSNR-vs.-bitrate for *Chain_link_fence_2*; (**b**) SSIM-vs.-bitrate for *Chain_link_fence_2*; (**c**) Y-PSNR-vs.-bitrate for *Flowers*; (**d**) SSIM-vs.-bitrate for *Flowers*; (**e**) Y-PSNR-vs.-bitrate for *Palais_du_Luxembourg*; (**f**) SSIM-vs.-bitrate for *Palais_du_Luxembourg*.

4.3. Visual Results

Figure 9 shows the pseudo-coloured image comparison between AEQE-CNN and HEVC [16] + DBF&SAO for two LF images in the test set, *Chain_link_fence_2* and *Flowers*. The green, blue, and red pixels mark the positions where AEQE-CNN provided an improved, similar, and worse performance, respectively, compared with HEVC [16] + DBF&SAO anchor. Green is the dominant color, which shows that AEQE-CNN enhanced the quality of almost all pixels in the LF image.

(a)

(b)

Figure 9. Pseudo-coloured image comparison between AEQE-CNN and HEVC [16] + DBF&SAO based on the absolute reconstruction error for the center SAI at position $(p, q) = (8, 8)$, and for $QP = 37$. Green marks the pixel positions where AEQE-CNN achieved better performance. Blue marks the pixel positions where the two methods had the same performance. Red marks pixels where HEVC [16] + DBF&SAO achieved better performance. The cyan rectangle marks an image area shown zoomed-in at the top-left corner and the corresponding Y channel in Figure 10. The results for two LF images in the test set: (**a**) *Chain_link_fence_2*; (**b**) *Flowers*.

Figure 10 shows the visual result comparison between AEQE-CNN and HEVC [16] + DBF&SAO for the corresponding Y channel of the two zoomed-in image areas marked by cyan rectangles in Figure 9. AEQE-CNN provided much sharper image edges and added more details to the image textures.

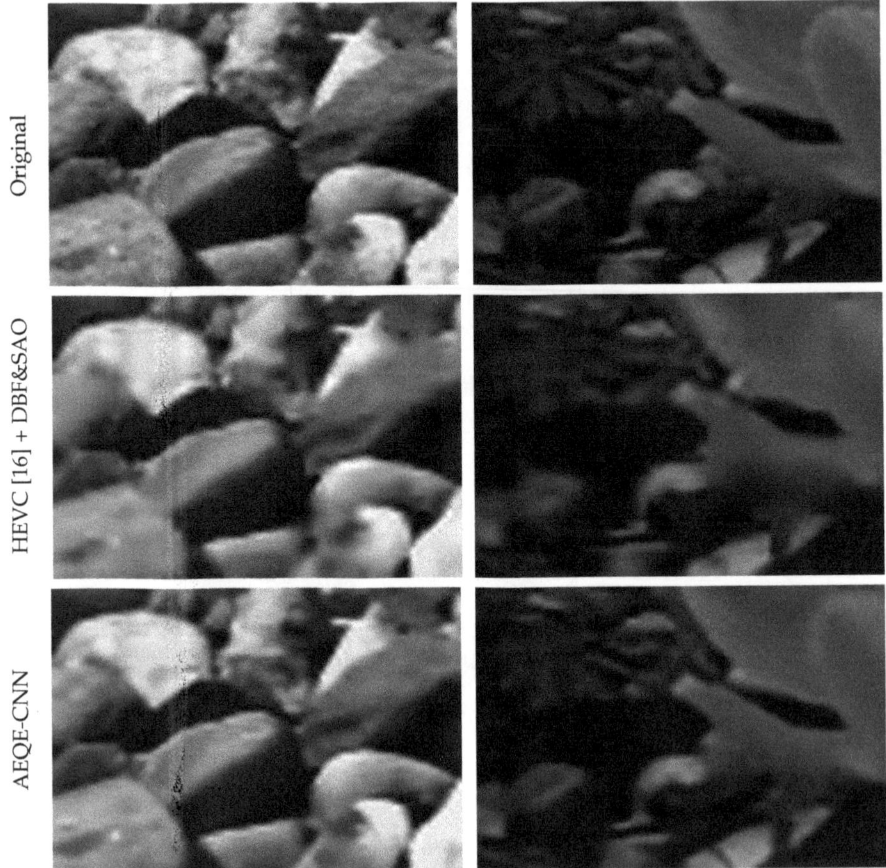

Figure 10. Visual comparison between AEQE-CNN and HEVC [16] + DBF&SAO for the Y channel of the zoomed-in image area marked by the cyan rectangle in Figure 9 above.

4.4. Ablation Study

In this work, we also studied the possibility to reduce the network complexity and runtime using two different approaches. In the first approach, an architecture variation of AEQE-CNN was generated by halving the number of channels used throughout the architecture by the 3D Convolution layers from $N = 32$ to $N = 16$. This first AEQE-CNN architecture variation is called AEQE-CNN [N=16]. In the second approach, the size of the MP neighbourhood, $\mathcal{N}_{x,y}$ (see Section 3.1), was reduced from 9×9 MPs (i.e., $b = 4$) to 3×3 MPs (i.e., $b = 1$).

More precisely, the same neighbourhood window as in [34] was used here with the goal of evaluating the influence of the size of the MP neighbourhood in the final enhancement results. In this case, the EPI volumes were of the size $15 \times 15 \times 3$; therefore, the CB_5 block in the AEQE-CNN architecture (see Figure 3) used a default stride of $s' = (1, 1, 1)$ instead of $s = (1, 1, 3)$. This second AEQE-CNN architecture variation is called AEQE-CNN [3×3].

Table 2 shows the average results obtained over the test set for the three AEQE-CNN architectures. The AEQE-CNN provided the best performance using the highest complexity and runtime. The network variations corresponding to the two approaches for complexity reduction still provided a better performance compared with the state-of-the-art methods and a close performance to AEQE-CNN. AEQE-CNN [N=16] offered a reduction of 44.6% in the inference runtime and a reduction of 74.7% in the network complexity, with a drop in the average performance of only 8.93% in Y-BD-PSNR and 3.59% in Y-BD-Rate.

Table 2. The average results obtained over the test set for the three AEQE-CNN network variations.

Method	Bjøntegaard Metric		Nr. of Trained	Inference Time
	Y-BD-PSNR	**Y-BD-Rate**	**Parameters**	**Per Img.**
AEQE-CNN [N=16]	2.0954 dB	−35.2581%	**197,661** (−74.7%)	**98 s** (−44.6%)
AEQE-CNN [3×3]	2.0799 dB	−35.0914%	782,661	105 s (−40.7%)
AEQE-CNN	**2.3006 dB**	**−36.5713%**	782,661	177 s

AEQE-CNN [3×3] offered a reduction of 40.7% in the inference runtime, with a drop in the average performance of only 9.6% in Y-BD-PSNR and of 4.05% in Y-BD-Rate. The ablation study demonstrate that AEQE-CNN [3×3] provided a large reduction in the network complexity and inference runtime while accepting a small performance drop compared with AEQE-CNN.

Figure 11 shows the rate-distortion curves computed over the test set for AEQE-CNN [N=16], AEQE-CNN [3×3], and AEQE-CNN. The results demonstrate again that the two network variations provided a close performance to AEQE-CNN. The performance dropped with less than 0.2 dB at low and high bitrates for the two architecture variations. The results obtained by AEQE-CNN [3×3] demonstrate that the proposed AEQE-CNN architecture, built using the PB and ARB blocks, provided an improved performance compared with the MPQE-CNN architecture [34] when operating on the same MP neighbourhood.

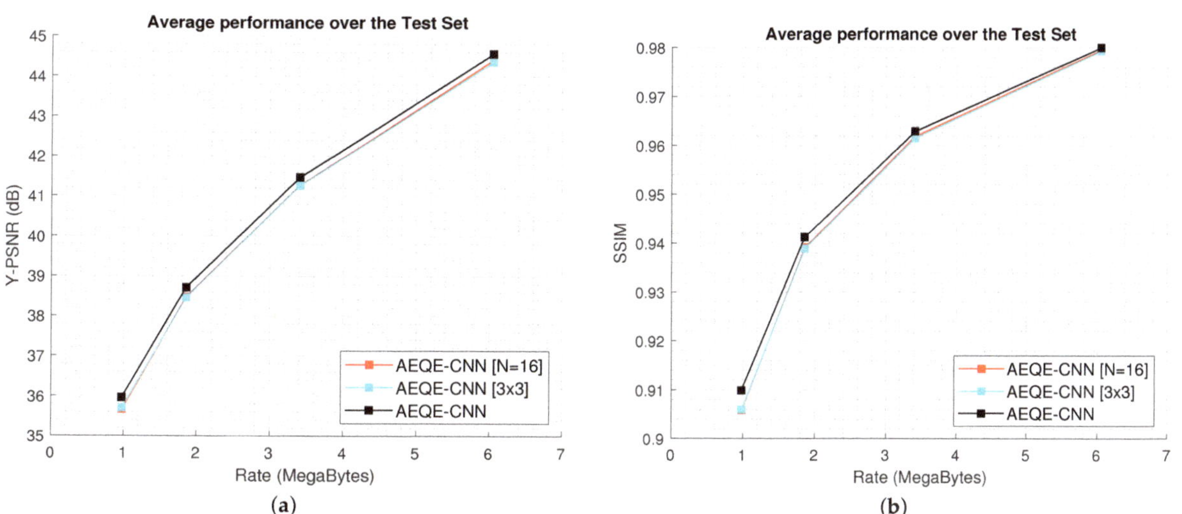

Figure 11. The Rate-Distortion results over the test set for the three network variations. (**a**) Y-PSNR-vs.-bitrate. (**b**) SSIM-vs.-bitrate.

Figure 12 shows the results of the Bjøntegaard metrics, Y-BD-PSNR and Y-BD-rate, computed for each LF image in the test set. The results demonstrate again that the two network variations provided a close performance to AEQE-CNN.

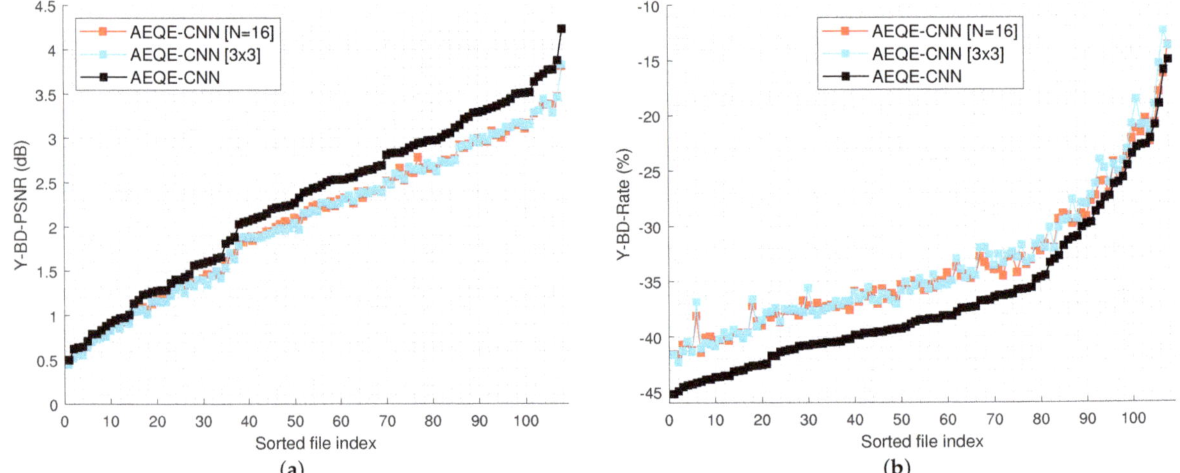

Figure 12. Bjøntegaard metrics results for every LF image in test set for the three network variations: (**a**) Y-BD-PSNR gains; (**b**) Y-BD-Rate savings.

5. Conclusions

In this paper, we proposed a novel CNN-based filtering method for the quality enhancement of LF images compressed by HEVC. The proposed architecture, AEQE-CNN, was built using novel layer structure blocks, such as complex processing blocks and attention-based residual blocks. AEQE-CNN operated on an EPI-based input patch extracted from an MP neighbourhood of 9×9 MPs and followed an MP-wise filtering approach that was specific to LF images. Similar to previous research works, the proposed AEQE-CNN filtering method provided an increased performance when the conventional HEVC built-in filtering methods were skipped. The results demonstrate the high potential of attention networks for the quality enhancement of LF images.

In our future work, we plan to study different strategies to reduce the inference runtime using lightweight neural network architectures, and to employ the CNN-based filtering method to enhance the quality of the light field images compressed using other video codecs, such as AV1 and VVC.

Author Contributions: Conceptualization, I.S.; methodology, I.S; software, I.S.; validation, I.S.; investigation, I.S.; resources, A.M.; writing—original draft preparation, I.S. and A.M.; writing—review and editing, I.S. and A.M.; visualization, I.S.; project administration, I.S. and A.M.; funding acquisition, I.S. and A.M. All authors have read and agreed to the published version of the manuscript.

Funding: This research work was funded by Innoviris within the research project DRIvINg, and by Ionut Schiopu's personal fund.

Conflicts of Interest: The authors declare no conflict of interest. The funders had no role in the design of the study; in the collection, analyses, or interpretation of data; in the writing of the manuscript; nor in the decision to publish the results. All authors read and approved the final manuscript.

References

1. Jeon, H.G.; Park, J.; Choe, G.; Park, J.; Bok, Y.; Tai, Y.W.; Kweon, I.S. Accurate depth map estimation from a lenslet light field camera. In Proceedings of the 2015 IEEE Conference on Computer Vision and Pattern Recognition (CVPR), Boston, MA, USA, 7–12 June 2015; pp. 1547–1555. [CrossRef]
2. Wang, T.C.; Efros, A.A.; Ramamoorthi, R. Depth Estimation with Occlusion Modeling Using Light-Field Cameras. *IEEE Trans. Pattern Anal. Mach. Intell.* **2016**, *38*, 2170–2181. [CrossRef]
3. Schiopu, I.; Munteanu, A. Deep-learning-based depth estimation from light field images. *Electron. Lett.* **2019**, *55*, 1086–1088. [CrossRef]

4. Rogge, S.; Schiopu, I.; Munteanu, A. Depth Estimation for Light-Field Images Using Stereo Matching and Convolutional Neural Networks. *Sensors* **2020**, *20*, 6188. [CrossRef]
5. Flynn, J.; Broxton, M.; Debevec, P.; DuVall, M.; Fyffe, G.; Overbeck, R.; Snavely, N.; Tucker, R. DeepView: View Synthesis With Learned Gradient Descent. In Proceedings of the 2019 IEEE/CVF Conference on Computer Vision and Pattern Recognition (CVPR), Long Beach, CA, USA, 16–20 June 2019; pp. 2362–2371. [CrossRef]
6. Peng, J.; Xiong, Z.; Zhang, Y.; Liu, D.; Wu, F. LF-fusion: Dense and accurate 3D reconstruction from light field images. In Proceedings of the 2017 IEEE Visual Communications and Image Processing (VCIP), St. Petersburg, FL, USA, 10–13 December 2017; pp. 1–4. [CrossRef]
7. Chen, M.; Tang, Y.; Zou, X.; Huang, K.; Li, L.; He, Y. High-accuracy multi-camera reconstruction enhanced by adaptive point cloud correction algorithm. *Opt. Lasers Eng.* **2019**, *122*, 170–183. [CrossRef]
8. Forman, M.C.; Aggoun, A.; McCormick, M. A novel coding scheme for full parallax 3D-TV pictures. In Proceedings of the 1997 IEEE International Conference on Acoustics, Speech, and Signal Processing, Munich, Germany, 21–24 April 1997; Volume 4, pp. 2945–2947. [CrossRef]
9. de Carvalho, M.B.; Pereira, M.P.; Alves, G.; da Silva, E.A.B.; Pagliari, C.L.; Pereira, F.; Testoni, V. A 4D DCT-Based Lenslet Light Field Codec. In Proceedings of the 2018 25th IEEE International Conference on Image Processing (ICIP), Athens, Greece, 7–10 October 2018; pp. 435–439. [CrossRef]
10. Chang, C.-L.; Zhu, X.; Ramanathan, P.; Girod, B. Light field compression using disparity-compensated lifting and shape adaptation. *IEEE Trans. Image Process.* **2006**, *15*, 793–806. [CrossRef]
11. Rüefenacht, D.; Naman, A.T.; Mathew, R.; Taubman, D. Base-Anchored Model for Highly Scalable and Accessible Compression of Multiview Imagery. *IEEE Trans. Image Process.* **2019**, *28*, 3205–3218. [CrossRef]
12. Jang, J.S.; Yeom, S.; Javidi, B. Compression of ray information in three-dimensional integral imaging. *Opt. Eng.* **2005**, *44*, 1–10. [CrossRef]
13. Kang, H.H.; Shin, D.H.; Kim, E.S. Compression scheme of sub-images using Karhunen-Loeve transform in three-dimensional integral imaging. *Opt. Commun.* **2008**, *281*, 3640–3647. [CrossRef]
14. Elias, V.; Martins, W. On the Use of Graph Fourier Transform for Light-Field Compression. *J. Commun. Inf. Syst.* **2018**, *33*. [CrossRef]
15. Hog, M.; Sabater, N.; Guillemot, C. Superrays for Efficient Light Field Processing. *IEEE J. Sel. Top. Signal Process.* **2017**, *11*, 1187–1199. [CrossRef]
16. Sullivan, G.; Ohm, J.; Han, W.; Wiegand, T. Overview of the High Efficiency Video Coding (HEVC) Standard. *IEEE Trans. Circuits Syst. Video Technol.* **2012**, *22*, 1649–1668. [CrossRef]
17. Ramanathan, P.; Flierl, M.; Girod, B. Multi-hypothesis prediction for disparity compensated light field compression. In Proceedings of the 2001 International Conference on Image Processing (Cat. No.01CH37205), Thessaloniki, Greece, 7–10 October 2001; Volume 2, pp. 101–104. [CrossRef]
18. Wang, G.; Xiang, W.; Pickering, M.; Chen, C.W. Light Field Multi-View Video Coding With Two-Directional Parallel Inter-View Prediction. *IEEE Trans. Image Process.* **2016**, *25*, 5104–5117. [CrossRef] [PubMed]
19. Conti, C.; Nunes, P.; Soares, L.D. New HEVC prediction modes for 3D holoscopic video coding. In Proceedings of the 2012 19th IEEE International Conference on Image Processing, Orlando, FL, USA, 30 September–3 October 2012; pp. 1325–1328. [CrossRef]
20. Zhong, R.; Schiopu, I.; Cornelis, B.; Lu, S.P.; Yuan, J.; Munteanu, A. Dictionary Learning-Based, Directional, and Optimized Prediction for Lenslet Image Coding. *IEEE Trans. Circuits Syst. Video Technol.* **2019**, *29*, 1116–1129. [CrossRef]
21. Dricot, A.; Jung, J.; Cagnazzo, M.; Pesquet, B.; Dufaux, F. Improved integral images compression based on multi-view extraction. In *Applications of Digital Image Processing XXXIX*; Tescher, A.G., Ed.; International Society for Optics and Photonics, SPIE: San Diego, CA, USA, 2016; Volume 9971, pp. 170–177. [CrossRef]
22. Astola, P.; Tabus, I. Coding of Light Fields Using Disparity-Based Sparse Prediction. *IEEE Access* **2019**, *7*, 176820–176837. [CrossRef]
23. Bahdanau, D.; Cho, K.; Bengio, Y. Neural Machine Translation by Jointly Learning to Align and Translate. *arXiv* **2016**, arXiv:1409.0473.
24. Zhu, M.; Jiao, L.; Liu, F.; Yang, S.; Wang, J. Residual Spectral–Spatial Attention Network for Hyperspectral Image Classification. *IEEE Trans. Geosci. Remote. Sens.* **2021**, *59*, 449–462. [CrossRef]
25. Wan, S.; Tang, S.; Xie, X.; Gu, J.; Huang, R.; Ma, B.; Luo, L. Deep Convolutional-Neural-Network-based Channel Attention for Single Image Dynamic Scene Blind Deblurring. *IEEE Trans. Circuits Syst. Video Technol.* **2020**. [CrossRef]
26. Fu, C.; Yin, Y. Edge-Enhanced with Feedback Attention Network for Image Super-Resolution. *Sensors* **2021**, *21*, 2064. [CrossRef]
27. Zhou, K.; Zhan, Y.; Fu, D. Learning Region-Based Attention Network for Traffic Sign Recognition. *Sensors* **2021**, *21*, 686. [CrossRef]
28. Lian, J.; Yin, Y.; Li, L.; Wang, Z.; Zhou, Y. Small Object Detection in Traffic Scenes Based on Attention Feature Fusion. *Sensors* **2021**, *21*, 3031. [CrossRef]
29. Schiopu, I.; Gabbouj, M.; Gotchev, A.; Hannuksela, M.M. Lossless compression of subaperture images using context modeling. In Proceedings of the 2017 3DTV Conf.: The True Vision—Capture, Transmission and Display of 3D Video (3DTV-CON), Copenhagen, Denmark, 7–9 June 2017; pp. 1–4. [CrossRef]

30. Schiopu, I.; Munteanu, A. Macro-Pixel Prediction Based on Convolutional Neural Networks for Lossless Compression of Light Field Images. In Proceedings of the 2018 25th IEEE International Conference on Image Processing (ICIP), Athene, Greece, 7–10 October 2018; pp. 445–449. [CrossRef]

31. Schiopu, I.; Munteanu, A. Deep-learning-based macro-pixel synthesis and lossless coding of light field images. *Apsipa Trans. Signal Inf. Process.* **2019**, *8*, e20. [CrossRef]

32. Schiopu, I.; Munteanu, A. Deep-Learning-Based Lossless Image Coding. *IEEE Trans. Circuits Syst. Video Technol.* **2020**, *30*, 1829–1842. [CrossRef]

33. Huang, H.; Schiopu, I.; Munteanu, A. Frame-wise CNN-based Filtering for Intra-Frame Quality Enhancement of HEVC Videos. *IEEE Trans. Circuits Syst. Video Technol.* **2020**. [CrossRef]

34. Huang, H.; Schiopu, I.; Munteanu, A. Macro-pixel-wise CNN-based filtering for quality enhancement of light field images. *Electron. Lett.* **2020**, *56*, 1413–1416. [CrossRef]

35. Rerabek, M.; Ebrahimi, T. New Light Field Image Dataset. Proc. Int. Conf. Qual. Multimedia Experience (QoMEX). 2016; pp. 1–2. Available online: https://infoscience.epfl.ch/record/218363/files/Qomex2016_shortpaper.pdf?version=1 (accessed on 1 July 2017).

36. Dong, C.; Deng, Y.; Loy, C.C.; Tang, X. Compression Artifacts Reduction by a Deep Convolutional Network. In Proceedings of the 2015 IEEE International Conference on Computer Vision (ICCV), Santiago, Chile, 7–13 December 2015; pp. 576–584. [CrossRef]

37. Cavigelli, L.; Hager, P.; Benini, L. CAS-CNN: A deep convolutional neural network for image compression artifact suppression. In Proceedings of the 2017 International Joint Conference on Neural Networks (IJCNN), Anchorage, AK, USA, 14–19 May 2017; pp. 752–759. [CrossRef]

38. Wang, Z.; Liu, D.; Chang, S.; Ling, Q.; Yang, Y.; Huang, T.S. D3: Deep Dual-Domain Based Fast Restoration of JPEG-Compressed Images. In Proceedings of the 2016 IEEE Conference on Computer Vision and Pattern Recognition (CVPR), Las Vegas, NV, USA, 27–30 June 2016; pp. 2764–2772. [CrossRef]

39. Galteri, L.; Seidenari, L.; Bertini, M.; Bimbo, A.D. Deep Generative Adversarial Compression Artifact Removal. In Proceedings of the 2017 IEEE International Conference on Computer Vision (ICCV), Venice, Italy, 22–29 October 2017; pp. 4836–4845. [CrossRef]

40. Ororbia, A.G.; Mali, A.; Wu, J.; O'Connell, S.; Dreese, W.; Miller, D.; Giles, C.L. Learned Neural Iterative Decoding for Lossy Image Compression Systems. In Proceedings of the 2019 Data Compression Conference (DCC), Snowbird, UT, USA, 26–29 March 2019; pp. 3–12. [CrossRef]

41. Dai, Y.; Liu, D.; Wu, F. A Convolutional Neural Network Approach for Post-Processing in HEVC Intra Coding. *Lect. Notes Comput. Sci.* **2016**, 28–39. [CrossRef]

42. Yang, R.; Xu, M.; Wang, Z.; Li, T. Multi-frame Quality Enhancement for Compressed Video. In Proceedings of the 2018 IEEE/CVF Conference on Computer Vision and Pattern Recognition, Salt Lake City, UT, USA, 18–22 June 2018. [CrossRef]

43. He, X.; Hu, Q.; Zhang, X.; Zhang, C.; Lin, W.; Han, X. Enhancing HEVC Compressed Videos with a Partition-Masked Convolutional Neural Network. In Proceedings of the 2018 25th IEEE International Conference on Image Processing (ICIP), Athene, Greece, 7–10 October 2018; pp. 216–220. [CrossRef]

44. Ma, C.; Liu, D.; Peng, X.; Wu, F. Convolutional Neural Network-Based Arithmetic Coding of DC Coefficients for HEVC Intra Coding. In Proceedings of the 2018 25th IEEE International Conference on Image Processing (ICIP), Athene, Greece, 7–10 October 2018; pp. 1772–1776. [CrossRef]

45. Song, X.; Yao, J.; Zhou, L.; Wang, L.; Wu, X.; Xie, D.; Pu, S. A Practical Convolutional Neural Network as Loop Filter for Intra Frame. In Proceedings of the 2018 25th IEEE International Conference on Image Processing (ICIP), Athene, Greece, 7–10 October 2018; pp. 1133–1137. [CrossRef]

46. Wan, S. CE13-Related: Integrated in-Loop Filter Based on CNN. JVET Document, JVET-N0133-v2, 2019. Available online: https://www.itu.int/wftp3/av-arch/jvet-site/2019_03_N_Geneva/JVET-N_Notes_d2.docx (accessed on 1 July 2020).

47. Norkin, A.; Bjontegaard, G.; Fuldseth, A.; Narroschke, M.; Ikeda, M.; Andersson, K.; Zhou, M.; Van der Auwera, G. HEVC Deblocking Filter. *IEEE Trans. Circuits Syst. Video Technol.* **2012**, *22*, 1746–1754. [CrossRef]

48. Fu, C.; Alshina, E.; Alshin, A.; Huang, Y.; Chen, C.; Tsai, C.; Hsu, C.; Lei, S.; Park, J.; Han, W. Sample Adaptive Offset in the HEVC Standard. *IEEE Trans. Circuits Syst. Video Technol.* **2012**, *22*, 1755–1764. [CrossRef]

49. Park, W.; Kim, M. CNN-based in-loop filtering for coding efficiency improvement. In Proceedings of the 2016 IEEE 12th Image, Video, and Multidimensional Signal Processing Workshop (IVMSP), Bordeaux, France, 11–12 July 2016; pp. 1–5. [CrossRef]

50. Zhang, Z.; Chen, Z.; Lin, J.; Li, W. Learned Scalable Image Compression with Bidirectional Context Disentanglement Network. In Proceedings of the 2019 IEEE International Conference on Multimedia and Expo (ICME), Shanghai, China, 8–12 July 2019; pp. 1438–1443. [CrossRef]

51. Li, F.; Tan, W.; Yan, B. Deep Residual Network for Enhancing Quality of the Decoded Intra Frames of Hevc. In Proceedings of the 2018 25th IEEE International Conference on Image Processing (ICIP), Athene, Greece, 7–10 October 2018; pp. 3918–3922. [CrossRef]

52. Lai, P.; Wang, J. Multi-stage Attention Convolutional Neural Networks for HEVC In-Loop Filtering. In Proceedings of the 2020 2nd IEEE International Conference on Artificial Intelligence Circuits and Systems (AICAS), Genova, Italy, 31 August–2 September 2020; pp. 173–177. [CrossRef]

53. Zhang, X.; Xiong, R.; Lin, W.; Zhang, J.; Wang, S.; Ma, S.; Gao, W. Low-Rank-Based Nonlocal Adaptive Loop Filter for High-Efficiency Video Compression. *IEEE Trans. Circuits Syst. Video Technol.* **2017**, *27*, 2177–2188. [CrossRef]

54. Zhang, Y.; Shen, T.; Ji, X.; Zhang, Y.; Xiong, R.; Dai, Q. Residual Highway Convolutional Neural Networks for in-loop Filtering in HEVC. *IEEE Trans. Image Process.* **2018**, *27*, 3827–3841. [CrossRef] [PubMed]

55. Jia, C.; Wang, S.; Zhang, X.; Wang, S.; Liu, J.; Pu, S.; Ma, S. Content-Aware Convolutional Neural Network for In-Loop Filtering in High Efficiency Video Coding. *IEEE Trans. Image Process.* **2019**, *28*, 3343–3356. [CrossRef]

56. Fraunhofer Institute for Telecommunications, Heinrich Hertz Institute (HHI). HEVC Reference Software. Available online: hevc.hhi.fraunhofer.de (accessed on 1 July 2019)

57. Bossen, F. Common HM Test Conditions and Software Reference Configurations. JCT-VC Document, JCTVC-G1100, 2012. Available online: https://www.itu.int/wftp3/av-arch/jctvc-site/2012_02_H_SanJose/JCTVC-H_Notes_dI.doc (accessed on 1 July 2017).

58. Ioffe, S.; Szegedy, C. Batch Normalization: Accelerating Deep Network Training by Reducing Internal Covariate Shift. *arXiv* **2015**, arxiv:1502.03167.

59. Woo, S.; Park, J.; Lee, J.Y.; Kweon, I.S. CBAM: Convolutional Block Attention Module. In Proceedings of the European Conference on Computer Vision (ECCV), Munich, Germany, 8–14 September 2018; pp. 3–19.

60. Kingma, D.; Ba, J. Adam: A Method for Stochastic Optimization. *arXiv* **2014**, arxiv:1412.6980.

61. FFmpeg. Libx265 Implementation of HEVC. Available online: http://ffmpeg.org (accessed on 1 April 2021).

62. Viitanen, M.; Koivula, A.; Lemmetti, A.; Ylä-Outinen, A.; Vanne, J.; Hämäläinen, T.D. Kvazaar: Open-Source HEVC/H.265 Encoder. In Proceedings of the 24th ACM International Conference on Multimedia, Amsterdam, The Netherlands, 15–19 October 2016; pp. 1179–1182. [CrossRef]

63. Hamidouche, W.; Raulet, M.; Déforges, O. 4K Real-Time and Parallel Software Video Decoder for Multilayer HEVC Extensions. *IEEE Trans. Circuits Syst. Video Technol.* **2016**, *26*, 169–180. [CrossRef]

64. Pescador, F.; Chavarrías, M.; Garrido, M.; Malagón, J.; Sanz, C. Real-time HEVC decoding with OpenHEVC and OpenMP. In Proceedings of the 2017 IEEE International Conference on Consumer Electronics (ICCE), Berlin, Germany, 3–6 September 2017; pp. 370–371. [CrossRef]

65. Wang, Z.; Bovik, A.C.; Sheikh, H.R.; Simoncelli, E.P. Image quality assessment: from error visibility to structural similarity. *IEEE Trans. Image Process.* **2004**, *13*, 600–612. [CrossRef] [PubMed]

66. Bjøntegaard, G. Calculation of average PSNR differences between RD-curves. In Proceedings of the ITU-T Video Coding Experts Group (VCEG) 13th Meeting, Austin, TX, USA, 2–4 April 2001; pp. 2–4.

 *sensors*

Article

DNet: Dynamic Neighborhood Feature Learning in Point Cloud

Fujing Tian, Zhidi Jiang and Gangyi Jiang *

Faculty of Information Science and Engineering, Ningbo University, Ningbo 315211, China;
tianfujing@foxmail.com (F.T.); jiangzhidi@nbu.edu.cn (Z.J.)
* Correspondence: jianggangyi@nbu.edu.cn

Abstract: Neighborhood selection is very important for local region feature learning in point cloud learning networks. Different neighborhood selection schemes may lead to quite different results for point cloud processing tasks. The existing point cloud learning networks mainly adopt the approach of customizing the neighborhood, without considering whether the selected neighborhood is reasonable or not. To solve this problem, this paper proposes a new point cloud learning network, denoted as Dynamic neighborhood Network (DNet), to dynamically select the neighborhood and learn the features of each point. The proposed DNet has a multi-head structure which has two important modules: the Feature Enhancement Layer (FELayer) and the masking mechanism. The FELayer enhances the manifold features of the point cloud, while the masking mechanism is used to remove the neighborhood points with low contribution. The DNet can learn the manifold features and spatial geometric features of point cloud, and obtain the relationship between each point and its effective neighborhood points through the masking mechanism, so that the dynamic neighborhood features of each point can be obtained. Experimental results on three public datasets demonstrate that compared with the state-of-the-art learning networks, the proposed DNet shows better superiority and competitiveness in point cloud processing task.

Keywords: point cloud; dynamic neighborhood; feature learning; attention mechanism; masking mechanism

Citation: Tian, F.; Jiang, Z.; Jiang, G. DNet: Dynamic Neighborhood Feature Learning in Point Cloud. *Sensors* **2021**, *21*, 2327. https://doi.org/10.3390/s21072327

Academic Editors: Kwong Tak Wu Sam, Yun Zhang, Xu Long and Tiesong Zhao

Received: 20 February 2021
Accepted: 23 March 2021
Published: 26 March 2021

Publisher's Note: MDPI stays neutral with regard to jurisdictional claims in published maps and institutional affiliations.

Copyright: © 2021 by the authors. Licensee MDPI, Basel, Switzerland. This article is an open access article distributed under the terms and conditions of the Creative Commons Attribution (CC BY) license (https://creativecommons.org/licenses/by/4.0/).

1. Introduction

With the rapid development of three dimensional (3D) sensing technologies, using deep learning to understand and analyze point clouds is becoming one of the important research topics [1–3]. As the output of 3D sensor, point cloud is composed of much number of points in 3D space. The neighborhood of point cloud is similar to the neighborhood of pixels in image, but point cloud does not have the regular grid structure as the image [4,5]. For learning-based point cloud processing, too large a neighborhood may lead to incorrect learning, but too small a neighborhood cannot ensure sufficient information being included for learning.

In recent years, deep learning has made great progress in point cloud classification and segmentation [6,7], and the existing methods can be roughly divided into the multi-view approach, the voxel approach, the graph convolution approach, and the point set approach. The multi-view approach projects point cloud to 2D plane from multiple angles to generate image data, then the traditional Convolutional Neural Network (CNN) is used for feature learning [8–10]. For this kind of approach, when the objects in the scene are obscured or the point density changes, the accuracy of object classification and segmentation will be reduced. The voxel approach converts point cloud into regular 3D meshes, and then processes the meshes with 3D convolutions [11,12]. However, the voxel approach is greatly limited because of the reduced resolution resulted from quantization, a large amount of data preprocessing and the computational complexity of 3D convolution. In addition, the voxels of point cloud will make 3D convolution away from the surface of the point cloud, leading to the loss of effective surface information. Riegler et al. [13] and Klokov et al. [14] used different spatial segmentation methods to solve the problem of computational complexity. However, these methods still rely on the accuracy of spatial segmentation and cannot effectively extract surface features of point cloud, thus they may lose some information of the fine-grained geometric manifolds.

Since the points in point cloud are similar to the nodes in graph, some works used graph convolution approaches to process the point cloud [15–19]. The graph convolution approach can be divided into the spectral convolution and the spatial convolution [7]. The spectral convolution method uses the eigenvector decomposition of the Laplacian matrix, and then obtains the global descriptor of the point cloud through network learning, based on which the classification and segmentation of the point cloud can be achieved [15]. Since the Laplacian matrix of each point cloud should be calculated, the computational cost is huge. At the same time, because the spectral convolution is associated with the Laplacian matrix, its generalization ability is weak. By contrast, the spatial convolution approach can directly perform convolution on the local neighborhoods of point cloud [16–19], and has high computational efficiency and strong generalization.

The point set approach can learn features from point cloud directly and efficiently. Qi et al. [20] designed the PointNet, which learns each point individually and uses a symmetric max-pooling function to maintain the permutation invariance of the points. In the PointNet, the network only considers every point itself, without combining its neighborhood information. To improve the PointNet, Qi et al. [21] further proposed the PointNet++ with a multi-scale mechanism to capture multi-scales local regions. Graham et al. [22] constrained the execution of volumetric convolution only along the input sparse set of active voxels of the grid. Hua et al. [23] put the points into a kernel unit, and then convolved the point cloud with the kernel weights. Su et al. [24] mapped the input data to a high-dimensional grid and processed it using bilateral convolution. Li et al. [25] proposed learning the X-transform from the input point cloud, and then obtained the invariant feature of point cloud permutation with traditional convolution. Huang et al. [26] designed the RSNet, which projects unordered points onto an ordered sequence of feature vectors through a slice pooling layer, and then used Recurrent Neural Network (RNN) to learn the sequence. Tchapmi et al. [27] combined trilinear interpolation and conditional random fields to perform segmentation on point clouds. Li et al. [28] simulated the spatial distribution of point cloud by establishing a self-organizing map (SOM), and then extracted the hierarchical features from SOM nodes. Huang et al. [29] used multi-scale point embedding, manifold learning and global graph-based optimization to deal with laser scanning point clouds.

For the point set approach, in order to learn the features of point cloud more effectively, many methods have been proposed. Wu et al. [6] regarded the convolution kernel as a non-linear function of local coordinates composed of a weight function and a density function, and then used it to convolve point cloud. Xu et al. [7] processed irregular data through the parameterized filters. Groh et al. [30] extended the traditional convolution to larger scale point cloud processing through exploring different parameterizations to generate the edge-dependent filters. Verma et al. [31] used soft-assignment matrices to extend traditional convolution into point cloud. Wang et al. [32] proposed a learnable operator to learn feature from non-grid structured data. Hermosilla et al. [33] proposed the density-based 3D convolution Markov approximation, which is used to learn the features of non-uniform point clouds. Shen et al. [34] defined the point set kernel as a set of learnable 3D points by measuring the geometric relationship between adjacent points, and then used the point set kernel to extract the feature of point cloud.

Although the methods mentioned above can be used to learn point clouds, most of them have the problem that feature extraction of local regions is rough because only simple regular range (such as the k-nearest neighborhood, spherical neighborhood, etc.) is defined as the neighborhood, without considering the semantics of the neighborhood. To solve this problem, this paper proposes a Dynamic neighborhood Network (DNet) with an adaptive selection strategy of the neighborhood. Firstly, the single-head structure is designed to obtain the attention weight of the neighborhood by learning the self-features, manifold features and neighborhood features of the point cloud. Then, the mask mechanism is used to remove some pseudo neighborhood points, and the dynamic neighborhood features are obtained. Finally, the multi-head structure is utilized to learn features in different neighborhood range so that multi-scale features can be obtained. The contributions of this paper are as follows:

- To learn the features of different scales of a point cloud, a multi-head structure is designed to effectively capture multi-scale features, and the Feature Enhancement Layer (FELayer) inside each head supplements the manifold features of local regions of the point cloud, so that each head can learn enough contextual information;
- An attention mechanism is proposed to obtain the contribution degree of each neighborhood point in a local region through learning the self-features, 2D manifold features and neighborhood features of the local region;
- A masking mechanism is designed to remove the pseudo neighborhood points that may mislead the neighborhood learning but keep the ones which are conducive to network understanding, so that the network can learn neighborhood features more reasonably and effectively.

The rest of this paper is organized as follows. Section 2 analyzes the motivation of this paper, and the proposed method is described in detail in Section 3. Section 4 gives the comparison results of the DNet and the state-of-the-art point cloud classification and segmentation networks. Section 5 concludes this paper.

2. Motivation

In this section, the works of point cloud neighborhood learning are reviewed. Then, the difference between the proposed attention mechanism and some traditional attention networks is introduced. Finally, the neighborhood problem worth thinking about and the motivation of this paper are put forward.

Local feature of point cloud is very important to understanding point cloud. For determining the neighborhood of a point in point cloud, most existing methods usually calculate the k-Nearest Neighbor (k-NN) points or use the spherical neighborhood with radius r, and then learn features on the neighborhood. For the neighborhood learning, PointNet++ [21] divided point cloud into multiple spherical neighborhoods to extract multi-scale context information. Wang et al. [35] proposed a dynamic graph CNN (DGCNN) to aggregate the features learned from local regions by calculating the k-NN points of each point. Thomas et al. [36] defined a new multi-scale neighborhood method of point cloud and maintained a reasonable point density in network learning. Weinmann et al. [37] defined the neighborhood of point cloud in advance, which is independent of network training. By contrast, the purpose of this paper is to select neighborhood points while training the network.

The non-adaptive neighborhood selection, such as the k-NN method and spherical neighborhood method, may result in pathological neighborhoods. Figure 1 shows two point cloud models with such pathological neighborhoods, where the k-NN method is used to find the neighborhood (marked as green points) of the red point, and the brown line indicates the geodesic distance from the red point to one of its pathological neighborhood points (the black point). For the red point at fishing rod in Figure 1a, its correct neighborhood points should also be points at the fishing rod, but not the points representing the fisherman. For the red point on a man's right knee in Figure 1b, the correct neighborhood points should be the points on the right knee, not the points on the left knee. Obviously, such pathological neighborhoods will lead to the network learning incorrect local information and further lead to pathological inferences. It is clear that discarding the pseudo neighborhood points with small Euclidean distance but large geodesic distance is helpful for the network to better understand the local surface information. Since surface-based geodesic topology is conducive to semantic analysis and geometric modeling of objects, He et al. [38] proposed deep geodesic networks for point cloud analysis.

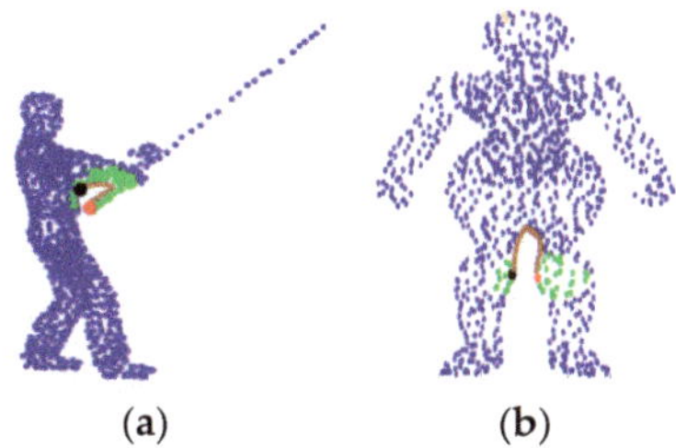

(a)   (b)

Figure 1. Examples of pathological neighborhood. (**a**) The pathological neighborhood of the fishing rod. (**b**) The pathological neighborhood of the knee.

Attention mechanism was used for weighting aggregation of point features in local regions [17,39–41], and it is also important for neighborhood learning. Chen et al. [17] used graph attention mechanism to learn local geometric representations of point clouds. Xie et al. [39] designed a self-attention module, which can realize the functions of feature transformation and feature aggregation. Feng et al. [40] proposed a Local Attention-Edge Convolution (LAE-Conv) to construct a local graph based on the neighborhood points searched in multi-directions. Xie et al. [41] used the local graph structure and the global graph structure to enhance the feature learning of point clouds. However, the traditional attention mechanism mainly focuses on using different features to obtain the weights of the neighborhood points, even for the pathological neighborhood as shown in Figure 1, such a kind of attention network also counts these pathological neighborhood points. By contrast, in this paper, the proposed attention mechanism will be used to evaluate the contribution degree of the neighborhood points, so as to filter out pseudo neighborhood points according to the evaluated contribution degree. Thus, it is necessary to consider which kind of features can be used to effectively obtain the contribution degree.

Figure 2 shows the neighborhoods obtained with two common methods, in which the green points are the neighborhood points of the red point. The two methods are the k-NN neighborhood, and the spherical neighborhood, respectively. As shown in Figure 2, for the red point at the wing of the aircraft, theoretically, the network is expected to learn the features of the edge of the aircraft wing, rather than the features of the plane of this region. Therefore, it is better to remove points on the plane of the wing as much as possible to reduce the impact of these points on the network, but retain points at the edge of the wing. This indicates that the following problems are worth to be considered:

(1) How to choose the number of points in a neighborhood, and whether the number of neighborhood points of all points in a point cloud should be equal.
(2) If the neighborhood is determined, whether all points in the neighborhood help to understand the point cloud.
(3) Do these neighborhood points contribute equally to the correct understanding of point clouds?

Considering the pathological neighborhood in Figure 1 and unreasonable neighborhood in Figure 2, the motivation of this paper starts from the following two points:

(1) When the point cloud has pathological neighborhood (as shown in Figure 1), the network is expected to have the ability of learning the correct neighborhood points and discarding the pseudo neighborhood point.
(2) When the center point is at the edge (as shown in the red point in Figure 2), the network is hoped to learn the edge features of the point cloud instead of the plane features.

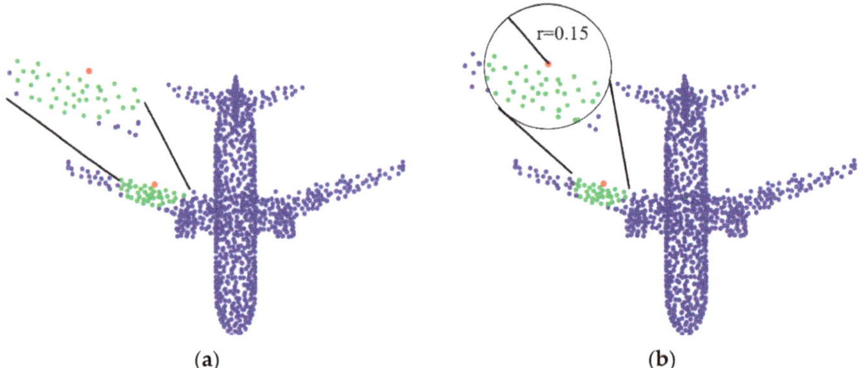

(a)   (b)

Figure 2. Neighborhoods obtained with two commonly used methods. (**a**) The *k*-NN neighborhood. (**b**) The spherical neighborhood with the radius *r* of 0.15.

3. The Proposed Network

Based on the above analyses, this paper propose a Dynamic neighborhood Network, denoted as DNet, to enhance neighborhood features learning for point cloud, so as to improve classification and segmentation of point cloud. Figure 3 shows the architecture of the DNet proposed in this paper, which has two branches: the classification sub-network and the segmentation sub-network. The core of the proposed DNet is a multi-head structure and its internal masking mechanism. Each head uses the attention mechanism to learn the contribution degree of each neighborhood point, and uses the masking mechanism to remove the neighborhood points with low contribution degree. Then, the weighted summation of the remaining neighborhood points is calculated to replace the maximum pooling of the neighborhood, so that the designed network has the ability to dynamically learn the effective neighborhood features of each point in the point cloud. Finally, multi-head structure composed of multiple single-head structures is used to learn multiple effective neighborhood features which are stacked as the final feature for subsequent point cloud classification and segmentation tasks.

Here, the neighborhood convolution of point cloud is first defined. Then, the multi-head structure in the proposed DNet is designed and its internal masking mechanism is described. Finally, the working principle and loss function of DNet are described.

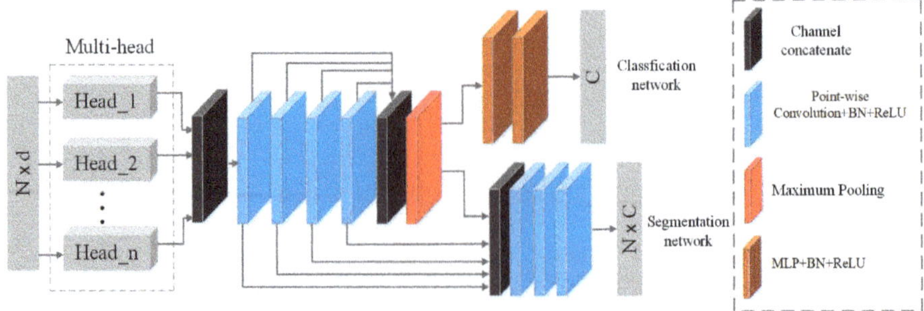

Figure 3. The architecture of the proposed Dynamic neighborhood Network (DNet).

3.1. Neighborhood Convolution

Given an unordered point set P in 3D space as a point cloud, where $P = \{P_i \mid i = 1, \ldots, n\}$, $P_i \in R^d$ (generally, $d = 3$), which is the coordinate of the i-th point, denoted as $P_i = \{x, y, z\}$, and n is the number of points in the point cloud. Then, let $N_{all}(P_i)$ denote the neighborhood of the point P_i, $N_{all}(P_i) = \left\{ P_i^j \middle| j = 1, \cdots, k \right\}$, where P_i^j is the j-th neighborhood point of P_i, and k is the number of neighborhood points of the point P_i. Since it is easy for the k-NN method to quickly construct a neighborhood graph, the k-NN neighborhood is used as the initial neighborhood in the proposed DNet. For the constructed neighborhood graph of P_i, neighborhood learning can be performed on all points of $N_{all}(P_i)$ to obtain the feature $F_{all}(P_i)$ with respect to the point P_i as follows

$$F_{all}(P_i) = Max(\sigma(h_\theta(P_i^j))), \quad \forall P_i^j \in N_{all}(P_i) \tag{1}$$

where $Max(\cdot)$ is the max-pooling operation, $\sigma(\cdot)$ is the activation function, and $h_\theta(\cdot)$ is point-wise convolution with a set of learnable parameters θ. For 2D image, $h_\theta(\cdot)$ can be a convolution kernel with the size of 3×3 and 5×5. However, for point cloud, since it is unstructured, $h_\theta(\cdot)$ is a convolution kernel with the size of 1×1, which is called as point-wise convolution [20].

In order to make Equation (1) more generalized, it is modified as follows

$$F_{all}(P_i) = A(\sigma(h_\theta(P_i^j, Oth))), \quad \forall P_i^j \in N_{all}(P_i) \tag{2}$$

where $A(\cdot)$ is the aggregation function (such as the max-pooling, summing, averaging, etc.). "*Oth*" represents some additional information such as the density of the local region, the 3D Euclidean distance from the neighborhood point to the center point P_i, etc. [35].

The traditional network only conducts neighborhood learning from all points of $N_{all}(P_i)$ in the local region, no matter whether the points in the neighborhood are suitable or not. Therefore, this work tries to remove some of the points in the neighborhood $N_{all}(P_i)$ through network learning, so as to adaptively obtain an effective neighborhood of the point P_i, namely $N_{eff}(P_i) = \left\{ P_i^j \middle| j = 1, \cdots, m \right\}$, $m \leq k$. Thus, the more effective feature $F_{eff}(P_i)$ of the point P_i can be learned as follows

$$F_{eff}(P_i) = A(\sigma(h_\theta(P_i^j))), \quad \forall P_i^j \in N_{eff}(P_i) \text{ and } N_{eff}(P_i) \subseteq N_{all}(P_i) \tag{3}$$

As an example, Figure 4 shows the feature learning with two different neighborhood methods, where the green and orange points mark the neighborhood of the red point. In the figure, since the red point is located at the edge of the airplane wing, the feature of the red point should reflect the characteristics of the wing edge. It is seen that for the $N_{all}(P_i)$, which is selected with k-NN method, some of the neighborhood points are not suitable for the feature learning of the wing edge. By contrast, the effective neighborhood $N_{eff}(P_i)$ marked as the orange is more helpful for learning the features of wing edge. In other words, $N_{eff}(P_i)$ is more expected for feature learning of the edge of the airplane wing compared with $N_{all}(P_i)$.

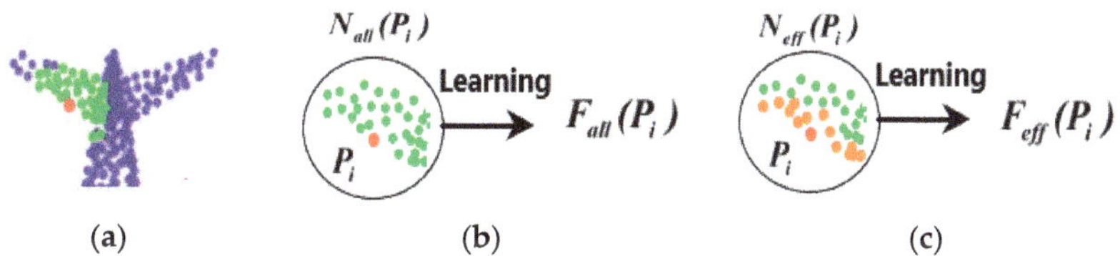

Figure 4. The diagram of the neighborhood learning. (**a**) Aircraft tail. (**b**) Feature learning on the k-NN neighborhood of the point P_i. (**c**) Feature learning on the effective neighborhood (the orange points) which is more appropriate for feature learning of the edge of the wing.

3.2. Multi-Head Structure

The proposed DNet utilizes the attention mechanism and masking mechanism to learn the more effective feature $F_{eff}(P_i)$. The main modules in the proposed DNet are the multi-head structure, which allows the network to learn information of different neighborhood ranges of the point clouds, that is, multi-scale features, so as to obtain sufficient context information and stabilize the network. Given a point cloud P, the effective feature $F(P)$ of the point cloud learned by the multi-head structure can be expressed as follows

$$F(P) = \overset{m}{\underset{t=1}{||}} F_{eff}(P)^{(t)} \tag{4}$$

where $||$ is the multi-channel cascade operation, m is the number of heads, and $m = 3$ in this paper, $F_{eff}(P)^{(t)}$ denotes the effective feature learned by the t-th head.

The proposed multi-head structure does not need to manually set multi-scale receptive fields as in [21]. For each head, as long as the number of initial neighborhood points in a neighborhood is set, an adaptive masking mechanism inside the heads will spontaneously filter out the neighborhood points with low contribution to obtain the features of different neighborhood ranges.

After designing the structure that captures multi-scale features, the next task is how to design the structure of each head so that it can select effective points in the neighborhood to promote network understanding of point cloud. Figure 5 shows the designed single-head structure. The attention mechanism can be used to obtain the feature of a point by weighted aggregation of features of the point's neighborhood points. The attention mechanism will be used to assign a contribution degree to each point in the neighborhood, which indicates the contribution of the point to the learning of this local region. Therefore, the contribution of the neighborhood points can be identified according to the attention mechanism, based on which an adaptive masking mechanism can be designed. For a point $P_i \in P$ with the neighborhood $N_{all}(P_i)$, the effective feature $F_{eff}(P_i)$ of the point P_i can be defined as follows

$$F_{eff}(P_i) = \sum_{j=1}^{k} M_i^j \cdot \alpha_i^j \cdot \overline{F}_i^j + b_i \tag{5}$$

where α_i^j is the contribution degree of the neighborhood point learned by the network, b_i is the bias term, and M_i^j denotes an adaptive mask determined by the contribution degrees of the neighborhood. $\overline{F}_i^j$ is the integration feature that needs to be multiplied with the mask, it is composed of neighborhood features and manifold features, and defined as follows

$$\overline{F}_i^j = h_\theta(F_i^j \oplus h_\theta(C(P_i^j))) \tag{6}$$

where $\oplus$ represents channel concatenate, $C(P_i^j)$ is the coding feature of P_i^j, and $h_\theta(C(P_i^j))$ is the manifold features of P_i^j. $h_\theta(C(P_i^j))$ is extracted from FELayer, which contains an autoencoding and point-wise convolution.

In order to establish the connection between different local regions, the covariance feature of the local region is added for each point P_i^j in the local region, and F_i^j can be expressed as follows

$$F_i^j = h_\theta(COV(N_{all}(P_i)) \oplus P_i^j) \tag{7}$$

In probability theory, covariance is used to measure the error between different variables, because it can well represent the statistical characteristics of the local regions. Therefore, the 3×3 covariance matrix of each region is calculated, and flattened to get a 9-dimensional covariance feature $COV(N_{all}(P_i))$, then it is concatenate with each point in the neighborhood to obtain the 12-dimensional data, which extends the neighborhood features of the point cloud.

The contribution degree α_i^j of the point P_i is obtained through the feature $\widetilde{F}_i^j$. $\widetilde{F}_i^j$ learned inside each head is composed of two parts: the self-features F_i and integration feature $\overline{F}_i^j$. Therefore, $\widetilde{F}_i^j$ can be denoted as follows

$$\widetilde{F}_i^j = F_i \oplus \overline{F}_i^j \tag{8}$$

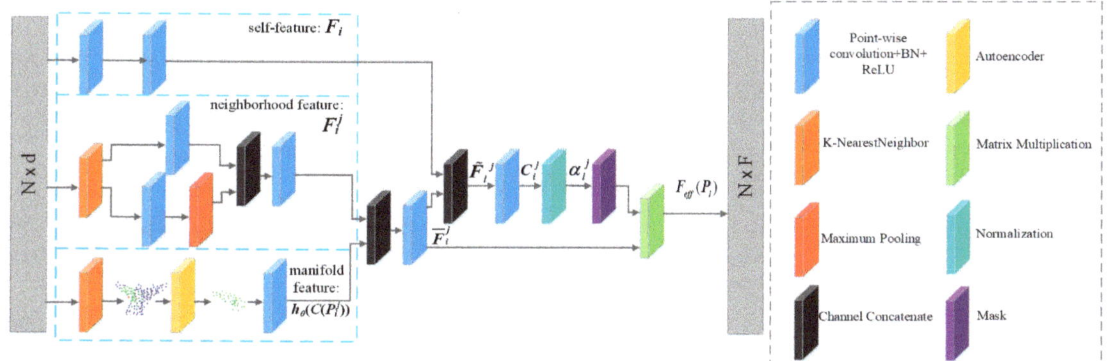

Figure 5. Single-head structure.

Then, for the point P_i and its neighborhood point P_i^j, the weight C_i^j of the neighborhood point P_i^j is learned through the single-head structure as follows

$$C_i^j = h_\theta(\widetilde{F}_i^j) \tag{9}$$

Finally, in order to better compare the attention coefficients C_i^j, it is normalized as the contribution degree of the neighborhood points, which is defined as follows

$$\alpha_i^j = \frac{\exp(C_i^j)}{\sum\limits_{l=1}^{k} \exp(C_i^l)} \tag{10}$$

where $exp(\cdot)$ is an exponential function, and k is the number of neighborhood points.

In order to better understand the multi-head structure, Figure 6 shows the contribution degree of neighborhood points when the center point (red point) is an edge point. The contribution degree indicates how much the network learns from the neighborhood points of the red point.

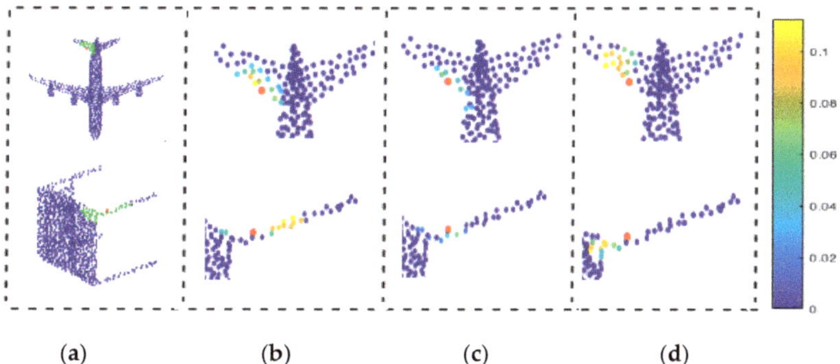

    (a)   (b)   (c)   (d)

Figure 6. The contribution degree output from the three heads when the center point is an edge point. (**a**) Two point cloud models. (**b**) The first head. (**c**) The second head. (**d**) The third head.

In Figure 6, as shown in the right colored bar, the closer the color of a neighborhood point is to yellow, the more features the network learns from the neighborhood point when processing the local region of the red point. Figure 6a shows the input models in which the green points indicate the initial neighborhood of the red point. Figure 6b–d show the contribution of the neighborhood points learned by the three heads to the center point. It is clear that the neighborhood range learned by each head is different. From the Figure 6, there are two points worth noting. Firstly, it is not that the closer the neighborhood point is to the red point, the more important it is; secondly, since the red point is at the edge of the airplane wing, the contribution degree of other edge points is significantly higher than that of the point on the wing plane. This indicates that the network is more willing to learn local features that are conducive to understand point clouds.

3.3. Masking Mechanism

As an important part of the multi-head structure, the masking mechanism is adopted to filter out the pseudo neighborhood points in the initial neighborhood so that the proposed network can learn neighborhood features more effectively. The adaptive mask M_i^j in Equation (5) can be expressed as follows

$$M_i^j = \begin{cases} 0, & if \quad \alpha_i^j < T \\ \alpha_i^j, & otherwise \end{cases} \tag{11}$$

where T is a threshold of the mask. The threshold can be obtained by different methods (e.g., the mean value of the weight of neighborhood points). If the contribution degree of a neighborhood point is less than the threshold, the point is regarded as the pseudo neighborhood point and will be removed from the neighborhood; otherwise, the neighborhood point is retained.

Assume that the dimension of the input point cloud is $(n, 3)$, where n is the number of points with 3D coordinate (x, y, z). Ideally, the network is expected to be able to select k_i neighborhood points of P_i for effective neighborhood learning, and k_i is different for the different center point P_i. However, because the shape of the convolution kernel is fixed, the network cannot handle irregular data. For example, if the first point has 10 neighborhood points with the shape of (1,10,3), while the second point has 20 neighborhood points with the shape of (1,20,3), the network cannot stack these two points for learning. However, if both of the shapes of the two points is (1,20,3), the network can stack the two points into the shape of (2,20,3). Therefore, in this paper, the number of initial neighborhood points is fixed to k, and the mask M_i^j is used to remove the pseudo neighborhood points from the neighborhood since these points are not conducive to the network learning of the local region.

The traditional neighborhood learning methods do not consider the geodesic information, which may result in pathological neighborhood, as shown in Figure 1. By contrast, GeoNet [38] learns the point cloud with geodesic information to avoid learning pathological neighborhood features. For the proposed DNet, it can use the mask M_i^j to remove the neighborhood points with low contribution so that more effective neighborhood features can be learned even if only coordinate information of point cloud is available. This can effectively prevent the network from learning pathological region features such as the body or another knee in Figure 7a, where the green points are the initial neighborhood points of the red point. Figure 7b–d show the neighborhood points selected by the first, second and third heads, respectively. It can be seen from Figure 7 that the masking mechanism shields many pseudo neighborhoods points with large geodesic distance, thereby it effectively summarizes the neighborhood. Instead, if the pseudo neighborhood is not shielded by mask, the point cloud learning network will learn the wrong neighborhood information, leading to a decrease in the accuracy of classification or segmentation.

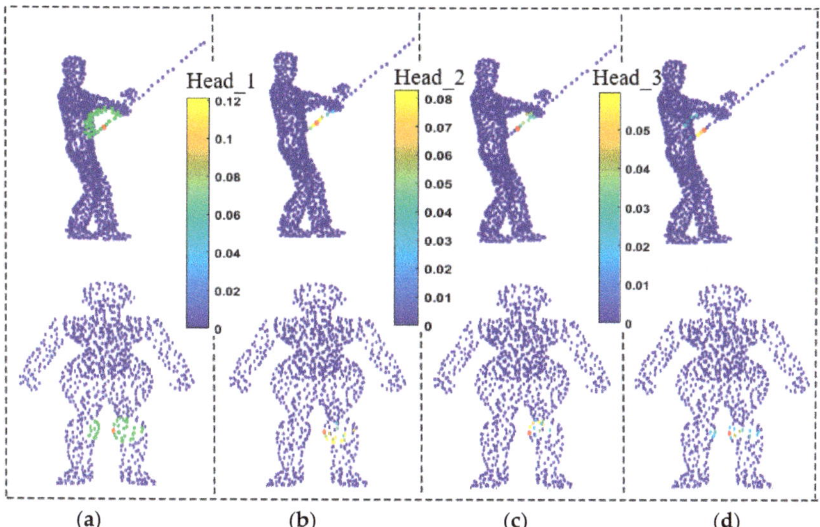

(**a**)   (**b**)   (**c**)   (**d**)

Figure 7. Neighborhood learning of multi-head structure under pathological conditions. (**a**) Two point cloud models. (**b**) The first head. (**c**) The second head. (**d**) The third head.

3.4. Learning with DNet

The architecture of the proposed DNet in Figure 3 can be used for point cloud classification (the upper branch) or segmentation (the lower branch). The point cloud classification sub-network in Figure 3 takes the coordinates of the whole point cloud as the input of the network, and after extracting multi-scale effective neighborhood features, it aggregates the point features through the max-pooling to output the classification results. The point cloud segmentation sub-network in Figure 3 concatenates global features with shallow features and outputs the segmentation results.

The core of the network consists of three heads, each of which can learn local information of different neighborhood ranges. Inside each head, the original local 3D space coordinates are used as the input, and the effective neighborhood features are learned as the output. The head obtains the attention weight of the neighborhood points by learning self-features, manifold features and neighborhood features. Then, the mask is used to remove some pseudo neighborhood points to obtain dynamic neighborhood features. Finally, a multi-head structure is used to learn multiple effective neighborhood features and stack them as the final feature for subsequent classification and segmentation tasks.

3.5. Loss Function

In this paper, an autoencoder is used to extract the 2D manifold features of the point clouds. Usually, for reconstruction networks whose purpose is to reconstruct the entire point cloud model, the complex loss of Chamfer Distance (CD) or Earth Mover's distance (EMD) are used as the loss function because of the disorder of point cloud. However, the task of this paper is not to reconstruct the entire point cloud model, but to roughly reconstruct the shape of the local neighborhood so as to extract the 2D manifold features of the point clouds. Therefore, since the local neighborhood is generally with simple topological structure, a simple L2 loss function is used in this work, and expressed as follows

$$L_{AE} = \sum_{i=1}^{n}\sum_{j=1}^{k}\left(P_i^j - \widetilde{P}_i^j\right)^2 \tag{12}$$

where $\widetilde{P}_i^j$ is the reconstructed point of P_i^j.

Figure 8 illustrates the effectiveness of the autoencoder with L2 loss function. We draw a grid in the figure to distinguish 3D points from 2D points. Figure 8a is the original input 3D point cloud, and Figure 8b enlarges the green local neighborhood in Figure 8a. Figure 8c depicts the result of using an autoencoder to compress Figure 8b to a 2D plane, and Figure 8d depicts the 3D points reconstructed from Figure 8c. It is clear that even though the simple L2 loss function is used instead of the more complex loss function in the autoencoder, the shape of the reconstructed 3D points is similar with that of the original shape of the local neighborhood.

Let y be the label of point cloud classification or segmentation, and $\hat{y}$ be the prediction result of DNet. The loss function of point cloud classification or segmentation is $L_{task} = -y \cdot \log(\hat{y})$, the final loss function of the proposed DNet is defined by

$$L_{total} = L_{task} + L_{AE} \tag{13}$$

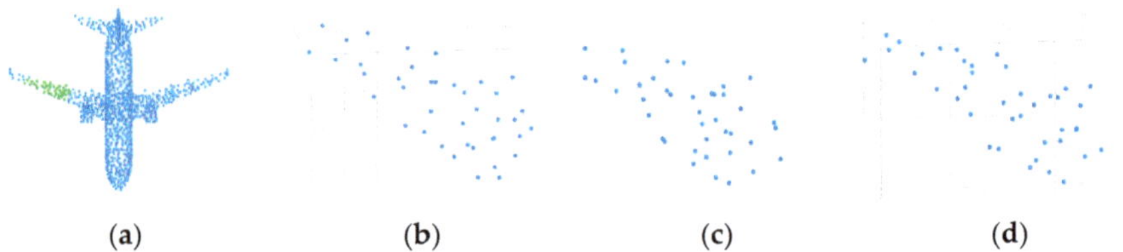

(a)   (b)   (c)   (d)

Figure 8. Compression and reconstruction of point cloud by using autoencoder with L2 loss function. (**a**) 3D point cloud. (**b**) 3D points at the green area of (**a**). (**c**) Compressed result of (**b**). (**d**) 3D points reconstructed from (**c**).

4. Experimental Results and Discussions

In this section, the training configuration of the networks is first introduced, and then the proposed DNet is tested on the benchmark dataset ModelNet40 [42] for point cloud classification, and on the benchmark datasets ShapeNet [43] and S3DIS [44] for point cloud segmentation, compared with other deep learning networks.

4.1. Network Training

The proposed DNet is constructed on Tensorflow, and the experiments are implemented on a computer with Intel Core I7-7820X CPU (3.6 GHz, 128GB memory) and GeForce RTX2080Ti GPUs. For the point cloud classification, 1024 points are uniformly sampled from the 3D grid of each point cloud as the network input, and the number of initial neighborhood points, that is, k, is set to 40. For part segmentation and indoor segmentation of point cloud, the number of input points of the DNet is 2048 and 4096, respectively, and k is set to 50. For the multi-head structure, in total three heads are used, and the output dimension of each head is 16. During the training phase, Adaptive Moment Estimation (ADAM) solver is used with the base learning rate of 0.001, the learning rate decay is executed every 40 epochs. ReLU and batch normalization are applied after each layer except the last fully connected layer. For the classification dataset, 200 epochs are trained with the batchsize of 32; while for the segmentation datasets, 100 epochs are trained with the batchsize of 16.

4.2. Point Cloud Classification

The performance of the proposed DNet on point cloud classification is tested on the ModelNet40 dataset [42]. This dataset contains 40 categories, including beds, chairs, airplanes, etc., with a total of 12,311 3D mesh models. In the experiments, 9843 models in the ModelNet40 dataset are used as the training set, while the remaining 2468 models constitute the testing set. For each model, 1024 points are uniformly sampled from the grid model and normalized into the unit circle. During the training, data augmentation techniques are used to scale point clouds in the range of [0.8, 1.25] and translate the point clouds in the range of [−0.1, 0.1].

Table 1 shows the classification results of the proposed DNet compared with the other sixteen advanced networks. As shown in the "input" column of Table 1, the methods, including the Spec-GCN [15], Pointconv [6], AGCN [41], PointNet++ [21], SpiderCNN [7] and SO-Net [28], require coordinates of point cloud as well as normal information as the input of their networks, while the other eleven comparison networks and the proposed DNet only need the coordinates of point cloud. Moreover, the networks listed in the last three (PointNet++ [21], SpiderCNN [7] and SO-Net [28]) for comparison use 5k points, rather than 1k points as other networks do. To evaluate the performance of different networks, the mean accuracy of each class of point cloud classification (mA) and the overall accuracy of point cloud classification (OA) are used, as shown in Table 1. It can be found that the proposed DNet has achieved good results. However, for most of the networks in Table 1, their focus is not on the effective neighborhood selection, which is emphasized by the proposed DNet. Therefore, in order to make a fairer comparison, the proposed DNet is mainly compared with DGCNN [35] and the PointNet++ [21] without normal information, because DGCNN also utilizes the k-NN neighborhood while PointNet++ adopts a spherical neighborhood. Table 1 shows that in terms of OA, the proposed DNet has 1.4% and 2.9% improvement over the DGCNN and the PointNet++ without normal information, respectively. It illustrates the importance of effective neighborhood selection for feature learning in the learning-based point cloud classification methods.

Table 1. Classification accuracy of different networks (%). (mA and OA denote the mean accuracy of each class of point cloud classification and the overall accuracy of point cloud classification, respectively.).

Method	Input	Points	mA	OA
Pointwise-CNN [23]	xyz	1k	81.4	86.1
ECC [18]	xyz	1k	-	87.4
PointNet [20]	xyz	1k	86.2	89.2
SCN [39]	xyz	1k	87.6	90.0
Kd-Net [14]	xyz	1k	86.3	90.6
PointNet++ [21]	xyz	1k	-	90.7
KCNet [34]	xyz	1k	-	91.0
Spec-GCN [15]	xyz	1k	-	91.5
PointCNN [25]	xyz	1k	88.1	92.2
DGCNN [35]	xyz	1k	90.2	92.2
GAPNet [17]	xyz	1k	89.7	92.4
Spec-GCN [15]	xyz+normal	1k	-	91.8
Pointconv [6]	xyz+normal	1k	-	92.5
AGCN [41]	xyz+normal	1k	90.7	92.6
PointNet++ [21]	xyz+normal	5k	-	91.9
SpiderCNN [7]	xyz+normal	5k	-	92.4
SO-Net [28]	xyz+normal	5k	90.8	93.4
DNet	xyz	1k	90.9	93.6

To test the influence of the number of initial neighborhood points k on the networks, GAPNet [17], DGCNN [35], and the proposed DNet are compared with each other, and all of them are the k-NN neighborhood-based networks. In the experiments, k is set to 10, 20, 30, 40, 50, and 60, respectively, and the networks are trained at each k separately, without using any data augmentation techniques. Figure 9 gives the corresponding OAs of the three networks with respect to each k. As shown in Figure 9, GAPNet and DGCNN achieve their highest accuracy when k is 20, and then the accuracy decreases with the increase in k. By contrast, the proposed DNet can achieve higher accuracy under more neighborhood points benefiting from the attention mechanism and masking mechanisms, and the highest accuracy is achieved when k is 40. On one hand, more initial neighborhood points can ensure that there are enough points describing the local region to be included in the network learning. On the other hand, the masking mechanism can filter out the pseudo neighborhood points with low contribution which are not conducive to the correct learning of the network. Therefore, the proposed DNet achieves higher classification accuracy.

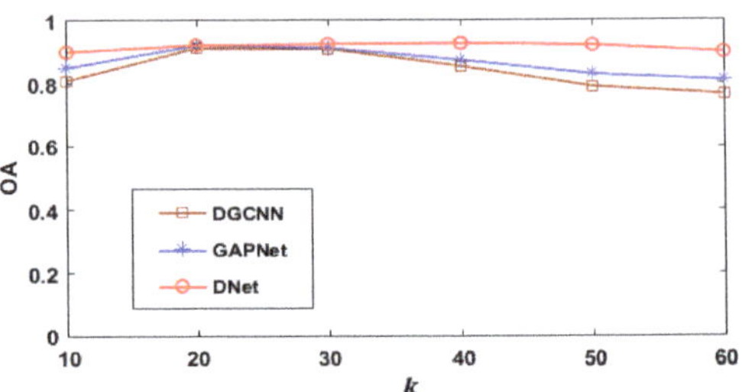

Figure 9. The influence of the number of initial neighborhood points on classification accuracy.

Additionally, in order to further analyze the influence of the number of initial neighborhood points on the performance of multi-head structure, the average numbers of the neighborhood points retained by the three heads of DNet are calculated, as shown in Figure 10, where all "airplane" models are used for the calculation. It should be noted that the neighborhood points retained by the three heads are the real learning content of the network. As shown in Figure 10, when the number of initial neighborhood points, k, is small, the average numbers of neighborhood points retained by the three heads are similar, and this will reduce the ability of the multi-head structure to capture multi-scale features. However, when k reaches 40, 50 or 60, the difference of the number among the three heads is obvious, indicating that the multi-head structure can capture multi-scale features. However, if k is too large, it will increase the burden of searching neighborhood and wash out high-frequency features [45], so k is set to 40 in this work.

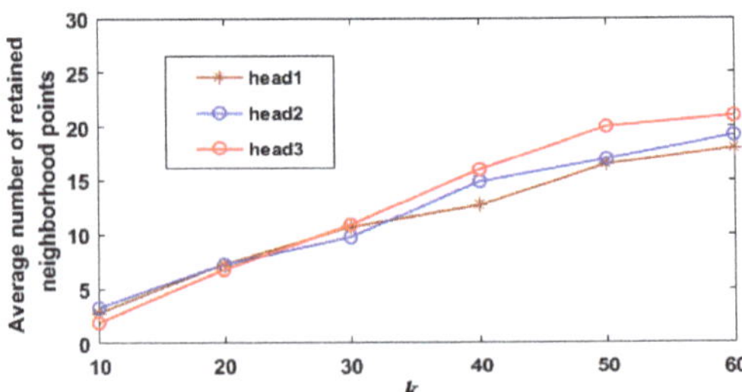

Figure 10. The average number of neighborhood points retained by the multi-head structure.

In the proposed DNet, the multi-head structure is utilized to learn multi-scale neighborhood features. However, too many heads will increase the complexity of the network. Therefore, to balance the complexity and accuracy, the number of head N is set to 3 in this paper. We have also tested the computational complexity of the proposed DNet with $N = 3$, compared with PointNet [20], PointNet++ [21] and DGCNN [35]. The comparison experimental results are given in Table 2. PointNet is not a neighborhood-based method, and it has the lowest complexity but also lowest classification accuracy in Table 2. PointNet++ and DGCNN are the representations of spherical neighborhood and k-NN neighborhood-based methods, respectively. In this experiment, for DGCNN, the number of neighborhood points k is 20, which is the default set by the author, while for the proposed DNet, k is set to 40. For PointNet++, the default parameters are used. It is seen that compared with the other networks, the proposed DNet is more lightweight, faster and more accurate.

Table 2. Comparison of different methods on model complexity, forward time, and classification accuracy.

Method	Model Size (MB)	Time (ms)	Accuracy (%)
PointNet [20]	40	6.7	89.2
PointNet++ [21]	12	21.3	90.7
DGCNN [35]	21	24.6	92.2
Proposed DNet	17	19.2	93.6

As a very important part of DNet, the masking mechanism can remove the pseudo neighborhood points in the initial neighborhood to achieve effective feature learning. There are some different kinds of masking mechanisms: for example, the mean masking and median masking mechanisms. The mean masking mechanism uses the average of contribution degrees of all the initial neighborhood points as the threshold to remove the pseudo neighborhood points. However, in the median masking mechanism, the median is used as the threshold instead of the average, and therefore the number of retained neighborhood points is fixed. Table 3 gives the point cloud classification results with respect to the two different masking mechanisms. The median masking mechanism is superior to the no masking scheme but inferior to the mean masking mechanism because of the fixed number of retained neighborhood points. Therefore, the mean masking mechanism is used in this paper. The experimental results indicate that not all points in a local region are helpful to network learning, in fact, some of them may weaken the learning and understanding ability of the network to point cloud processing.

Table 3. Effect of different masking mechanisms on point cloud classification (%).

Mask	mA	OA
No mask	92.9	89.2
Median mask	93.3	90.1
Mean mask	93.6	90.9

4.3. Point Cloud Segmentation

Point cloud segmentation is a fine-grained recognition task that requires understanding the role of each point playing in its respective category, so it is one of the challenging point cloud processing tasks.

4.3.1. Part Segmentation of Point Cloud

The part segmentation is tested on a ShapeNet dataset [43], which has 16,881 models in 16 categories, with 50 annotated parts in total. In the experiments, for each model in the ShapeNet dataset, 2048 points are extracted as the input of the networks. On the premise that the model category is known, the one-hot encoding of the category is concatenated to the last feature layer as the input of the fully connected layer in DNet, and finally the prediction result is obtained.

Intersection over Union (IoU) is used to evaluate the performance of the proposed DNet and other comparison networks. The IoU of a class refers to the average of all IoUs with respect to such kind of objects, denoted as class mean IoU (cIoU). The average of cIoU of all classes is denoted as mcIoU. The average IoU of all classes refers to the average of the IoU of all test objects, denoted as instance mean IoU (mIoU). Table 4 gives the cIoU, mcIoU and mIoU results of several different networks implemented on ShapeNet dataset, and the best results are shown in bold. Compared to the PointCNN [25] which is not a neighborhood-based method, the proposed DNet has demonstrated its potential, surpassing in several categories. For the sake of fairness, the proposed DNet is further compared in detail with the two representative neighborhood-based learning networks, that is, PointNet++ and DGCNN. PointNet++ does not consider how to learn effective regional features, but simply stacks features in multiple ranges; its mcIoU and mIoU are of 81.9% and 85.1%, respectively. Although DGCNN considers the neighborhood information of both the spatial and feature spaces, it does not consider which features of the neighborhood points are effective, its performance of mcIoU and mIoU is 82.3% and 85.2%, respectively. By contrast, the proposed DNet can reasonably learn the effective neighborhood information to achieve better results. We also carried out a qualitative analysis, and the visualization results of the components were visualized in Figure 11.

Table 4. Comparison of part segmentation of point cloud (%).

Method	mcIoU	mIoU	cIoU															
			Air Plane	Bag	Cap	Car	Chair	Ear Phone	Guitar	Knife	Lamp	Laptop	Motor Bike	Mug	Pistol	Rocket	Skate Ball	Table
Kd-Net [14]	77.4	82.3	80.1	74.6	74.3	70.3	88.6	73.5	90.2	87.2	81.0	94.9	57.4	86.7	78.1	51.8	69.9	80.3
PointNet [20]	80.4	83.7	83.4	78.7	82.5	74.9	89.6	73.0	91.5	85.9	80.8	95.3	65.2	93.0	81.2	57.9	72.8	80.6
SPLATNet [24]	82.0	84.6	81.9	83.9	88.6	79.5	90.1	73.5	91.3	84.7	84.5	96.3	69.7	95.0	81.7	59.2	70.4	81.3
KCNet [34]	82.2	84.7	82.8	81.5	86.4	77.6	90.3	76.8	91.0	87.2	84.5	95.5	69.2	94.4	81.6	60.1	75.2	81.3
GAPNet [17]	82.0	84.7	84.2	84.1	88.8	78.1	90.7	70.1	91.0	87.3	83.1	96.2	65.9	95.0	81.7	60.7	74.9	80.8
RSNet [26]	81.4	84.9	82.7	86.4	84.1	78.2	90.4	69.3	91.4	87.0	83.5	95.4	66.0	92.6	81.8	56.1	75.8	82.2
SpiderCNN [7]	82.4	85.3	83.5	81.0	87.2	77.5	90.7	76.8	91.1	87.3	83.3	95.8	70.2	93.5	82.7	59.7	75.8	82.8
AGCN [41]	82.6	85.4	83.3	79.3	87.5	78.5	90.7	76.5	91.7	87.8	84.7	95.7	72.4	93.2	84.0	63.7	76.4	82.5
SCN [39]	-	84.6	83.8	80.8	83.5	79.3	90.5	69.8	91.7	86.5	82.9	96.0	69.2	93.8	82.5	62.9	74.4	80.8
PointCNN [25]	84.6	86.1	84.1	86.5	86.0	80.8	90.6	79.7	92.3	88.4	85.3	96.1	77.2	95.3	84.2	64.2	80.0	83.0
PointNet++ [21]	81.9	85.1	82.4	79.0	87.7	77.3	90.8	71.8	91.0	85.9	83.7	95.3	71.6	94.1	81.3	58.7	76.4	82.6
DGCNN [35]	82.3	85.2	84.0	83.4	86.7	77.8	90.6	74.7	91.2	87.5	82.8	95.7	66.3	94.9	81.1	63.5	74.5	82.6
DNet	83.8	86.1	84.5	85.2	88.6	79.3	91.7	77.8	91.5	88.7	84.7	95.7	73.4	95.3	82.3	62.8	76.8	82.1

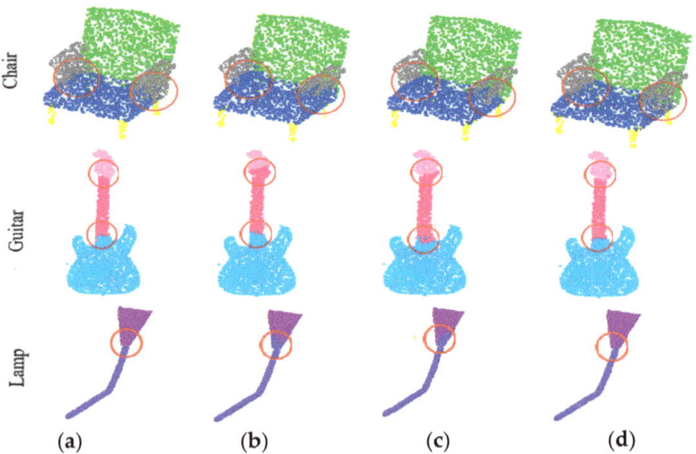

Figure 11. Part segmentation results of three models in ShapeNet dataset. (**a**) Ground Truth. (**b**)PointNet++. (**c**) Dynamic Graph Convolutional Neural Network (DGCNN). (**d**) the proposed DNet.

Figure 11 shows some of the part segmentation results, where Figure 11a shows the ground truth of the part segmentation. In Figure 11, the parts marked with red circles are segmented incorrectly by PointNet++ and DGCNN, while the segmentation results achieved by the proposed DNet are consistent with the ground truth. The segmentation results of PointNet++ and DGCNN at some of the parts of the connection are incorrect, while the DNet can predict these parts better. From the perspective of an effective neighborhood, the proposed DNet assigns lower contribution degree to the neighborhood point whose label is different from that of the central point, thereby the segmentation accuracy of the proposed DNet is improved.

4.3.2. Scene Segmentation of Point Cloud

For scene segmentation, comparative experiments are implemented on S3DIS dataset [44]. The dataset has six areas, including 271 indoor scenes (for example conference room, hallway, office etc.) with a total of 13 types of objects (such as chair, table, floor, wall and so on). In S3DIS dataset, each point has nine attributes: XYZ space coordinates, RGB color information, and a normalized location in the room. In the experiments, the same training strategy as in PointNet [20] is adopted, and 4096 points are randomly sampled from the scene as the network input.

In the experiments, 6-fold cross validation is adopted to verify the performance of the comparison networks. In this case, five areas of S3DIS dataset are used for training while the remaining one area is for testing. Then, the average results of the six tests are reported as the indicators of the performance of the networks, as shown in Table 5. In this table, the experimental results of the comparison networks also come from the corresponding literature. Considering that some of the networks only provided the experimental results of the segmentation of Area 5, that is, only Area 5 is used for testing while the other five areas are used for training, we also show such experimental results in Table 6. In Tables 5 and 6, the best results are in bold. It is seen that the proposed DNet achieves better results compared with other networks except the PointCNN and PCCN. Figure 12 shows the scene segmentation results obtained with different learning networks. It is seen that for the points in red circles, the segmentation achieved by the proposed DNet is closer to the label compared with the DGCNN.

PointCNN transforms the point cloud into the feature space by learning an X-matrix, and then weights and sums it using traditional convolution. This method maintains the invariance of the displacement of the point cloud in the feature space. When the point cloud is rotated or translated, PointCNN can still capture the fine-grained information of each point, so it achieves better results in point cloud segmentation. By contrast, the proposed DNet learns the point cloud from the perspective of the neighborhood and also shows its competitive performance. Compared with PointNet++ and DGCNN, which are also neighborhood-based learning networks, DNet achieves better performance in classification and segmentation of point cloud. This indicates that both of point cloud permutation invariance and effective neighborhood learning are indispensable for deep learning-based point cloud processing.

Table 5. Scene segmentation results on S3DIS dataset evaluated with 6-fold cross validation (%).

Method	OA	mA	mIOU
PointNet [20]	78.5	66.2	47.6
SCN [39]	81.6	-	52.7
DGCNN [35]	84.1	-	56.1
RSNet [26]	-	66.4	56.4
AGCN [41]	84.1	-	56.6
SPGraph [19]	85.5	73.0	62.1
PointCNN [25]	88.1	75.6	65.3
DNet	86.3	75.3	66.7

Table 6. Scene segmentation results of Area 5 in S3DIS dataset (%).

Method	OA	mA	mIOU
PointNet [20]	-	49.0	41.1
SegCloud [27]	-	57.4	48.9
PointCNN [25]	85.9	63.9	57.3
SPGraph [19]	86.4	66.5	58.0
PCCN [32]	-	67.0	58.3
DNet	86.5	66.3	59.7

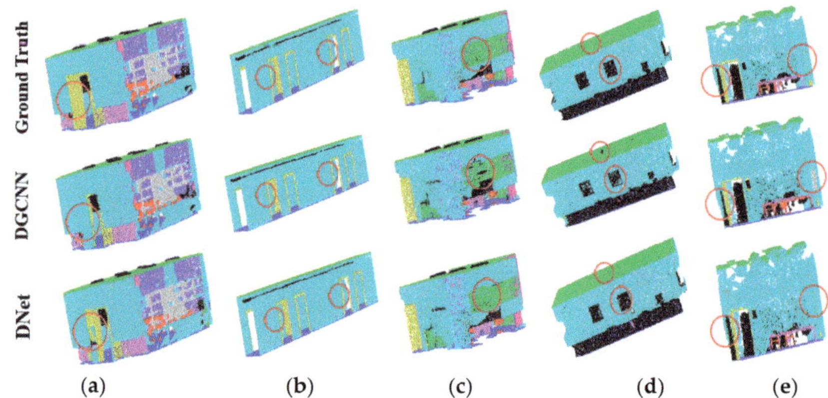

Figure 12. Comparison of point cloud segmentation with indoor scenes in the S3DIS dataset. (**a**) ConferenceRoom. (**b**) Hallway. (**c**) Office. (**d**) Pantry. (**e**) Storage.

4.4. Ablation Experiments

To clearly show the effect of the three different kinds of features in DNet, ablation experiments are implemented, and the results are given in Table 7. It is seen that if the neighborhood features are absent, the classification accuracy of DNet is significantly reduced, implying that the neighborhood features are very important for the network to understand the point cloud. Figure 13 gives the visualized results of the neighborhood points selected by the proposed DNet in the absence of some features. In Figure 13, the self-features have relatively less influence on neighborhood point selection, while neighborhood features, manifold features and neighborhood features can improve the performance of the DNet.

Table 7. Classification accuracy of DNet using different features on ModelNet40 dataset.

DNet Using Different Features	OA
without self-features	93.0
without manifold features	92.3
without neighborhood features	90.2
all features	93.6

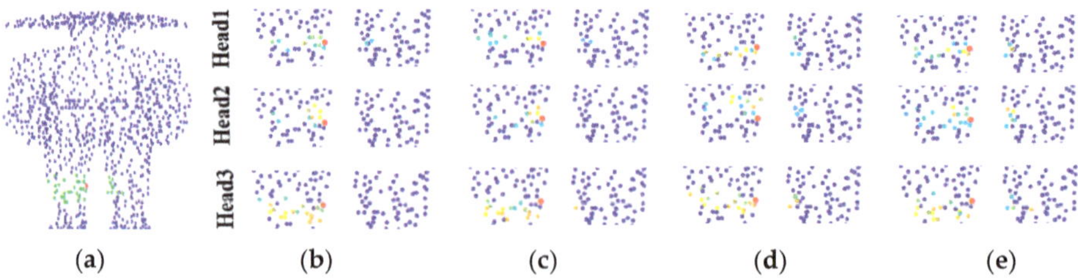

Figure 13. Neighborhood points selected by multi-head structure with different features. (**a**) Model. (**b**) All features. (**c**) Without self-features. (**d**) Without manifold features. (**e**) Without neighborhood features.

4.5. Robustness Analysis

In order to verify the robustness of the proposed DNet, uniform noise is added to the point cloud models in the testing set of the ModelNet40 dataset, and the number of noise points is set to 10, 50, 100 and 200, respectively, as shown in Figure 14a–d. Since the input points of networks are uniformly sampled from the point cloud model and normalized into the unit circle, the coordinates of the added noise points are also limited to the range of $[-1, 1]$. The training set is noise-free, and the data augmentation is not used in the training process. The final result is shown in Figure 14e, where the abscissa is the number of noise points, and the ordinate denotes the overall accuracy of classification of a network. For the four comparison networks, it is seen that the classification accuracy decreases at different rate with the increase in the number of noise points. PointNet does not consider the neighborhood, so it is most affected by noise points. PointNet++ and DGCNN are relatively better than PointNet. By contrast, the proposed DNet further considers the dynamic neighborhood, so it has strong robustness to noise compared with the other three networks.

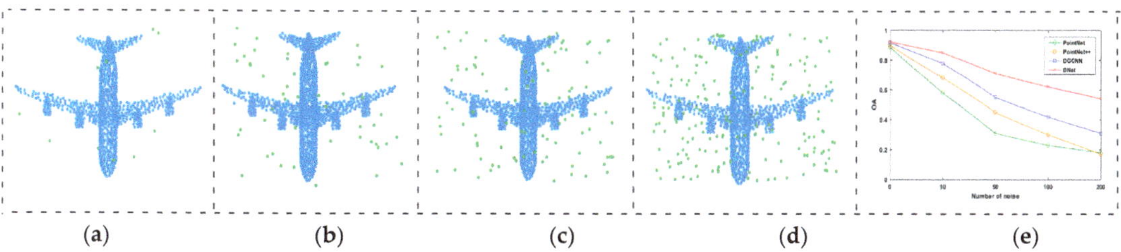

(a)   (b)   (c)   (d)   (e)

Figure 14. The influence of noise on the classification accuracy of different networks. (**a**) 10 noise points. (**b**) 50 noise points. (**c**) 100 noise points. (**d**) 200 noise points. (**e**) Classification accuracy.

5. Conclusions

In view of the lack of an effective learning network for point cloud neighborhood selection, a new Dynamic neighborhood Network, known as DNet, has been proposed to extract effective neighborhood features in this paper. The proposed DNet has a multi-head structure with two important modules: the Feature Enhancement Layer (FELayer) and the masking mechanism. The FELayer enhances the manifold features of the point cloud, while the masking mechanism can suppress the effects of some pseudo neighborhood points, so that the network can learn features that are conducive to understanding the local geometric information of the point cloud. In order to obtain sufficient contextual information in the proposed DNet, the multi-head structure is designed to allow the network to autonomously learn multi-scale features of a local region. The experimental results on three benchmark datasets have proved the effectiveness of the proposed DNet. The visualization results also show that the proposed DNet can capture more effective neighborhood features that are easy to understand.

Author Contributions: Conceptualization, F.T. and G.J.; methodology, F.T.; validation, G.J.; formal analysis, F.T and Z.J.; visualization, F.T and Z.J.; paper writing, F.T., Z.J., and G.J.; supervision, G.J. All authors have read and agreed to the published version of the manuscript.

Funding: This work was supported in part by the National Natural Science Foundation of China under Grant no. 61871247.

Institutional Review Board Statement: Not applicable.

Informed Consent Statement: Not applicable.

Data Availability Statement: Not applicable.

Conflicts of Interest: The authors declare no conflict of interest.

References

1. Tsai, C.; Lai, Y.; Sun, Y.; Chung, Y.; Perng, J. Multi-dimensional underwater point cloud detection based on deep learning. *Sensors* **2021**, *21*, 884. [CrossRef] [PubMed]
2. Yang, Y.; Ma, Y.; Zhang, J.; Gao, X.; Xu, M. AttPNet: Attention-based deep neural network for 3D point set analysis. *Sensors* **2020**, *20*, 5455. [CrossRef] [PubMed]

3.	Yuan, Y.; Borrmann, D.; Hou, J.; Schwertfeger, S. Self-supervised point set local descriptors for point cloud registration. *Sensors* **2021**, *21*, 486. [CrossRef] [PubMed]
4.	Liu, Y.; Fan, B.; Xiang, S.; Pan, C. Relation-shape convolutional neural network for point cloud analysis. In Proceedings of the IEEE Conference on Computer Vision and Pattern Recognition, Long Beach, CA, USA, 16–20 June 2019; pp. 8895–8904.
5.	Li, F.; Jin, W.; Fan, C.; Zou, L.; Chen, Q.; Li, X.; Jiang, H.; Liu, Y. PSANet: Pyramid splitting and aggregation network for 3D object detection in point cloud. *Sensors* **2021**, *21*, 136. [CrossRef] [PubMed]
6.	Wu, W.; Qi, Z.; Fuxin, L. Pointconv: Deep convolutional networks on 3D point clouds. In Proceedings of the IEEE Conference on Computer Vision and Pattern Recognition, Long Beach, CA, USA, 16–20 June 2019; pp. 9621–9630.
7.	Xu, Y.; Fan, T.; Xu, M.; Zeng, L.; Qiao, Y. SpiderCNN: Deep learning on point sets with parameterized convolutional filters. In Proceedings of the European Conference on Computer Vision, Munich, Germany, 8–14 September 2018; pp. 87–102.
8.	Su, H.; Maji, S.; Kalogerakis, E.; Learned-Miller, E. Multi-view convolutional neural networks for 3D shaperecognition. In Proceedings of the IEEE International Conference on Computer Vision, Santiago, Chile, 7–13 December 2015; pp. 945–953.
9.	Feng, Y.; Zhang, Z.; Zhao, X.; Ji, R.; Gao, Y. GVCNN: Group-view convolutional neural networks for 3D shape recognition. In Proceedings of the IEEE Conference on Computer Vision and Pattern Recognition, Salt Lake City, UT, USA, 18–22 June 2018; pp. 264–272.
10.	Guo, H.; Wang, J.; Li, J.; Lu, H. Multi-view 3D object retrieval with deep embedding network. *IEEE Trans. Image Proces.* **2016**, *25*, 5526–5537. [CrossRef] [PubMed]
11.	Maturana, D.; Scherer, S. VoxNet: A 3D Convolutional Neural Network for real-time object recognition. In Proceedings of the IEEE International Conference on Intelligent Robots and Systems, Hamburg, Germany, 28 September–2 October 2015; pp. 922–928.
12.	Qi, C.R.; Su, H.; Nießner, M.; Dai, A.; Yan, M.; Guibas, L.J. Volumetric and multi-view cnns for object classification on 3D data. In Proceedings of the IEEE Conference on Computer Vision and Pattern Recognition, Las Vegas, NV, USA, 27–30 June 2016; pp. 5648–5656.
13.	Riegler, G.; Osman Ulusoy, A.; Geiger, A. Octnet: Learning deep 3D representations at high resolutions. In Proceedings of the IEEE Conference on Computer Vision and Pattern Recognition, Honolulu, HI, USA, 21–26 July 2017; pp. 3577–3586.
14.	Klokov, R.; Lempitsky, V. Escape from cells: Deep Kd-networks for the recognition of 3D point cloud models. In Proceedings of the IEEE International Conference on Computer Vision, Venice, Italy, 22–29 October 2017; pp. 863–872.
15.	Wang, C.; Samari, B.; Siddiqi, K. Local spectral graph convolution for point set feature learning. In Proceedings of the European Conference on Computer Vision, Amsterdam, The Netherlands, 8–16 October 2016; pp. 52–66.
16.	Wang, L.; Huang, Y.; Hou, Y.; Zhang, S.; Shan, J. Graph Attention Convolution for Point Cloud Semantic Segmentation. In Proceedings of the IEEE Conference on Computer Vision and Pattern Recognition, Long Beach, CA, USA, 16–20 June 2019; pp. 10296–10305.
17.	Chen, C.; Fragonara, L.Z.; Tsourdos, A. GAPNet: Graph attention based point neural network for exploiting local feature of point cloud. *arXiv* **2019**, arXiv:1905.08705.
18.	Simonovsky, M.; Komodakis, N. Dynamic edge-conditioned filters in convolutional neural networks on graphs. In Proceedings of the IEEE Conference on Computer Vision and Pattern Recognition, Honolulu, HI, USA, 21–26 July 2017; pp. 3693–3702.
19.	Landrieu, L.; Simonovsky, M. Large-scale point cloud semantic segmentation with superpoint graphs. In Proceedings of the IEEE Conference on Computer Vision and Pattern Recognition, Salt Lake City, UT, USA, 18–22 June 2018; pp. 4558–4567.
20.	Qi, C.R.; Su, H.; Mo, K.; Guibas, L.J. Pointnet: Deep learning on point sets for 3D classification and segmentation. In Proceedings of the IEEE Conference on Computer Vision and Pattern Recognition, Honolulu, HI, USA, 21–26 July 2017; pp. 652–660.
21.	Qi, C.R.; Yi, L.; Su, H.; Guibas, L.J. Pointnet++: Deep hierarchical feature learning on point sets in a metric space. In Proceedings of the Advances in Neural Information Processing Systems, Curran Associates, Long Beach, CA, USA, 4–9 December 2017; pp. 5099–5108.
22.	Graham, B.; van der Maaten, L. Submanifold sparse convolutional networks. *arXiv* **2017**, arXiv:1706.01307.
23.	Hua, B.; Tran, M.; Yeung, S.-K. Pointwise convolutional neural networks. In Proceedings of the IEEE Conference on Computer Vision and Pattern Recognition, Salt Lake City, UT, USA, 18–22 June 2018; pp. 984–993.
24.	Su, H.; Jampani, V.; Sun, D.; Maji, S.; Kalogerakis, E.; Yang, M.H.; Kautz, J. Splatnet: Sparse lattice networks for point cloud processing. In Proceedings of the IEEE Conference on Computer Vision and Pattern Recognition, Salt Lake City, UT, USA, 18–22 June 2018; pp. 2530–2539.
25.	Li, Y.; Bu, R.; Sun, M.; Wu, W.; Di, X.; Chen, B. Pointcnn: Convolution on x-transformed points. In Proceedings of the Advances in Neural Information Processing Systems, Montréal, QC, Canada, 3–8 December 2018; pp. 820–830.
26.	Huang, Q.; Wang, W.; Neumann, U. Recurrent slice networks for 3D segmentation of point clouds. In Proceedings of the IEEE Conference on Computer Vision and Pattern Recognition, Salt Lake City, UT, USA, 18–22 June 2018; pp. 2626–2635.
27.	Tchapmi, L.; Choy, C.; Armeni, I.; Gwak, J.; Savarese, S. Segcloud: Semantic segmentation of 3D point clouds. In Proceedings of the International Conference on 3D Vision, Qingdao, China, 10–12 October 2017; pp. 537–547.
28.	Li, J.; Chen, B.M.; Hee Lee, G. So-net: Self-organizing network for point cloud analysis. In Proceedings of the IEEE Conference on Computer Vision and Pattern Recognition, Salt Lake City, UT, USA, 18–22 June 2018; pp. 9397–9406.
29.	Huang, R.; Xu, Y.; Hong, D.; Yao, W.; Ghamisi, P.; Stilla, U. Deep point embedding for urban classification using ALS point clouds: A new perspective from local to global. *ISPRS J. Photogramm. Remote Sens.* **2020**, *163*, 62–81. [CrossRef]

30. Groh, F.; Wieschollek, P.; Lensch, H.P. Flex-Convolution. In Proceedings of the Asian Conference on Computer Vision, Perth, Australia, 2–6 December 2018; pp. 105–122.
31. Verma, N.; Boyer, E.; Verbeek, J. Feastnet: Feature-steered graph convolutions for 3D shape analysis. In Proceedings of the IEEE Conference on Computer Vision and Pattern Recognition, Salt Lake City, UT, USA, 18–22 June 2018; pp. 2598–2606.
32. Wang, S.; Suo, S.; Ma, W.-C.; Pokrovsky, A.; Urtasun, R. Deep parametric continuous convolutional neural networks. In Proceedings of the IEEE Conference on Computer Vision and Pattern Recognition, Salt Lake City, UT, USA, 18–22 June 2018; pp. 2589–2597.
33. Hermosilla, P.; Ritschel, T.; Vázquez, P.-P.; Vinacua, À.; Ropinski, T. Monte Carlo convolution for learning on non-uniformly sampled point clouds. *ACM Tran. Graph.* **2019**, *37*, 6. [CrossRef]
34. Shen, Y.; Feng, C.; Yang, Y.; Tian, D. Mining Point Cloud Local Structures by Kernel Correlation and Graph Pooling. In Proceedings of the IEEE Conference on Computer Vision and Pattern Recognition, Salt Lake City, UT, USA, 18–22 June 2018; pp. 4548–4557.
35. Wang, Y.; Sun, Y.; Liu, Z.; Sarma, S.E.; Bronstein, M.M.; Solomon, J.M. Dynamic Graph CNN for Learning on Point Clouds. *ACM Tran. Graph.* **2019**, *38*, 5. [CrossRef]
36. Thomas, H.; Deschaud, J.-E.; Marcotegui, B. Semantic classification of 3D point clouds with multiscale spherical neighborhoods. In Proceedings of the International Conference on 3D Vision, Verona, Italy, 5–8 September 2018; pp. 390–398.
37. Weinmann, M.; Jutzi, B.; Hinz, S.; Mallet, C. Semantic point cloud interpretation based on optimal neighborhoods, relevant features and efficient classifiers. *ISPRS J. Photogramm. Remote Sens.* **2015**, *105*, 286–304. [CrossRef]
38. He, T.; Huang, H.; Yi, L. GeoNet: Deep geodesic networks for point cloud analysis. In Proceedings of the IEEE Conference on Computer Vision and Pattern Recognition, Long Beach, CA, USA, 16–20 June 2019; pp. 6888–6897.
39. Xie, S.; Liu, S.; Chen, Z.; Tu, Z. Attentional ShapeContextNet for Point Cloud Recognition. In Proceedings of the 2018 IEEE Conference on Computer Vision and Pattern Recognition, Salt Lake City, UT, USA, 18–22 June 2018; pp. 4606–4615.
40. Feng, M.; Zhang, L.; Lin, X.; Gilani, S.Z.; Mian, A. Point Attention Network for Semantic Segmentation of 3D Point Clouds. *Pat. Recog.* **2020**, *107*, 107446. [CrossRef]
41. Xie, Z.; Chen, J.; Peng, B. Point clouds learning with attention-based graph convolution networks. *Neurocomputing* **2020**, *402*, 245–255. [CrossRef]
42. Wu, Z.; Song, S.; Khosla, A.; Yu, F.; Zhang, L.; Tang, X.; Xiao, J. 3D ShapeNets: A deep representation for volumetric shapes. In Proceedings of the 2015 IEEE Conference on Computer Vision and Pattern Recognition, Boston, MA, USA, 7–12 June 2015; pp. 1912–1920.
43. Yi, L.; Kim, V.G.; Ceylan, D.; Shen, I.-C.; Yan, M.; Su, H.; Lu, C.; Huang, Q.; Sheffer, A.; Guibas, L. A scalable active framework for region annotation in 3D shape collections. *ACM Tran. Graph.* **2019**, *35*, 6. [CrossRef]
44. Armeni, I.; Sener, O.; Zamir, A.R.; Jiang, H.; Brilakis, I.; Fischer, M.; Savarese, S. 3D semantic parsing of large-scale indoor spaces. In Proceedings of the IEEE Conference on Computer Vision and Pattern Recognition, Las Vegas, NV, USA, 27–30 June 2016; pp. 1534–1543.
45. Mel, B.W.; Omohundro, S.M. How receptive field parameters affect neural learning. In Proceedings of the Advances in Neural Information Processing Systems, Denver, CO, USA, 2–5 December 1991; pp. 757–763.

Article

Compressed Video Quality Index Based on Saliency-Aware Artifact Detection

Liqun Lin, Jing Yang, Zheng Wang, Liping Zhou, Weiling Chen and Yiwen Xu *

Fujian Key Lab for Intelligent Processing and Wireless Transmission of Media Information, College of Physics and Information Engineering, Fuzhou University, Fuzhou 350002, China; lin_liqun@fzu.edu.cn (L.L.); 201127079@fzu.edu.cn (J.Y.); n191120078@fzu.edu.cn (Z.W.); n181120080@fzu.edu.cn (L.Z.); weiling.chen@fzu.edu.cn (W.C.)
* Correspondence: xu_yiwen@fzu.edu.cn

Abstract: Video coding technology makes the required storage and transmission bandwidth of video services decrease by reducing the bitrate of the video stream. However, the compressed video signals may involve perceivable information loss, especially when the video is overcompressed. In such cases, the viewers can observe visually annoying artifacts, namely, Perceivable Encoding Artifacts (PEAs), which degrade their perceived video quality. To monitor and measure these PEAs (including blurring, blocking, ringing and color bleeding), we propose an objective video quality metric named Saliency-Aware Artifact Measurement (SAAM) without any reference information. The SAAM metric first introduces video saliency detection to extract interested regions and further splits these regions into a finite number of image patches. For each image patch, the data-driven model is utilized to evaluate intensities of PEAs. Finally, these intensities are fused into an overall metric using Support Vector Regression (SVR). In experiment section, we compared the SAAM metric with other popular video quality metrics on four publicly available databases: LIVE, CSIQ, IVP and FERIT-RTRK. The results reveal the promising quality prediction performance of the SAAM metric, which is superior to most of the popular compressed video quality evaluation models.

Keywords: video quality assessment; saliency detection; perceivable encoding artifacts; Dense Convolutional Network (DenseNet)

Citation: Lin, L.; Yang, J.; Wang, Z.; Zhou, L.; Chen, W.; Xu, Y. Compressed Video Quality Index Based on Saliency-Aware Artifact Detection. *Sensors* **2021**, *21*, 6429. https://doi.org/10.3390/s21196429

Academic Editors: Yun Zhang, Kwong Tak Wu Sam, Xu Long and Tiesong Zhao

Received: 30 July 2021
Accepted: 22 September 2021
Published: 26 September 2021

Publisher's Note: MDPI stays neutral with regard to jurisdictional claims in published maps and institutional affiliations.

Copyright: © 2021 by the authors. Licensee MDPI, Basel, Switzerland. This article is an open access article distributed under the terms and conditions of the Creative Commons Attribution (CC BY) license (https://creativecommons.org/licenses/by/4.0/).

1. Introduction

Video coding technology largely reduces storage capacity and transmission bandwidth. However, lossy compression and transmission via changeable channel inevitably cause various distortions. Thus, compressed video often shows visually annoying distortions, named Perceivable Encoding Artifacts (PEAs), which greatly affect video perceived quality [1].

For effective analysis and improvement of user experience, it is necessary to accurately evaluate visual quality of video. Subjective Video Quality Assessment (VQA) is the most accurate and reliable reflection of human perception, because it is the quality scored by viewers. At present, only the results of subjective quality evaluation are used as a benchmark to measure the accuracy of objective quality evaluation methods. According to the standard given by International Telecommunications Union (ITU) [2], Mean Opinion Score (MOS) and Different Mean Opinion Score (DMOS) are employed to expressed video subjective quality. Therefore, MOS and DMOS are the most reliable quality indicators and are used to assessment objective quality of videos. However, subjective experiments are tedious, time-consuming and expensive. Consequently, it is imperative to establish reliable objective VQA index.

According to the availability of reference, the objective VQA metrics can be categorized into Full-Reference (FR), Reduced-Reference (RR) and No-Reference (NR) metrics. Typical FR metrics such as Peak Signal to Noise Ratio (PSNR) and Structural SIMilarity (SSIM) [3] have been extensively applied. RR-VQA metrics, such as Spatio-Temporal RR Entropic Differences (STRRED) [4] and Spatial Efficient Entropic Differencing for Quality Assessment (SpEED-QA) [5], also show good performance. In real-life video display, the unimpaired original video source is inaccessible to end users, thus NR metric is highly desirable. It is also the most difficult one among three types of VQA metrics due to the lack of prior knowledge from reference video. However, it is the most widely used in different applications.

As mentioned above, NR-VQA metrics have a wide range of applications, but they require that the extracted features are not sensitive to the video content and highly related to the degree of distortion. Furthermore, they have high computational complexity and still have room for improvement in the accuracy. With the development of the Natural Video Statistic (NVS) model, researchers extracted features from natural scenes. These features can describe the temporal and spatial statistical characteristics of video, and were fed into the regression model (RM) to realize the evaluation of video quality in the transform domain [6–8]. Motivated by the effort of unsupervised feature learning for NR image quality assessment [9], Xu et al. [10] presented a NR-VQA algorithm named video COdebook Representation for No-reference Image quality Assessment (CORNIA), where a linear Support Vector Regression (SVR) is utilized to predict the video quality based on frame-level features. In [11], a blind NR-VQA model was developed by using the statistical properties in natural videos. The model employs the output data to directly predict the video quality, without any external information about the reference video such as subjective quality score. Zhu et al. [12] presented a blind VQA method considering the characteristics of human visual system (HVS). Reddy et al. [13] proposed a NR-VQA metric utilizing an asymmetric generalized gaussian distribution model, which performs the statistics of the characteristic parameters in natural videos.

The distortion-specific NR-VQA approaches assess video quality under the premise that distortion types of video are known. In [14], a method was proposed to measure the perceived strength of blocking artifact in decoded video at the position of the non-fixed grid. Next, it was combined with the entropy measurement to predict video quality. Amor et al. [15] proposed a NR-VQA index based on blocking artifact estimation in spatial domain by calculating the difference of gray-level conversion between adjacent blocks. Xue et al. [16] proposed a VQA metric method to evaluate the impact of frame freezing caused by packet loss or delay on perceived quality. In [17], a NR-VQA model based on discrete cosine transform was developed to measure distortion, such as blocking, clearness and noise, in which a multilayer neural network was used to obtain the prediction scores of videos. A model was built by Men et al. [18] to achieve the prediction of video quality scores by combining features including blurring artifact, contrast and color bleeding. In [19], blocking, packet-loss and freezing artifacts were obtained to predict video quality. Rohil et al. [20] developed a holistic NR-VQA model based on quantifying certain distortions in video frames, such as ringing and contrast distortion. Next, the intensity values of various distortions were input to the neural network to evaluate the quality of videos. In summary, most of the existing NR-VQA algorithms are aimed at traditional videos. Some algorithms involve the transmission distortion caused by channel error, such as packet loss and frame freezing. However, there are few NR-VQA researches on compressed videos, and most of existing works only detect a single type of compression artifact. These methods can not abundantly reflect the impact of PEAs on HVS. Therefore, it is necessary to develop a NR compressed video quality evaluation algorithm combined with two or more PEAs, which shows highly correlated with subjective perception quality.

To further improve the NR-VQA performance, it is feasible to detect more artifacts. In this paper, we propose a NR-VQA metric named Saliency-Aware Artifact Measurement (SAAM) to estimate video quality by analyzing four typical types of spatial PEAs including blocking, blurring, color bleeding and ringing. We also exploit visual saliency detection and patch segmentation of interested regions to map the PEA intensities to objective score with reduced complexity. We sum up major contributions of this work as follows.

(1) Proposed a NR-VQA method based on PEA detection. The PEA detection module accurately identifies four typical types of PEAs (i.e., blurring, blocking, ringing and color bleeding). Based on the PEA detection module, the PEA intensities are obtained to analyze video quality.

(2) Introduced visual saliency detection and patch segmentation for high VQA accuracy and reduced complexity. The visual saliency detection can make useful information of videos and maximize utilization of computing resources, as well as help to eliminate the impact of redundant visual information on subjective evaluation.

(3) Achieved the superior performance of our method in terms of compressed videos. Compared with multiple typical VQA metrics, our index has the highest overall correlation coefficient with the subjective quality score. In addition, our algorithm can achieve reasonable performance in cross-database verification, which shows that our algorithm has good generalization and robustness.

2. PEA-Based Video Quality Index

The overall architecture of the SAAM metric is shown in Figure 1. It consists of four steps: video saliency detection with Attentive CNN-LSTM Network (ACLNet) [21] (input video frame F_i, output saliency map S_i), image patch segmentation (enter saliency map S_i to guide the generation of 72×72 image patches P_{ij}), PEA detection (input image patches P_{ij}, output PEA intensities I_{ij} of patches) and SVR prediction (input PEA intensities I_V of video, output predicted quality Q_V). These detail contents are elaborated as follows.

2.1. Perceivable Encoding Artifacts

In this section, we review PEA classification in [1] and select typical PEAs (including blocking, blurring, ringing and color bleeding) to develop our SAAM algorithm.

2.1.1. Four Typical PEAs

The causes and manifestations of four typical types of PEAs are summarized as follows.

(1) Blurring: Modern video compression techniques involve a frequency transformation step followed by a quantization process, which usually eliminates small amplitude transformation coefficients. Because the energy of natural visual signals is concentrated at low frequencies, quantification reduces high frequency energy in such signals. It results in a significant blurring artifact in the reconstructed signal. Besides, de-blocking filter also leads to blurring artifact. Visually, blurring typically manifests itself as damage to spatial structure or reduced sharpness at edge or texture areas in images [22]. A blurring artifact example is shown in Figure 2b, which exhibits the spatial loss of the building field.

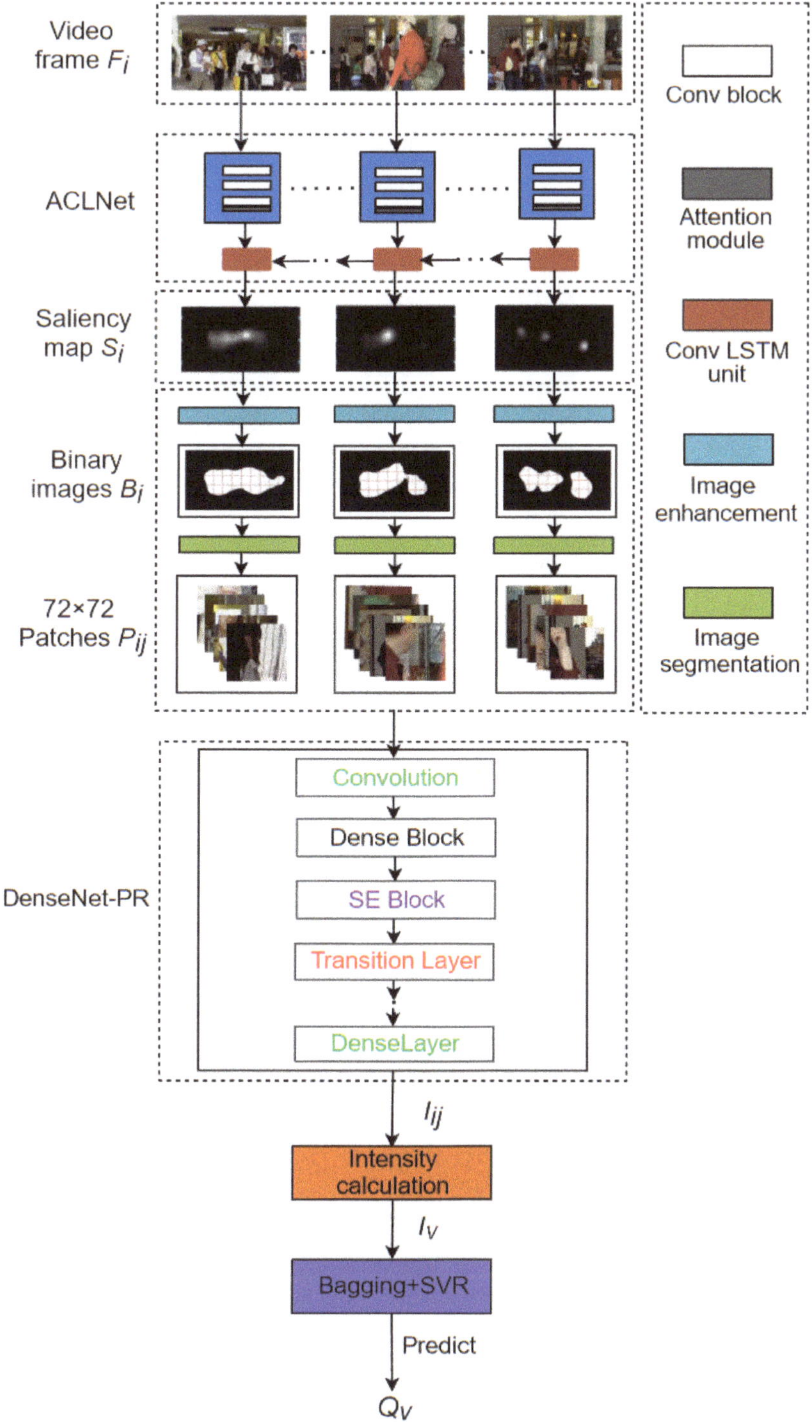

Figure 1. The overall architecture of SAAM.

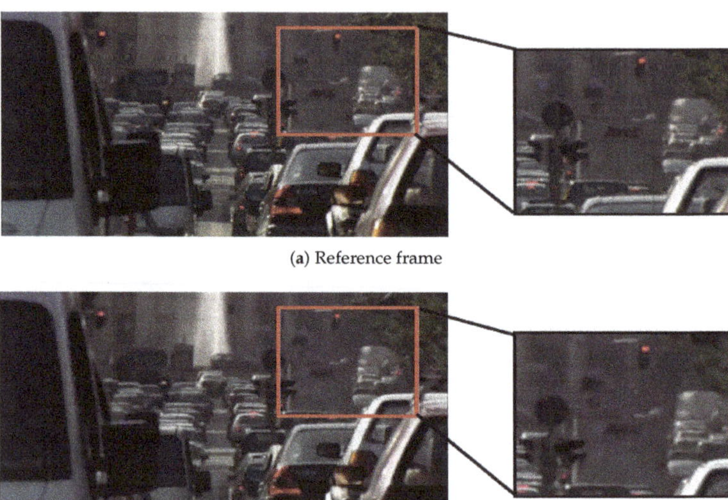

(**a**) Reference frame

(**b**) Compressed frame with blurring artifact

Figure 2. An example of blurring [23–25].

(2) Blocking: Standard video compression techniques utilize blocks of different sizes as basic units for frequency transformation and quantization. The quantization errors introduced in each block are presented in different forms, which result in discontinuities on block boundaries. In decoded videos, different forms of the blocking artifact are demonstrated, such as mosaic artifact, ladder artifact and pseudo-edge artifact. Blocking refers to discontinuities on the boundaries of adjacent blocks. The visual shape of blocking depends on the region where blocking occurs [26]. A blocking artifact example is shown in Figure 3b, which demonstrates the visual blocks of the face field.

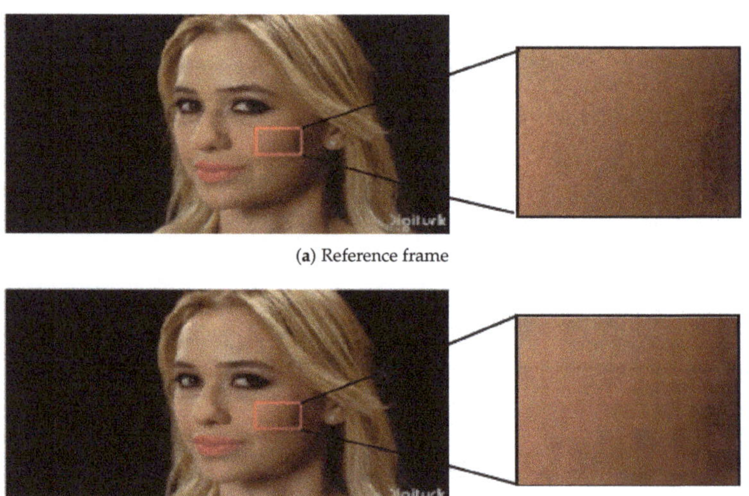

(**a**) Reference frame

(**b**) Compressed frame with blocking artifact

Figure 3. An example of blocking [27].

(3) Ringing: When the quantization error of the high-frequency component corresponding to strong edges of image occurs, a corrugated pseudo-boundary will appear

near the strong edge. With high-contrast edges, the ringing artifact is the most obvious in areas with smoother textures during the reconstruction process. Ringing shows ripple or vibration structures near the strong edge [20]. A ringing artifact example is shown in Figure 4b, in which the marked letters show the phenomenon of boundary ripples.

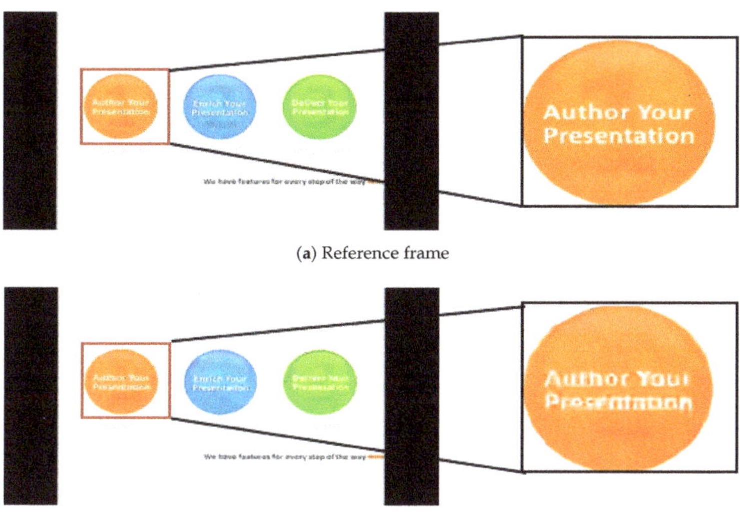

(**a**) Reference frame

(**b**) Compressed frame with ringing artifact

Figure 4. An example of ringing.

(4) Color bleeding: Color bleeding is the result of coarse quantification of chroma information. Color diffusion occurs in areas with very large chroma variation, resulting in blurring of chroma. After compression, due to low resolution of the color channel, interpolation operation is inevitably involved in the rendering process, which results in additional inconsistent color diffusion. Color bleeding is the result of inconsistent image rendering between brightness and chromaticity channels. A color bleeding artifact example is shown in Figure 5b, which exhibits a color diffusion in the marked rectangular field.

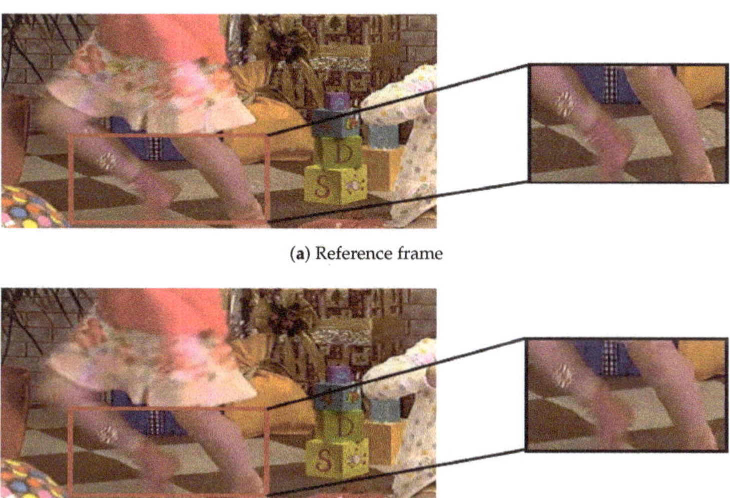

(**a**) Reference frame

(**b**) Compressed frame with color bleeding artifact

Figure 5. An example of color bleeding.

2.1.2. Correlation between PEAs and Visual Quality

To verify the effects of four typical types of PEAs on visual quality, the correlation between PEAs and visual quality is studied. The sensitivity of human eyes to different types of PEAs is different. In [28], the blocking and blurring artifacts were observed to show significant impacts on visual quality of compressed videos. To explore the correlation between PEAs and compressed video quality [29], it is necessary to adopt a PEA detection algorithm. In this work, a PEA recognition model [1] is adopted, which can detect different types of PEAs with superior performance. The intensities of color bleeding, blocking, blurring and ringing are measured on the LIVE Video Quality Database [30], which consists of 10 reference videos and 40 compressed videos. The scatterplot of each PEA intensity value and video subjective quality score DMOS is shown in Figure 6, where the abscissa is PEA intensity, the ordinate represents DMOS value and each legend denotes compressed videos.

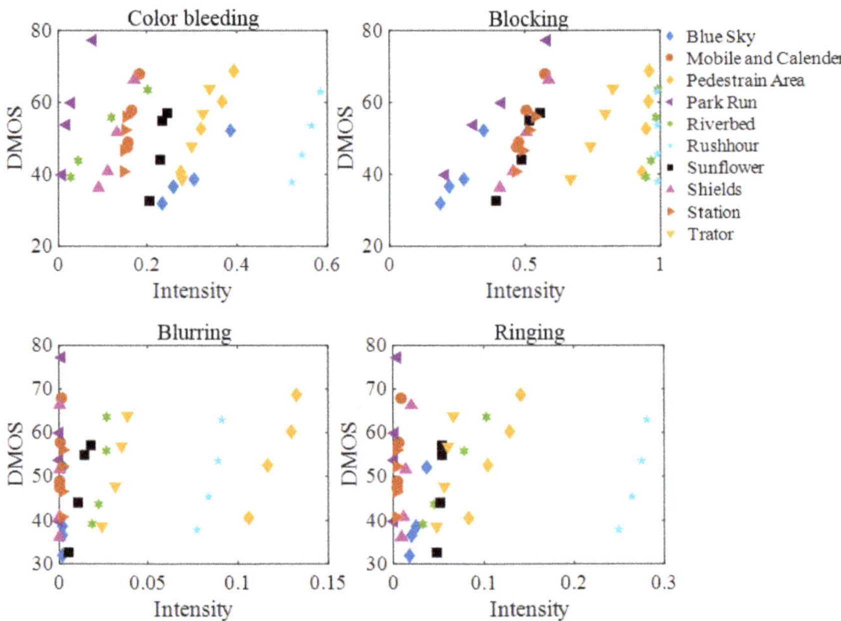

Figure 6. The scatterplot of each PEA intensity and DMOS on the LIVE Video Quality Database.

In addition, to study the influence of PEAs on video quality, it is necessary to analyze compressed videos with different distortion degrees. As can be seen from Figure 6, for different compressed videos with same content, there is a positive correlation between four PEA intensity values and their subjective quality scores, respectively, that is, the higher the PEA intensity value, the higher the DMOS value, the worse the video quality will be. Therefore, the existence of PEAs will reduce the quality of compressed videos. In order to further study the overall correlation between four types of PEAs and compressed video quality, the Pearson Linear Correlation Coefficient (PLCC) and Spearman Rank Correlation Coefficient (SRCC) of four PEA intensity values and their DMOS values are listed in Table 1, where the optimal and suboptimal correlation are represented in bold.

Table 1. Correlation between the PEA intensity and DMOS on the LIVE Video Quality Database.

Correlation	Blocking	Blurring	Ringing	Color Bleeding
PLCC	**0.5551**	**0.1921**	0.1900	0.1902
SRCC	**0.4208**	**0.1848**	0.1340	0.1109

The above results show that the PLCC and SRCC of blocking artifact are the best, and that of blurring artifact are second best. Therefore, the blocking and blurring artifacts are the most important factors leading to the deterioration of compressed video quality. It also confirms the finding that the human eye shows the highest sensitivity to the blocking and blurring artifacts.

2.2. Video Saliency Detection with ACLNet

Visual saliency is an inherent attribute of HVS and is also a key factor affecting video perceptual quality [31]. The advantages of introducing visual saliency into video quality assessment are primarily reflected in two aspects: first, it allocates constrained hardware resources to more significant regions, and second, video quality analyze considering visual saliency is more consistent with human visual perception. Therefore, we select ACLNet as our video saliency model based on comprehensive comparison and analysis of popular video saliency models. ACLNet has strong applicability and its high real-time processing speed.

ACLNet combines attention module to improve features extracted by CNN [32], and utilizes a convLSTM [33] to obtain temporal characteristic. Then, convLSTM [33] is employed to model the temporal characteristic of this sequential issue, which is completed by merging memory units with gated operations. Finally, saliency maps of all frames are summarized as video saliency map. In addition, ACLNet uses the first five convolution blocks of VGG16 [34] and removes pool4 and pool5 layers to preserve more spatial details [22]. Saliency map can be expressed as

$$S_i = f_s(F_i), \tag{1}$$

where S_i refers to saliency map of the i-th frame. $f_s(\cdot)$ represents saliency algorithm.

2.3. Image Patch Segmentation

To make ease of PEA detection, saliency regions are segmented into image patches, which is shown in Figure 7. First, the saliency map of each frame is binarized. In this work, we adopt grayscale transformation to obtain a appropriate threshold. The saliency regions below the threshold are ignored. All bright regions in enhanced map are clipped into patches. We utilize 72×72 as patch size, in accordance with the input patch size of our PEA detection module. To minimize the number of image patches, the connected regions in the binary image are framed by smallest circumscribed rectangles. R_i represents the binary images with minimum rectangle. The binary image marked with 72×72 square is denoted as B_i. All clipped image patches are grouped for PEA detection. The relationship between the original image marked with 72×72 square and image patches is calculated as follows:

$$\{O_{i_j}, j \in [1...N]\} = P_{ij}, \tag{2}$$

where O_i represents the original images marked with 72×72 square. O_{i_j} denotes the j-th original image marked with 72×72 square, respectively. N ($j \in [1...N]$) is the total number of original images marked with 72×72 square. P_{ij} refers to all the clipped image patches of the video and is grouped for PEA detection.

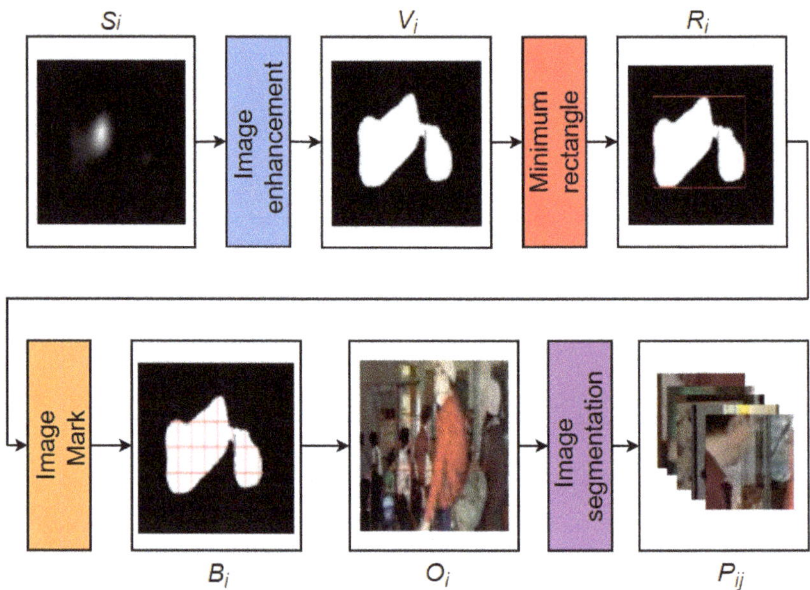

Figure 7. The process of image patch segmentation.

2.4. PEA Detection

To detect four typical types of PEAs, we utilize our DenseNet for PEA Recognition (Dense Net-PR) [1]. The specific structure of the DenseNet-PR is shown in Figure 8. First, Dense Block contains many layers, where the size of feature maps of each layer is the same. A feedforward fashion is utilized to establish connections between layers. The input of each layer is the feature maps of all previous layers, and the output feature maps of each layer are delivered to all subsequent layers. The nonlinear transformation function between layers is composed of Batch Normalization (BN), a rectified linear unit and a 3×3 convolution. Low computational complexity of the algorithm can be achieved by inserting a 1×1 convolution as the bottleneck layer before 3×3 convolution. The essence of this operation is to decrease the number of input feature maps. Additionally, we integrated the 3×3 convolution into a 3×3 and a 1×1 pointwise convolution to learn deeper features of feature channel. Second, we inserted a Squeeze and Excitation (SE) Block between each Dense Block and the transition layer to highlight vital characteristics of training set. Because this process also reused critical features of the transition layer, recognition accuracy is improved. Third, the transition layers are composed of a BN layer, a 1×1 convolutional layer and a 2×2 average pooling layer, where the 1×1 convolutional layer can decrease the number of feature maps. Finally, we utilize softmax classifier to return a list of probabilities. The label with the largest probability is chosen as the final classification.

The DenseNet-PR alleviates the vanishing-gradient problem, enhances feature propagation and greatly reduces the number of parameters. Based on the DenseNet-PR architecture, we randomly choose 50,000 ground-truth PEA samples to individually train four types of PEA recognition models from our subject-labeled database, which is composed of 324 compressed videos containing various PEAs [1]. The ratio of training sets and testing sets is 75:25 in these samples. Stochastic Gradient Descent (SGD) is adopted and the batch size is 256. The momentum is set to 0.9000. 0.0001 is the value of weight decay. The learning rate is adjusted following the schedule in [35] and its initial value is 0.1. The weight is initialized according to [35]. The depth and width of the DenseNet-PR network are set to 46 and 10, respectively. Based on the DenseNet-PR network, we individually trained four types of PEA recognition models to detect the presence of PEAs in image patches. It is worth mentioning that multi-objective classification is not utilized here because different types of PEAs may be coexist in one patch. Finally, based on the above models, a list of probabilities of each 72×72 patch are obtained to measure the PEA intensities, namely, I_{ij}. Then, we can calculate the intensity of each PEA for a video sequence. We calculate the PEA intensity value of each patch and assume that the intensity of each pixel in the patch is equal to the intensity of the patch. For a few pixels that belong to overlapping patches, we use their average intensity values as the intensity values of these pixels. Finally, the intensity of each PEA of each video is calculated as follows:

$$I_{\text{frame}_k} = \frac{1}{N_{\text{pixel}}} \sum_{n=1}^{N_{\text{pixel}}} I_{ij}, \tag{3}$$

$$I_{V_k} = \frac{1}{N_{\text{frame}}} \sum_{n=1}^{N_{\text{frame}}} I_{\text{frame}_k}, \tag{4}$$

where N_{pixel} refers to the total number of pixels in the saliency region of each frame. I_{frame_k} denotes the intensity value of the k-th type of PEAs per frame. N_{frame} is the total number of video frames. I_{V_k} represents the intensity values of the k-th type of PEAs of each video, respectively.

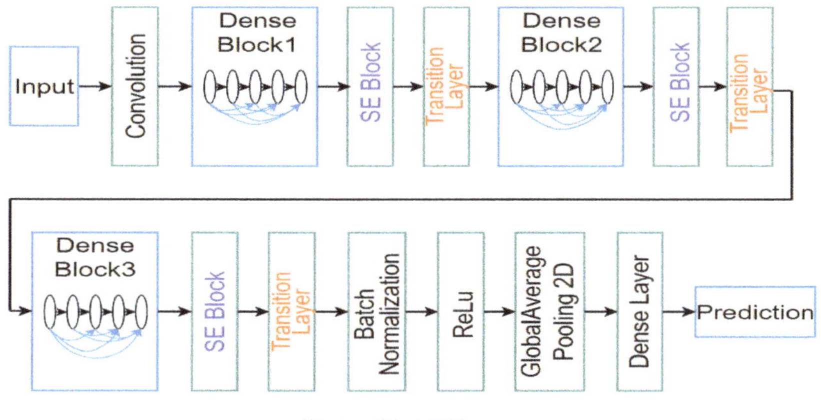

Figure 8. The structure of the DenseNet-PR [1]. (© 2020 IEEE)

2.5. Video Quality Prediction

To improve the generalization ability of our proposed SAAM metric, we design an ensemble model using Support Vector Regression (SVR) model based on Boostrap Aggregating (Bagging) as shown in Figure 9.

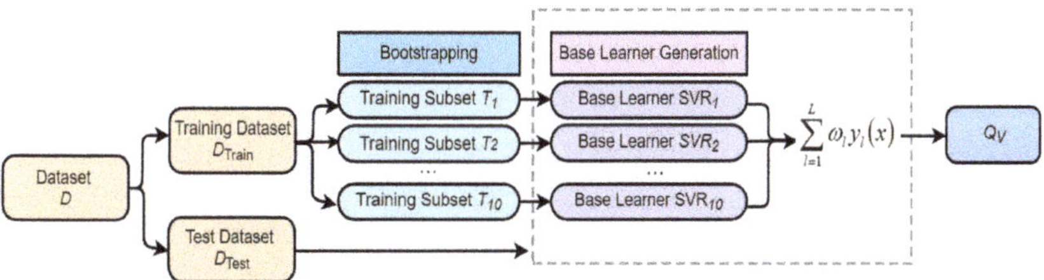

Figure 9. SVR model based on bagging.

After obtaining the intensity values of PEAs, ensemble learning model is adopted to map the intensity values of four types of artifacts to MOS|DMOS values. First, for anyone selected VQA database, we form a complete data set D by matching the intensity values of four types of artifacts to MOS $|$ DMOS, which can be expressed as

$$D = \{(I_{V_1}, \text{MOS}_1|\text{DMOS}_1), (I_{V_2}, \text{MOS}_2|\text{DMOS}_2), ..., (I_{V_k}, \text{MOS}_m|\text{DMOS}_m)\}, \tag{5}$$

where $\text{MOS}_m|\text{DMOS}_m$ represents the quality score MOS $|$ DMOS of the m-th compressed video. Second, the date set is randomly split into training set D_{Train} and testing set D_{Test} at a ratio of 80:20, and 10 sub-training sets $[T_1, T_2, ..., T_{10}]$ are resampled from the D_{Train}. Note that the sub-training set $[T_1, T_2, ..., T_{10}]$ and D_{Train} contain the same number of samples. Then, we train 10 SVR models as the base learners through $[T_1, T_2, ..., T_{10}]$, that is, the intensity values of four types of PEAs for video sequences are fed into SVR, and then SVR output a predicted value for each video. In this work, we chose the radial basis function as the kernel of SVR due to its better performance. Next, D_{Test} is utilized to evaluate the performance of these base learners by the PLCC between the predicted quality scores and the true quality scores. Finally, the final prediction result Q_V is obtained as follows:

$$Q_V = f(\sum \sum I_{ij}) = \sum_{l=1}^{L} \omega_l y_l(x), \tag{6}$$

$$\sum_{l=1}^{L} \omega_l = 1 \quad \omega_l \geq 0, \tag{7}$$

where $f(\cdot)$ refers to summation operation. x represents D_{Test}. $y_l(x)$ refers to the prediction output of the l-th base learner. L is the number of base learners, which value is 10. ω_l denotes the weight of the l-th base learner. In this work, we set the weights of based learners with the top three PLCC to $1/3$, and the weights of the remaining base learners are set to 0.

3. Experiments and Discussions

To evaluate the performance of our proposed algorithm, it is examined on four publicly and widely used Video Quality Databases (VQD): LIVE, CSIQ, IVP and FERIT-RTRK. Among them, the compressed videos generated by H.264 encoder are utilized here to evaluate the performance of SAAM. The LIVE VQD contains 40 compressed videos with a resolution of 768×432 [30]. The CSIQ VQD contains 36 compressed videos with a resolution of 832×480 [36]. The IVP VQD contains 40 compressed videos with a resolution of 1920×1088 [37]. The FERIT-RTRK VQD consists 30 compressed videos with a resolution of 1920×1080 [38]. Based on the compressed videos from the four VQA databases, we individually form four complete data sets by matching the intensity values of four types of artifacts to their MOS $|$ DMOS as described in Section 2.5.

To show the superiority of our method, it is compared with typical video quality metrics including PSNR, SSIM [3], MS-SSIM [39], STRRED [4], SpEED-QA [5], BRISQUE [40], NIQE [41] and VIIDEO [11]. Among them, PSNR, SSIM and MS-SSIM are FR metrics. STRRED and SpEED-QA are RR metrics, and BRISQUE, NIQE and VIIDEO are NR metrics. All methods are compared in terms of the PLCC and SRCC, which characterize the correlation between VQA results and MOS $|$ DMOS values. The results are reported in Tables 2 and 3. Among them, as our SAAM metric is based on machine learning, to fairly verify its performance, the result of our metric is the median value of 15 repeated processes. In addition, the overall performance of each VQA algorithm on the four video databases is listed in the last column of the tables, expressed by weighted PLCC and SRCC. The weight of each database depends on the number of distorted videos in the database, and the optimal performance is given in bold.

Table 2. Performance comparison in terms of PLCC.

Methods	LIVE	CSIQ	IVP	FERIT-RTRK	Overall
PSNR	0.5735	0.8220	0.7998	0.7756	0.7383
SSIM	0.6072	0.8454	0.8197	0.6870	0.7406
MS-SSIM	0.6855	0.8782	0.8282	0.8724	0.8105
STRRED	0.8392	0.8772	0.5947	0.8425	0.7823
SpEED-QA	0.7933	0.8554	0.6822	0.6978	0.7586
BRISQUE	0.2154	0.5526	0.2956	0.7653	0.4335
NIQE	0.3311	0.5350	0.3955	0.5817	0.4505
VIIDEO	0.6829	0.7211	0.4358	0.3933	0.5651
SAAM	**0.9023**	**0.9244**	**0.8717**	**0.9499**	**0.9091**

Table 3. Performance comparison in terms of SRCC.

Methods	LIVE	CSIQ	IVP	FERIT-RTRK	Overall
PSNR	0.4146	0.8028	0.8154	0.7685	0.6928
SSIM	0.5677	0.8440	0.8049	0.7236	0.7328
MS-SSIM	0.6773	0.9465	0.7917	0.8508	0.8107
STRRED	0.8358	**0.9770**	0.8595	0.8310	0.8761
SpEED-QA	0.7895	0.9639	**0.8812**	0.7945	0.8587
BRISQUE	0.2638	0.5655	0.1051	0.7574	0.3961
NIQE	0.1769	0.5012	0.2351	0.4855	0.3362
VIIDEO	0.6593	0.7153	0.1621	0.3177	0.4667
SAAM	**0.8691**	0.8810	0.8413	**0.9429**	**0.8796**

From the tables, our algorithm delivers strong competitive performance on these datasets. First, the PLCC of the proposed SAAM approach outperforms all of compared methods on the four databases. Second, on the CSIQ database, the SRCC of the SAAM outperforms that of PSNR, SSIM, BRISQUE, NIQE and VIIDEO and is competitive with the performance of MS-SSIM, STRRED and SpEED-QA. Finally, the overall performance of the SAAM is better than all of compared VQA methods. The experimental results also show that there is a strong correlation between the PEA intensity and subjective quality of a compressed video, and the PEAs affect the viewing experience of end users.

To further verify the generalization of the proposed algorithm, we also studied cross-dataset evaluation in Table 4. It can be observed that when CSIQ VQD is used as the testing set, the performance of using LIVE VQD as the training set is relatively better than that of utilizing FERIT-RTRK VQD as the training set. The most likely reason may be that the resolutions of LIVE and CSIQ databases are very close, but the difference of resolutions between FERIT-RTRK and CSIQ VQD is relatively bigger. In addition, the subjective quality scores provided by LIVE and CSIQ VQD are MOS values, while the scores provided by FERIT-RTRK VQD are DMOS values. Therefore, different scoring standards of the subjective quality scores may also cause differences in cross-database performance. Generally, the performances of image processing algorithms are usually not very good in cross-database performance verification, especially considering that the video resolutions and contents of various databases are different. At this point, our performance of cross-database experiment is acceptable. It also shows that our proposed model has good generalization and robustness. In our algorithm, we can adjust the PEA recognition models to further improve the recognition accuracy of artifacts and increase the correlation between SAAM results and MOS | DMOS, which will become our next work.

Table 4. Cross-database validation.

Training Set	Testing Set	PLCC	SRCC
LIVE	CSIQ	0.7107	0.7290
FERIT-RTRK	CSIQ	0.4437	0.5302

Besides, we also perform ablation experiments on CSIQ VQD to verify the advantages of saliency detection. We crop the whole frame into patches for PEA detection, and the size of data is 1.11 GB. We only retain the patches containing saliency regions for PEA detection, the size of data is reduced to 0.37 GB and the storage is saved by 66.49%. The time consumption of PEA detection is reduced from 6.47 h to 1.50 h, saving by 76.82%, as shown in Table 5.

Table 5. Ablation experiments on the CSIQ VQD database.

	Data Size	Time of PEAs Detection	PLCC	SRCC
Without Saliency	1.11 GB	6.47h	0.9557	0.8929
With Saliency	0.37 GB	0.50h	0.9244	0.8810

4. Conclusions

In this paper, we propose a NR-VQA metric called SAAM, based on the intensity values of four typical types of artifacts (i.e., blurring, blocking, ringing and color bleeding). To the best of our knowledge, this is the first work combining video saliency with artifacts detection to predict the quality of compressed video. The experimental results demonstrate that the proposed algorithm delivers competitive performance with common video quality metrics in different datasets. As future work, we plan to design a NR-VQA algorithm based on natural video statistics, which can detect more types of video PEAs.

Author Contributions: Conceptualization, L.L. and W.C.; methodology, J.Y. and Z.W.; software, Z.W. and L.Z.; validation, L.L., J.Y. and Z.W.; investigation, J.Y. and L.Z.; resources, L.L.; data curation, L.Z.; writing—original draft preparation, J.Y.; writing—review and editing, L.L. and W.C.; visualization, J.Y. and Z.W.; supervision, Y.X.; project administration, L.L.; funding acquisition, Y.X. All authors have read and agreed to the published version of the manuscript.

Funding: This research was funded by Natural Science Foundation of China No. 62171134 and No. 61901119 and Natural Science Foundation of Fujian Province No. 2019J01222.

Institutional Review Board Statement: Not applicable.

Informed Consent Statement: Informed consent was obtained from all subjects involved in the study.

Data Availability Statement: Data available on request from the authors.

Conflicts of Interest: The authors declare no conflict of interest.

References

1. Lin, L.; Yu, S.; Zhou, L.; Chen, W.; Zhao, T.; Wang, Z. PEA265: Perceptual assessment of video compression artifacts. *IEEE Trans. Circuits Syst. Video Technol.* **2020**, *28*, 3898–3910. [CrossRef]
2. International Telecommunication Union. *Methodology for the Subjective Assessment of the Quality of Television Pictures*; Recommendation ITU-R BT.500-13; International Telecommunication Union: Geneva, Switzerland, 2012.
3. Wang, Z.; Bovik, A.C.; Sheikh, H.R.; Simoncelli, E.P. Image quality assessment: From error visibility to structural similarity. *IEEE Trans. Image Process.* **2004**, *13*, 600–612. [CrossRef]
4. Soundararajan, R.; Bovik, A.C. Video quality assessment by reduced reference spatiotemporal entropic differencing. *IEEE Trans. Circuits Syst. Video Technol.* **2013**, *23*, 684–694. [CrossRef]
5. Bampis, C.G.; Gupta, P.; Soundararajan, R.; Bovik, A.C. SpEED-QA: Spatial efficient entropic differencing for image and video quality. *IEEE Signal Process. Lett.* **2017**, *24*, 1333–1337. [CrossRef]
6. Saad, M.A.; Bovik, A.C.; Charrier, C. Blind prediction of natural video quality. *IEEE Trans. Image Process.* **2014**, *23*, 1352–1365. [CrossRef] [PubMed]
7. Li, Y.; Po, L.; Cheung, C.; Xu, X.; Feng, L.; Yuan, F.; Cheung, K. No-reference video quality assessment with 3D shearlet transform and convolutional neural networks. *IEEE Trans. Circuits Syst. Video Technol.* **2016**, *26*, 1044–1057. [CrossRef]
8. Wang, S.; Gu, K.; Zhang, X.; Lin, W.; Zhang, L.; Ma, S.; Gao, W. Subjective and Objective Quality Assessment of Compressed Screen Content Images. *IEEE J. Emerg. Sel. Top. Circuits Syst.* **2016**, *6*, 532–543. [CrossRef]
9. Ye, P.; Kumar, J.; Kang, L.; Doermann, D. Unsupervised feature learning framework for no-reference image quality assessment. In Proceedings of the IEEE Conference on Computer Vision and Pattern Recognition (CVPR), Providence, RI, USA, 16–21 June 2012; pp. 1098–1105.

10. Xu, J.; Ye, P.; Liu, Y.; Doermann, D. No-reference video quality assessment via feature learning. In Proceedings of the IEEE Conference on Image Processing (ICIP), Paris, France, 27–30 October 2014; pp. 491–495.
11. Mittal, A.; Saad, M.A.; Bovik, A.C. A completely blind video integrity oracle. *IEEE Trans. Image Process.* **2016**, *25*, 289–300. [CrossRef]
12. Zhu, Y.; Wang, Y.; Shuai, Y. Blind video quality assessment based on spatio-temporal internal generative mechanism. In Proceedings of the IEEE Conference on Image Processing (ICIP), Beijing, China, 17–20 September 2017; pp. 305–309.
13. Reddy, D.S.V.; Channappayya, S.S. No-reference video quality assessment using natural spatiotemporal scene statistics. *IEEE Trans. Image Process.* **2020**, *29*, 5612–5624. [CrossRef]
14. Abate, L.; Ramponi, G.; Stessen, J. Detection and measurement of the blocking artifact in decoded video frames. *Signal Image Video Process.* **2013**, *7*, 453–466. [CrossRef]
15. Amor, M.B.; Larabi, M.C.; Kammoun, F.; Masmoudi, N. A no reference quality metric to measure the blocking artefacts for video sequences. *J. Photogr. Sci.* **2016**, *64*, 408–417.
16. Xue, Y.; Erkin, B.; Wang, Y. A novel no-reference video quality metric for evaluating temporal jerkiness due to frame freezing. *IEEE Trans. Multimedia* **2014**, *17*, 134–139. [CrossRef]
17. Zhu, K.; Li, C.; Asari, V.; Saupe, D. No-reference video quality assessment based on artifact measurement and statistical analysis. *IEEE Trans. Circuits Syst. Video Technol.* **2015**, *25*, 533–546. [CrossRef]
18. Men, H.; Lin, H.; Saupe, D. Empirical evaluation of no-reference VQA methods on a natural video quality database. In Proceedings of the International Conference on Quality of Multimedia Experience (QoMEX), Erfurt, Germany, 31 May–2 June 2017; pp. 1–3.
19. Vranje, M.; Bajinovci, V.; Grbi, R.; Vajak, D. No-reference artifacts measurements based video quality metric. *Signal Process Image Commun.* **2019**, *78*, 345–358. [CrossRef]
20. Rohil, M.K.; Gupta, N.; Yadav, P. An improved model for no-reference image quality assessment and a no-reference video quality assessment model based on frame analysis. *Signal Image Video Process.* **2020**, *14*, 205–213. [CrossRef]
21. Wang, W.; Shen, J.; Xie, J.; Chen, M.; Ling, H.; Borji, A. Revisiting video saliency prediction in the deep learning era. *IEEE Trans. Pattern Anal. Mach. Intell.* **2019**, *43*, 220–237. [CrossRef] [PubMed]
22. Kai, Z.; Zhao, T.; Rehman, A.; Zhou, W. Characterizing perceptual artifacts in compressed video streams. In Proceedings of the Human Vision and Electronic Imaging XIX, San Francisco, CA, USA, 3–6 February 2014; pp. 173–182.
23. LIVE Video Quality Database. Available online: http://live.ece.utexas.edu/research/quality/live_video.html (accessed on 20 September 2021).
24. Seshadrinathan, K.; Soundararajan, R.; Bovik, A.C.; Cormack, L. Study of subjective and objective quality assessment of video. *IEEE Trans. Image Process.* **2010**, *19*, 1427–1441. [CrossRef]
25. Seshadrinathan, K.; Soundararajan, R.; Bovik, A.C.; Cormack, L. A Subjective Study to Evaluate Video Quality Assessment Algorithms. In Proceedings of the Human Vision and Electronic Imaging XV, San Jose, CA, USA, 18–21 January 2010.
26. Li, L.; Lin, W.; Zhu, H. Learning structural regularity for evaluating blocking artifacts in JPEG images. *IEEE Signal Process. Lett.* **2014**, *21*, 918–922. [CrossRef]
27. Ultra Video Group. Available online: http://ultravideo.cs.tut.fi/#main (accessed on 20 September 2021).
28. Wu, J.; Liu, Y.; Dong, W.; Shi, G.; Lin, W. Quality Assessment for Video With Degradation Along Salient Trajectories. *IEEE Trans. Multimedia* **2019**, *21*, 2738–2749. [CrossRef]
29. Duanmu, Z.; Zeng, K.; Ma, K.; Rehman, A.; Wang, Z. A Quality-of-Experience Index for Streaming Video. *IEEE J. Sel. Top. Signal Process.* **2017**, *11*, 154–166. [CrossRef]
30. Hu, S.; Jin, L. Wang, H.; Zhang, Y.; Kwong, S.; Kuo, C. Objective Video Quality Assessment Based on Perceptually Weighted Mean Squared Error. *IEEE Trans. Circuits Syst. Video Technol.* **2017**, *27*, 1844–1855. [CrossRef]
31. Yang, J.; Ji, C.; Jiang, B.; Lu, W.; Meng, Q. No reference quality assessment of stereo video based on saliency and sparsity. *IEEE Trans. Broadcast.* **2018**, *64*, 341–353. [CrossRef]
32. Feng, W.; Li, X.; Gao, G.; Chen, X.; Liu, Q. Multi-Scale Global Contrast CNN for Salient Object Detection. *Sensors* **2020**, *20*, 2656. [CrossRef] [PubMed]
33. Shi, X.; Chen, Z.; Wang, H.; Yeung, D.; Wong, W.; Woo, W. Convolutional LSTM network: A machine learning approach for precipitation nowcasting. In Proceedings of the 28th International Conference on Neural Information Processing Systems, Montreal, QC, Canada, 7–12 December 2015; pp. 802–810.
34. Xu, K.; Ba, J.; Kiros, R.; Cho, K.; Courville, A.; Salakhutdinov, R.; Zemel, R.; Bengio, Y. Show, attend and tell: Neural image caption generation with visual attention. In Proceedings of the International Conference on Machine Learning, Lille, France, 7–9 July 2015; pp. 2048–2057.
35. He, K.; Zhang, X.; Ren, S.; Sun, J. Deep residual learning for image recognition. In Proceedings of the IEEE Conference on Computer Vision and Pattern Recognition (CVPR), Las Vegas, NV, USA, 27–30 June 2016; pp. 770–778.
36. Vu, P.V.; Chandler, D.M. ViS3: An algorithm for video quality assessment via analysis of spatial and spatiotemporal slices. *J. Electron. Imaging* **2014**, *23*, 1–25. [CrossRef]
37. Zhang, F.; Li, S.; Ma, L.; Wong, Y.; Ngan, K. IVP Subjective Quality Video Database. Available online: http://ivp.ee.cuhk.edu.hk/research/database/subjective (accessed on 20 September 2021).
38. Bajčinovci, V.; Vranješ, M.; Babić, D.; Kovačević, B. Subjective and objective quality assessment of MPEG-2, H.264 and H.265 videos. In Proceedings of the International Symposium ELMAR, Zadar, Croatia, 18–20 September 2017; pp. 73–77.

39. Sun, W.; Liao, Q.; Xue, J.; Zhou, F. SPSIM: A Superpixel-Based Similarity Index for Full-Reference Image Quality Assessment. *IEEE Trans. Image Process.* **2018**, *27*, 4232–4244. [CrossRef]
40. Mittal, A.; Moorthy, A.K.; Bovik, A.C. No-reference image quality assessment in the spatial domain. *IEEE Trans. Image Process.* **2012**, *21*, 4695–4708. [CrossRef] [PubMed]
41. Mittal, A.; Soundararajan, R.; Bovik, A.C. Making a completely blind image quality analyzer. *IEEE Signal Process. Lett.* **2013**, *20*, 209–212. [CrossRef]

Article

Wheat Ear Recognition Based on RetinaNet and Transfer Learning

Jingbo Li, Changchun Li *, Shuaipeng Fei, Chunyan Ma, Weinan Chen, Fan Ding, Yilin Wang, Yacong Li, Jinjin Shi and Zhen Xiao

School of Surveying and Mapping Land Information Engineering, Henan Polytechnic University, Jiaozuo 454000, China; lijingbo1024@163.com (J.L.); feishuaipeng@163.com (S.F.); mayan@hpu.edu.cn (C.M.); 211904020046@home.hpu.edu.cn (W.C.); 212004020067@home.hpu.edu.cn (F.D.); 211904010019@home.hpu.edu.cn (Y.W.); 211904020026@home.hpu.edu.cn (Y.L.); 211804010022@home.hpu.edu.cn (J.S.); 212004020068@home.hpu.edu.cn (Z.X.)
* Correspondence: lichangchun610@126.com

Abstract: The number of wheat ears is an essential indicator for wheat production and yield estimation, but accurately obtaining wheat ears requires expensive manual cost and labor time. Meanwhile, the characteristics of wheat ears provide less information, and the color is consistent with the background, which can be challenging to obtain the number of wheat ears required. In this paper, the performance of Faster regions with convolutional neural networks (Faster R-CNN) and RetinaNet to predict the number of wheat ears for wheat at different growth stages under different conditions is investigated. The results show that using the Global WHEAT dataset for recognition, the RetinaNet method, and the Faster R-CNN method achieve an average accuracy of 0.82 and 0.72, with the RetinaNet method obtaining the highest recognition accuracy. Secondly, using the collected image data for recognition, the R^2 of RetinaNet and Faster R-CNN after transfer learning is 0.9722 and 0.8702, respectively, indicating that the recognition accuracy of the RetinaNet method is higher on different data sets. We also tested wheat ears at both the filling and maturity stages; our proposed method has proven to be very robust (the R^2 is above 90). This study provides technical support and a reference for automatic wheat ear recognition and yield estimation.

Keywords: RetinaNet; deep learning; transfer learning; wheat ears; Global WHEAT

Citation: Li, J.; Li, C.; Fei, S.; Ma, C.; Chen, W.; Ding, F.; Wang, Y.; Li, Y.; Shi, J.; Xiao, Z. Wheat Ear Recognition Based on RetinaNet and Transfer Learning. *Sensors* **2021**, *21*, 4845. https://doi.org/10.3390/s21144845

Academic Editors: Yun Zhang, KWONG Tak Wu Sam, Xu Long and Tiesong Zhao

Received: 22 June 2021
Accepted: 12 July 2021
Published: 16 July 2021

Publisher's Note: MDPI stays neutral with regard to jurisdictional claims in published maps and institutional affiliations.

Copyright: © 2021 by the authors. Licensee MDPI, Basel, Switzerland. This article is an open access article distributed under the terms and conditions of the Creative Commons Attribution (CC BY) license (https://creativecommons.org/licenses/by/4.0/).

1. Introduction

Wheat is the largest grain crop in world trade [1,2]. Recently, with population growth and social and economic development, the demand for wheat has increased. However, due to extreme weather, pests, crop diseases, and yield, the wheat supply is unstable [3]. Therefore, maintaining a high and stable wheat yield is essential to improving people's living standards, maintaining social stability, and promoting the development of the national economy [4]. The number of wheat ears is an important factor that directly affects wheat yield [5–7]. Therefore, rapid and accurate identification and statistics of wheat ears are fundamental for crop growth monitoring and yield estimation [8]. Traditional counting methods rely on field surveys, sampling, and weighing. These methods are inefficient, costly, and difficult to determine accurate yield estimation for large areas, severely limiting their application for breeding, monitoring plant performance in crop management, or predicting grain yield. No model has been able to perform consistently across different wheat reproductive stages and identify the derived wheat spikes well. Additionally, some spike counting methods are based on wheat spike data collected at maturity and other traits, which are not suitable for early yield prediction [9]. In recent years, many studies that apply deep learning techniques [10] to unmanned aerial systems [11,12] for wheat spikelet detection under field conditions have received much attention.

In recent years, with the development of artificial intelligence [13], the target detection models built using deep learning, e.g., Faster regions with convolutional neural networks (R-CNN), has far surpassed traditional target detection techniques in feature representation, reaching top performance in terms of detection accuracy and speed [14]. However, deep learning techniques are limited by a large number of training datasets and training equipment. Since the color, shape, and awn of wheat ears change with growth, the same wheat varieties differ in performance in different growing regions. Thus, it is crucial to study a deep learning model for regional wheat detection. The current Faster R-CNN method achieves better detection accuracy in wheat ear detection [15], but its detection speed cannot meet the integration requirements in the unmanned aerial vehicle (UAV) system. Therefore, it is of great significance and application value to develop a wheat detection model with high detection accuracy, migration ability and be integrated into the UAV system.

The target detection models built using deep learning are mainly divided into two categories: two-stage and one-stage. The most representative target detection algorithms for two-stage detectors are Fast R-CNN [16], Faster R-CNN [14], and the most representative target detection algorithms for one-stage detectors are YOLO (You Only Look Once) [17,18] and RetinaNet [19]. Two-stage detectors are usually slower than one-stage detectors. In two-stage detectors, the first step determines the regions (regions of proposals) that might contain a target to be detected (location). The second step performs a detailed identification of the target contained in each candidate region (classification) [14]. RetinaNet combines the advantages of multiple target recognition methods, especially the"anchor" concept introduced by Region Proposal Network (RPN) [14], and the use of feature pyramids in Single Shot Multibox Detector (SSD) [20] and Feature Pyramid Networks (FPN) [21]. Retinanet has a wide range of applications, such as ship detection in remote sensing images of different resolutions [22], identification of storm drains and manholes in urban areas [23], fly identification [24], and rail surface crack detection [25].

The future development trend of the image-based automatic recognition method is to obtain wheat ears and yield data in real-time over a large area. Using machine learning techniques, the image-based recognition method exploits the color, texture, and shape of the target image to encode and represent the image. Thus, image feature representation is essential [26]. However, these features are different in different environmental conditions, limiting the effectiveness of these features. For example, [27] exploited color consistency coefficient, gray symbiosis matrix, and edge histogram to construct wheat ear feature matrix. The authors of [28] used scale-invariant feature transform (SIFT), and Fisher vector (FV) features to identify the wheat ears at the heading stage accurately. The authors of [9] created a binary image through the local maximum value of the pixel and the variance of the nearest neighbor pixel to calculate the number of wheat ears in the image. The experimental results of these studies may suffer from degradation due to different recognition angles of the wheat ears being photographed, the period of growth of the ears, and the field environment in which the wheat ears were photographed. As a branch of machine learning, current deep learning techniques can solve this problem for wheat recognition. Deep Learning [29] exploits a perceptron containing multiple hidden layers, and it transforms the features of samples from the original space to a new feature space based on the principle of learning hierarchical data. In this process, the hierarchical feature representation was automatically learned and obtained, and the accuracy of recognition improved [29]. The main advantage of deep learning techniques is that the characteristics of the input data are automatically learned, overcoming the bottleneck in many intelligent applications. Thus, the use of deep learning techniques has become the frontier in the field of crop phenotypes.

The current application of drones combined with deep learning technology has greatly promoted the development of precision agriculture. In recent years, some meaningful research [7–9,15,27,28,30–42] has emerged. These studies have used RGB (red, green, blue), multispectral, hyperspectral, and thermal infrared data acquired by UAV and CNN to evaluate the phenotypic characteristics of citrus crops [38], obtain key points of plants/plant leaves [39], plant stress analysis and plant disease identification [40,41]. The research on automatic recognition and counting wheat ears with deep learning technology has made great progress [8,15,37]. Hasan et al. [8] used R-CNN to identify and count wheat ears. Madec et al. [15] used Faster R-CNN to identify wheat ears in RGB images with different spatial resolutions. Sadeghi–Tehran et al. [37] proposed a visual recognition method based on linear iterative clustering and deep CNN to automatically identify and count wheat ears in the images obtained under natural field conditions. The research works [8,15,37] achieved a lower detection speed than [30,42], and cannot be integrated into UAV systems for real-time detection.

To our knowledge, there are no other systematic, quantitative assessments of how training sample size and sample selection methods affect the results of wheat identification models. This paper aims to: (1) Obtain a model with high recognition accuracy at different growth stages; (2) evaluate the impact of training samples on Faster-RCNN and RetinaNet in combination with transfer learning; (3) evaluate the detection speed of Faster-RCNN and RetinaNet. To achieve these goals, there are many different types of wheat ears in different growth environments considered in this paper, and a model suitable for regional wheat identification is proposed. The idea of transfer learning is integrated into the proposed model to explore its performance in different training samples, wheat fertility training samples, and obtain the recognition performance of the model in wheat ear images. Specifically, images of wheat ears in different fertility periods collected in the field are used in combination with the wheat ears recognition image database (Global WHEAT). Different deep learning models based on different data samples of wheat ears in different fertility periods are trained by labeling wheat ears. The recognition accuracy and detection speed of these models are then compared and analyzed. The model proposed in this paper achieves high detection accuracy and migration capability, and it can be integrated into UAV systems.

2. Materials and Methods

2.1. Data Acquisition and Processing

2.1.1. Global Wheat Data Acquisition

Global Wheat Head Detection (Global WHEAT) [43] obtains data from the wheat ear recognition image database (data source: https://www.kaggle.com/c/global-wheat-detection/data, accessed on 12 June 2020). The database was carefully produced by seven countries, nine research institutes, and more than ten phenotyping experts in a year. This database is composed of train.zip, test.zip, sample_ssubmission.csv, train.csv, and other files. The information contained in each file is listed in Table 1.

Table 1. The information of the data files in the Global WHEAT database.

Files	Information
train.csv	the training data
sample_submission.csv	a sample submission file in the correct format
train.zip	training images
test.zip	test images

The Global WHEAT data set contains a total of 3422 images, and each image has a size of 1024 × 1024 pixels. The Ground Sampling Distance (GSD) of the GWHD dataset ranges from 0.28 to 0.55 mm [43]. Part of the image data is shown in Figure 1.

Figure 1. Part of the image data in the Global WHEAT database.

2.1.2. Digital Image Data Acquisition

The digital image of wheat in the filling stage and mature stage are obtained by a high-definition digital camera. In clear and windless conditions, the backlit hand-held Sony DSC-H9 digital camera was used for vertical shooting. The shooting height was approximately 1 m higher than the top of the wheat canopy, and the shooting area was approximately 0.75 M^2 (5 rows of wheat with a spacing of 15 cm). Each digital image has 3088 × 2056 pixels, and the horizontal and vertical resolution is 72 dpi. A total of 715 images are taken. Among these 715 images, 365 images present the wheat ears in the filling stage and 350 in the mature stage with an approximate ratio of 1:1. The digital images of wheat ears in the partial filling stage and mature stage are illustrated in Figure 2.

According to the number of images, three groups of digital images of wheat in the filling and mature stages are set as the training data set, and the number of images in each group is 50, 100, and 150, respectively. The denotations of the three situations are listed in Table 2, where Filling Stage Model (FSM) and Mature Stage Model (MSM) represent a wheat data set at the grain filling stage and mature stage.

According to the number of wheat ears in each image, the test data set is divided into three groups, and the number of images in each group is 30. The detailed information of the training data set and the test data set is listed in Tables 2 and 3. Testing the same number of images ensures that each model has the same evaluation benchmark, and the test set is 180 images in total. The ratio of 90 images for each fertility period is 1:1, which guarantees the reliability of the results.

(a)   (b)

Figure 2. Digital images of wheat in key growth stages. (**a**) Partial Filling stage; (**b**) Mature stage.

Table 2. Training database.

Growth Period	Database	Number of Images Per Group	Number of Wheat Ears Per Piece
Filling stage	FSM50	50	6409
	FSM100	100	12,733
	FSM150	150	19,275
Mature stage	MSM50	50	6684
	MSM100	100	13,404
	MSM150	150	20,132

Table 3. Testing database.

Growth Period	Number of Wheat Ears in Each Image	Number of Images Per Group	Number of Wheat Ears Per Piece	Total Number of Wheat Ears Per Piece
Filling stage	less than 50	30	1125	
	50–100	30	2458	6814
	more than 100	30	3231	
Mature stage	less than 50	30	1128	
	50–100	30	2116	6699
	more than 100	30	3455	

2.1.3. Data Processing

(1) Image marking

To obtain better recognition results, many deep learning models require annotated training data sets. Although image-based high-throughput crop phenotyping systems already exist, such as Field Scanalyzer [44], which generates a large amount of image data every day, the annotated images with ground truth values are not available among the obtained crop image data. Therefore, the obtained images must be labeled to generate a training data set.

LABELIMG [45] is a free and open-source graphic image annotation tool (https://github.com/tzutalin/labelImg, accessed on 13 June 2020) that grants simultaneous access to different users and is available to all institutions. The LABELIMG tool outputs an annotation file with an interactive drawing of a bounding box containing all the pixels of the wheat ears. After the digital image is obtained, theannotation tool of LABELIMG can be used to draw bounding boxes around each identified wheat ear in the images. Specific information for each image is shown in Figure 3; Figure 3a exhibits the position of wheat ears in the image; Figure 3b illustrates the name of the target, and Figure 3c indicates the shape of the image. The bounding boxes contain all the pixels of the wheat ears, however, sometimes the bounding box can be too large and includes background lawns. If possible, the boxes also contain a small portion of the wheat stem. When the identification results from one of the developed models were compared, it was found that a few wheat ears were forgotten by the operator during the interactive labeling process. Therefore, the images were reprocessed with greater care.

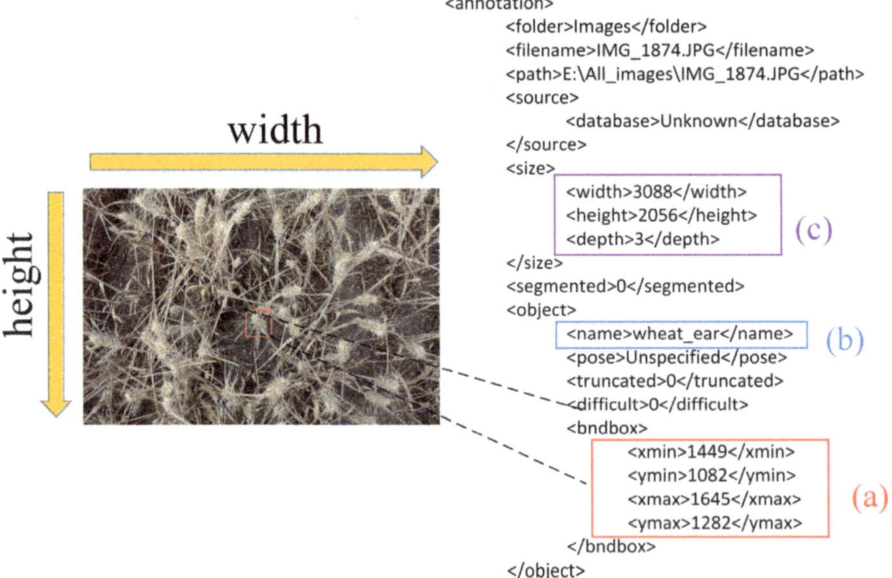

Figure 3. Wheat ear label image data. (**a**) the position of wheat ears in the image; (**b**) the name of the target; (**c**) the shape of the image.

Each labeled image has an additional text file containing the coordinates of the annotated bounding boxes. In this file, the boxes are stored as a 4-tuple $(x_{min}, y_{min}, x_{max}, y_{max})$, where (x_{min}, y_{min}) and (x_{max}, y_{max}) denote the top left corner and the lower right corner of the box, respectively.

(2) Denoising and enhancement

During the shooting process, the wheat image is easily affected by the changes in natural light, growth environment, shaking of the shooting equipment, and the unstable focus of the lens. Meanwhile, the obtained image may contain some noise caused by random signals in the process of transmission [34]. Therefore, the method of data denoising is exploited to remove the noise points in the obtained image and reduce the influence of noise on the recognition results. Firstly, the median filtering method with a kernel of 5 is used to remove the noise in the wheat image. The specific denoising process exploits the Python language to call the medianBlur function provided by the OpenCV library, and the parameter ksize is set to 5 [34].

In the training process, the training samples cannot always reflect all the information for each real target. Thus, the image data is enhanced by different transformations to improve the training data and generalization ability. The diversity allows the model to be applied to various situations and has more robustness. Additionally, most deep learning algorithms require a large amount of training data to obtain accurate recognition results. Although this work obtained approximately 3000 images, there were not enough to train and validate the model, emphasizing the need for data enhancement. In order to address this issue, programs have been written in Python language to shrink, enlarge, and flip the original image [34]. In order to simulate the change of light, the HSV (hue, saturation, and value) [46], color space, and various conversions are exploited, such as linear change of hue, linear change of saturation, and linear change of brightness. PIL (Python Image Library) [47] is a third-party image processing library provided by the python (https://python-pillow.org/, accessed on 10 September 2020) language. This library is featured with extensive file format support, efficient internal representation, and strong image processing capabilities. It provides a solid foundation for general image processing tools. The function FLIP_LEFT_RIGHT can flip an image horizontally. Specifically, a Python program is written to call the transpose function provided by the PIL library, and the parameter of FLIP_LEFT_RIGHT is used. Based on this, each HSV channel's value is changed linearly and randomly; the rand function provided by the Numpy library randomly returns a value between 0 and 1 for multiplication. The image denoising and data enhancement effects are illustrated in Figure 4.

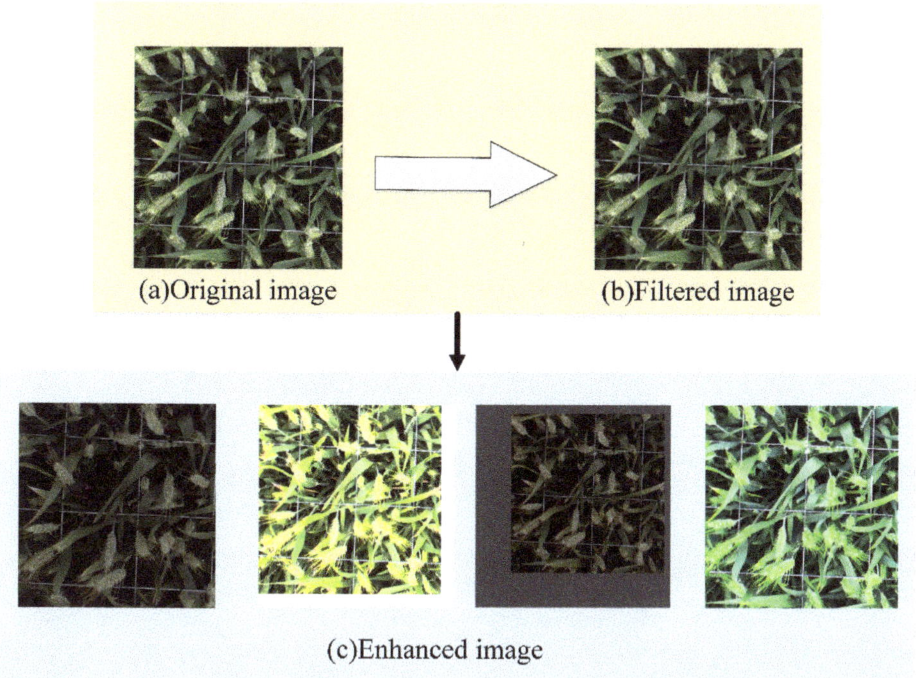

Figure 4. Image denoising and data enhancement. (**a**) original image; (**b**) filtered image; (**c**) enhanced image.

2.2. Method

The application of deep learning techniques has greatly contributed to the development of precision agriculture. Faster R-CNN has been widely applied and used in maize tassels detection [35] and wheat ears recognition [15]. RetinaNet combines the advantages of multiple target recognition methods, especially the "anchor" concept introduced by Region Proposal Network (RPN) [14], and the use of feature pyramids in Single Shot Multibox Detector (SSD) [20] and Feature Pyramid Networks (FPN) [21]. Retinanet has a wide range of applications, such as ship detection in remote sensing images of different resolutions [22], identification of storm drains and manholes in urban areas [23], fly identification [24], and rail surface crack detection [25]. The experiment described in this paper was conducted on a computer equipped with Intel® Xen(R) W-2145 CPU and NVIDIA GeForceRTX 2080Ti.

The Keras library based on the Tensorflow environment in Windows was employed. Additionally, the Python language was employed to realize the automatic recognition of wheat ears based on Faster R-CNN and RetinaNet and verify the recognition accuracy.

2.2.1. Faster R-CNN

As a typical two-stage target recognition algorithm, Faster regions with convolutional neural networks (Faster R-CNN) [14] has been widely applied to many fields since it was proposed. Faster R-CNN is an improved version of Fast-RCNN [16], which uses RPN network (Region Proposal Network) instead of Selective Search to generate candidate boxes. Additionally, the anchor concept is introduced and can be used in future target recognition models.

As shown in Figure 5, Faster R-CNN consists of four parts:

(1)   Convolution layer.

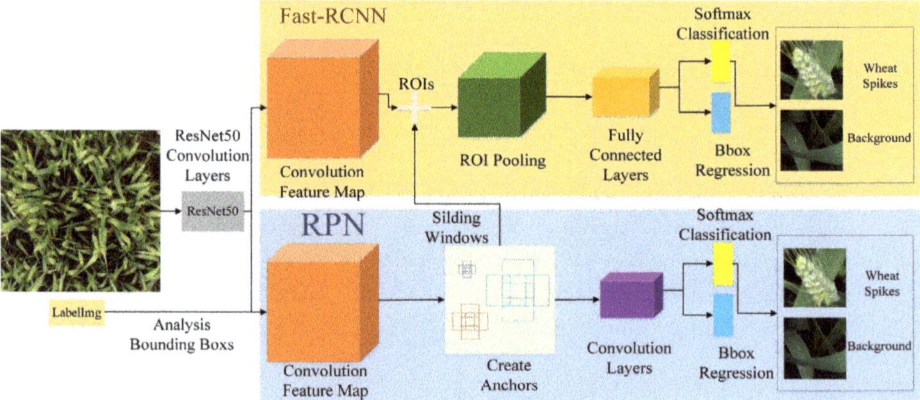

Figure 5. The diagram of the Faster R-CNN framework.

As a CNN network for target recognition, the convolution layer in Faster R-CNN uses ResNet50 as the feature extraction network. The network in this layer extracts the feature map of the image, which is passed to the subsequent RPN layer and the fully connected layer;

(2)   RPN layer.

This layer is used to generate target candidate regions, eliminating the time consumed by the process of Selective Search (pre-SS) [48] that generates candidate frames. Faster R-CNN uses an RPN network that shares part of the weight with the recognizer to generate candidate frames for the image directly, and then perform classification and position regression based on the candidate frames obtained by RPN;

(3)   Region of Interest (ROI) pooling layer [14].

This layer uses the feature map and suggestion box information output by the RPN layer to map to feature maps of the same size;

(4)   Recognition.

The feature map of the candidate target area is used to calculate the category of the candidate target area, and the coordinate frame position of the target is regressed again to obtain the final precise position of the target [49]. The score threshold of 0.5 was used to determine whether the bounding box contains wheat ears. In order to limit the overlap between bounding boxes containing the same wheat ear, the Intersection-over-Union (IOU) threshold was set to 0.5 so that only one bounding box was selected [28].

2.2.2. RetinaNet

Retinanet combines the advantages of multiple target recognition methods, especially the "anchor" concept introduced by RPN and the use of feature pyramids in Single Shot Multibox Detector (SSD) [20] and Feature Pyramid Networks (FPN) [21]. The structure of RetinaNet is composed of three parts: a convolutional neural network for feature extraction and two sub-networks for classification and box regression [19]. The structure is shown in Figure 6, where Figure 6a represents the backbone network, i.e., ResNet50; Figure 6b illustrates that FPN is used as a decoder to generate a multi-scale convolutional feature pyramid, and Figure 6c shows that two subnets are used for classification and bounding box regression. Based on feature mapping, two sub-networks of classification and box regression are constructed through simple convolution operations. Specifically, the classification sub-network performs object classification, and the box regression sub-network is used to return the position of the bounding box. The advantage of FPN is that the hierarchical structure of the deep convolutional network can be used to represent multi-scale objects to help the recognizer create a better prediction of the position.

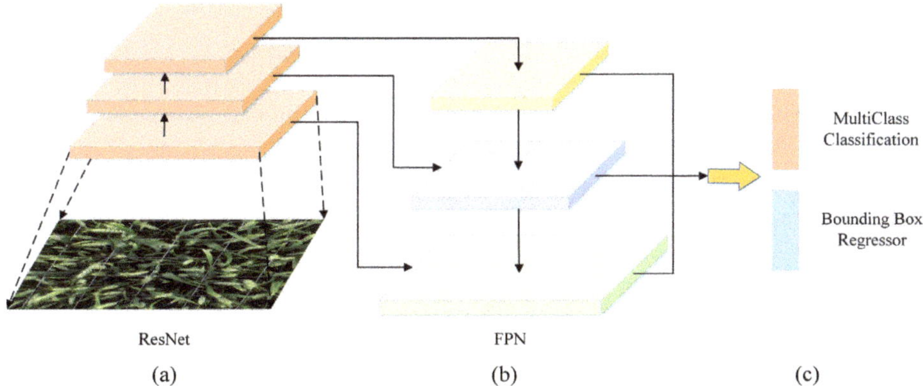

Figure 6. The structure of RetinaNet. (**a**) Backbone network; (**b**) decoder; (**c**) subnet).

This paper uses ResNet50 to extract image features [50]. Compared with the two-stage recognition method, the low accuracy of the one-stage target recognition is mainly caused by the extreme imbalance between the foreground and the background during the training process of the dense recognizer, which creates a large number of negative samples during the training process [19]. The focus loss is used to solve the problem of extreme imbalance of categories; it is implemented by modifying the standard cross-entropy, reducing the loss assigned to well-classified examples [19]. Under the supervision of focus loss, the retina can achieve significant improvements on the universal object recognition benchmark. It is expressed as Equation (1) and has been used to improve detection accuracy [19]. The definition of an a-balanced variant of the focus loss is:

$$FL(p_t) = -\alpha_t(1 - p_t)^y \log(p_t) \tag{1}$$

where α_t and γ are hypermeters. $\alpha_t \in [0,1]$ is the weighting factor to address class imbalance; parameter γ smoothly adjusts the rate at which easy examples are down-weighted [19]. For a convenient notation, p_t is defined as follows.

$$p_t = \begin{cases} p \ if \ y = 1 \\ 1 - p \ otherwise \end{cases} \tag{2}$$

where $p \in [0,1]$ is the probability estimated by the model, and $y = 1$ specifies the ground truth.

2.2.3. Recognition Accuracy Evaluation Index

The accuracy of wheat ear recognition is evaluated by precision and recall [51,52]. Precision measures the accuracy of the algorithm, recall measures the integrity of recognition, and *F1-score* is used to balance precision and recall. The classifier with a high *F-score* is now shown to have good recall and accuracy. The three indicators can be calculated as follows:

$$Precision = \frac{TP}{FP + TP} \tag{3}$$

$$Recall = \frac{TP}{FN + TP} \tag{4}$$

$$F1 - score = 2 \times \frac{Precision \times Recall}{Precision + Recall} \times 100\% \tag{5}$$

If the predicted bounding box overlaps with the marked ear bounding box and exceeds the IOU threshold (set to 0.5 in this paper), then the predicted bounding box represents the wheat ear sample; otherwise, it is the background sample. *TP* indicates the number of correctly classified wheat ear samples, and *FN* indicates the number of wrongly classified wheat ear samples. *FP* indicates the number of background samples that are wrongly classified, and TN indicates the number of background samples that are correctly classified.

Average Precision (*AP*) balances the precision and recall values, reflecting the model's performance [53]. Considering the accuracy as the ordinate and the recall as the abscissa, a Precision and Recall (*PR*) curve can be obtained; the area under the curve is *AP*.

$$AP = \int_0^1 P(R) \tag{6}$$

Additionally, indicators such as the mean absolute error (*MAE*), the root mean squared error (*MSE*), the relative *RMSE* (*rRMSE*), bias (*BIAS*), and coefficient of determination (R^2) are used to evaluate the result of wheat ear recognition. *MAE* and *rRMSE* represent the accuracy of recognition, and *MSE* represents the robustness of the recognition model. The lower the scores of *RMSE*, *rRMSE*, and *MAE*, the better the performance of the model. These indicators can be calculated as follows:

$$RMSE = \sqrt{\frac{1}{N} \sum_{k=1}^{n} (Truth_k - Predicted_k)^2} \tag{7}$$

$$rRMSE = \sqrt{\frac{1}{N} \sum_{k=1}^{n} \left(\frac{Truth_k - Predicted_k}{Truth_k}\right)^2} \tag{8}$$

$$BIAS = \frac{1}{N} \sum_{k=1}^{n} (Truth_k - Predicted_k) \tag{9}$$

$$MAE = \frac{1}{N} \sum_{k=1}^{n} |Truth_k - Predicted_k| \tag{10}$$

$$R^2 = 1 - \frac{\sum_{k=1}^{n} (Truth_k - Predicted_k)^2}{\sum_{k=1}^{n} (Truth_k - \overline{Truth_k})^2} \tag{11}$$

In Equations (7)–(10), N represents the number of test images for the model, the actual number of wheat ears, the number of identified wheat ears of the k-th image, and the average actual number of wheat ears, respectively.

3. Results

3.1. Analysis of the Recognition Results Obtained by Different Methods on the Global WHEAT Dataset

In order to evaluate the performance of the method used to identify wheat ears in this paper, two target recognition algorithms, Faster R-CNN and RetinaNet, as shown in Figures 5 and 6, are used. These three models are trained on the same data set (Global WHEAT data set), and the mean average precision (mAP) results of the test data set are shown in Figure 7. Since the wheat ear is the only identification target, mAP is equal to average precision (AP).

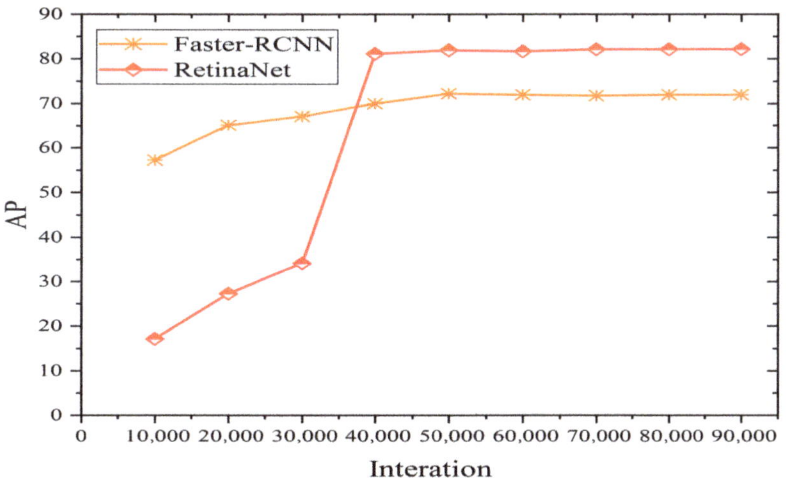

Figure 7. The AP value of Faster R-CNN and RetinaNet for identifying wheat ears.

A total of 90,000 iterations were executed to train the model, where 10,000 iterations were executed to calculate the AP value of the Faster R-CNN and RetinaNet (IOU set to 0.5). Each model uses the VOC data set for pre-training and initialization. The AP value and loss of Faster R-CNN and RetinaNet for identifying wheat ears are shown in Figure 7. As the number of iterations increases, the accuracy of the model gradually increases. When the number of iterations reaches 40,000, the accuracy of the model reaches its maximum. The Faster R-CNN model achieves high accuracy from the beginning, and the AP value does not increase significantly as the number of iterations increases. The AP value of the RetinaNet increases significantly between 30,000 to 40,000 iterations and then becomes stable to the maximum. This result is related to the two-part detection of Faster R-CNN. In Faster R-CNN, the first step generates region proposals that may contain a target to be localized, and the second step performs a fine distinction between the specific targets contained in each candidate region. Therefore, high accuracy is achieved at the beginning of the iterative process. By contrast, RetinaNet performs one-stage detection, and it directly generates the position and category information of the target that derived from the object. This method is prone to category classification errors and inaccurate target location information at the beginning. Through continuous iterative training, the focal loss mechanism of RetinaNet continuously and rapidly reduces the loss value, and a stable detection result can be obtained after 35,000 iterations. Additionally, it can be seen from Figure 7 that the AP value of the RetinaNet is higher than that of the Faster R-CNN, indicating that the RetinaNet achieves the best AP; this is because RetinaNet extracts multi-scale semantic features, which greatly improve the AP value. The RetinaNet model has a strong advantage in wheat ear classification and box regression using derivatives.

Though the Global WHEAT data set contains many types of wheat ear data, there are many wheat varieties globally, and even the same wheat varieties show great differences due to different growth environments. To better evaluate the performance of the Faster

R-CNN and RetinaNet models trained on the Global WHEAT data set, the wheat images collected in the field during the grain filling phase and the mature stage were used as the test set. Using the number of wheat ears identified by different models and obtained by manual methods to calculate the RMSE, rRMSE, MAE, Bias, and R^2, the results are shown in Figure 8.

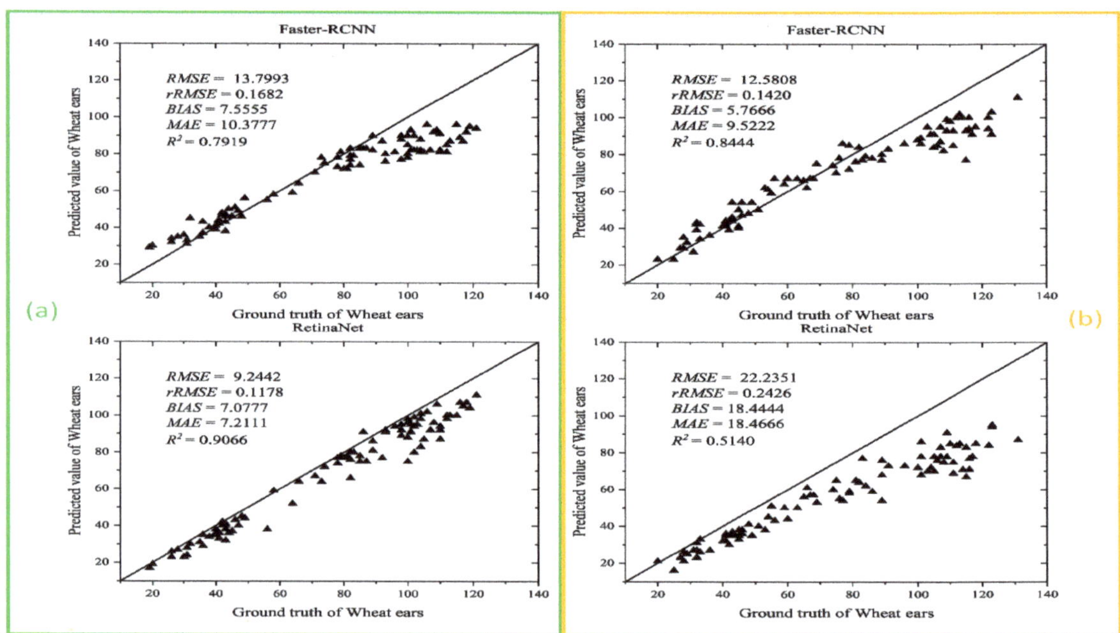

Figure 8. Wheat ear recognition accuracy of Faster R-CNN and RetinaNet models on the test set with images collected in (**a**) in filling stage; (**b**) in mature stage.

The results in Figure 8 show that R^2 of the Faster R-CNN and RetinaNet models on the test data set with images collected in the filling stage are 0.792 and 0.907, respectively. The R^2 of the RetinaNet is 14.5% higher than that of the Faster R-CNN, indicating that the RetinaNet model achieves the highest recognition accuracy for wheat ears in the filling stage. As for recognizing the wheat ear in the mature stage, the Faster R-CNN model achieves the R^2 of 0.844, and the RetinaNet model achieves the R^2 of 0.514. The number under each image represents the number of identified wheat ears, showing that the predicted value differs from the true value by more than 10 wheat ears. Figures 8 and 9 indicate that Faster R-CNN and RetinaNet models trained on the Global WHEAT dataset do not transfer well to the field for wheat ears identification. According to [43], most of the images from the Global WHEAT dataset are acquired before the appearance of head senescence. It also demonstrates the limitations of the Global WHEAT dataset. Therefore, it is important to obtain a model that can migrate to the field and perform with high accuracy for wheat ears of different fertility stages.

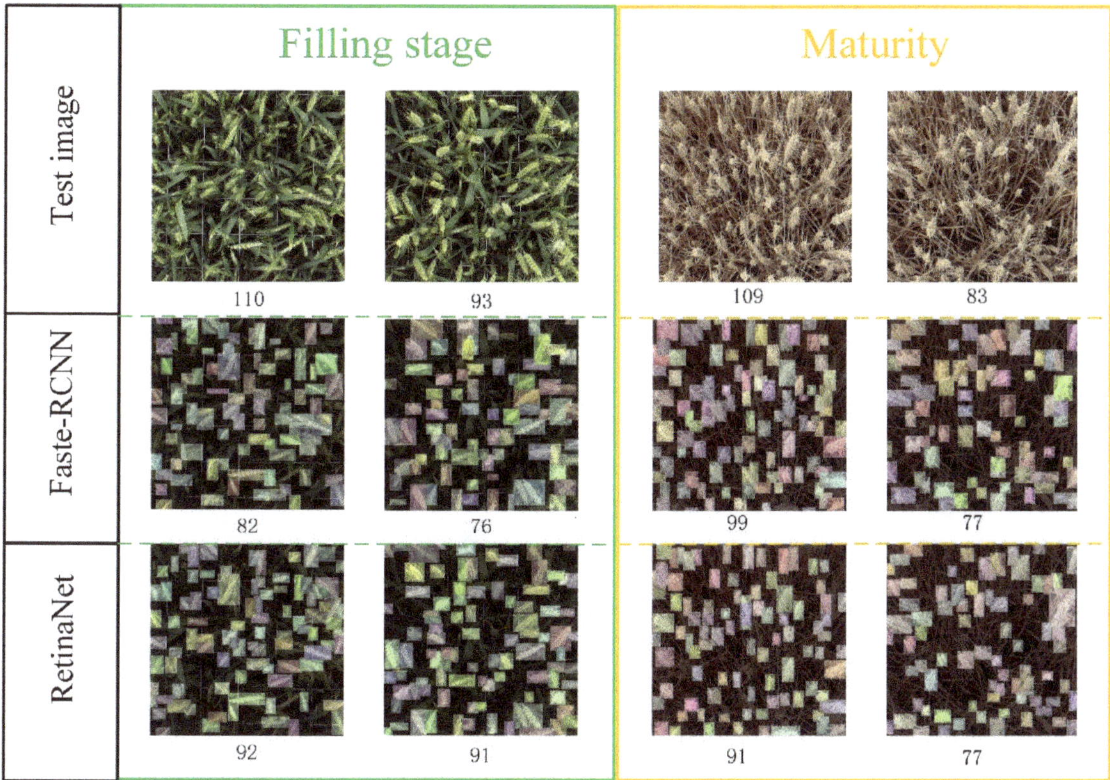

Figure 9. The predicted number of wheat ears (partial) by the Faster R-CNN and RetinaNet models on the test data set.

3.2. Results and Analysis of Wheat Ear Recognition Based on Transfer Learning

3.2.1. Recognition Results and Analysis of Different Numbers of Training Samples after Transfer Learning

In order to study the influence of training samples on the model's performance in the process of transfer learning, the Faster R-CNN and RetinaNet models trained on the Global WHEAT dataset were used as the initial models for transfer learning. Each model was trained 90,000 times. The samples in the test set were used to verify the accuracy of these models, and the results are shown in Figure 10. In this figure, FFSM indicates that the Faster R-CNN model is trained on wheat images collected in the filling stage, and FMSM indicates that the Faster R-CNN model is trained on wheat images collected in the mature stage. Similarly, RFSM and RMSM represent the training of the RetinaNet model on wheat images collected in the filling stage and the mature stage, respectively. Meanwhile, 50, 100, and 150, respectively, represent the number of training sample images.

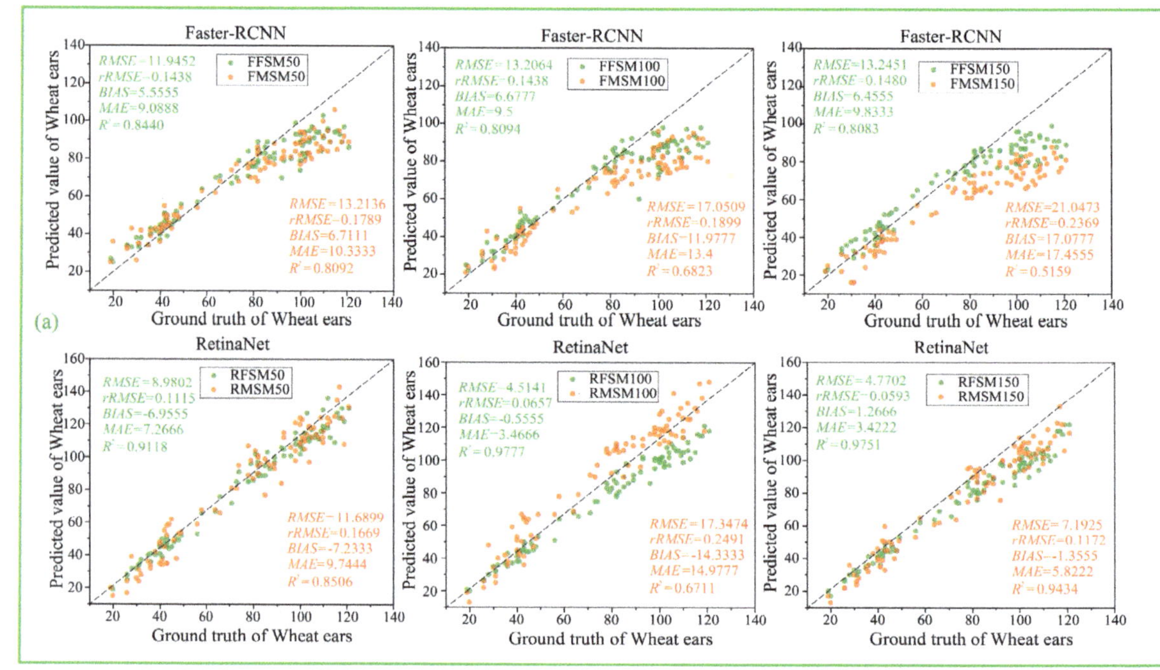

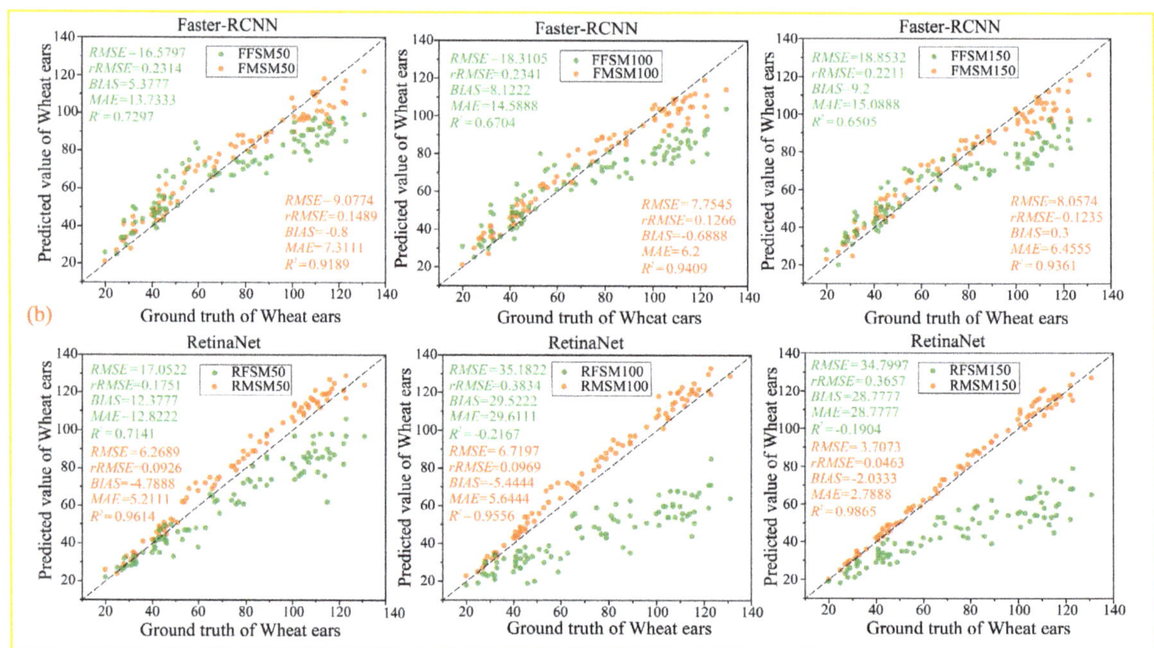

Figure 10. The performance of the models after transfer learning on the test data set with images collected in (**a**) filling stage; (**b**) mature stage.

For different numbers of training samples, manual counting and recognition results have a strong positive correlation. The comprehensive analysis of Figures 8 and 10 indicates that the recognition capabilities of Faster R-CNN and RetinaNet models have greatly improved by transfer learning, and the accuracy of the RetinaNet for recognizing the wheat ear in the filling stage and the mature stage is 97.77% and 98.65%, respectively. Meanwhile, the highest accuracy of the Faster R-CNN model for recognizing the wheat ear in the filling stage and the mature stage is 94.09% and 84.40%, respectively. Therefore, the use of the transfer learning method to place the wheat ear images collected in the field for model training can improve the recognition performance of the model. According to the overall analysis results shown in Figure 10, after 50 training samples were used for training, the recognition performance of RetinaNet and Faster R-CNN models greatly improved compared with the initial model. However, as the number of training samples increased, the recognition accuracy o the two models slowly improved, especially the accuracy of the Faster R-CNN model for recognizing that the wheat in the filling stage was slowly decreasing.

3.2.2. Recognition Results and Analysis of Transfer Learning in Different Growth Stages

The deep learning model consumes a lot of time and equipment (such as GPU) for model training. Therefore, a recognition model with high identification accuracy for wheat ears at different fertility stages is essential for crop yield estimation and a better understanding of the wheat ears and canopy. This paper is based on transfer learning to study the recognition performance of the Faster R-CNN and RetinaNet models in different growth stages of wheat ears. For the filling state and the mature state, the R^2 of the Faster R-CNN and RetinaNet models are calculated and the results are shown in Figure 11.

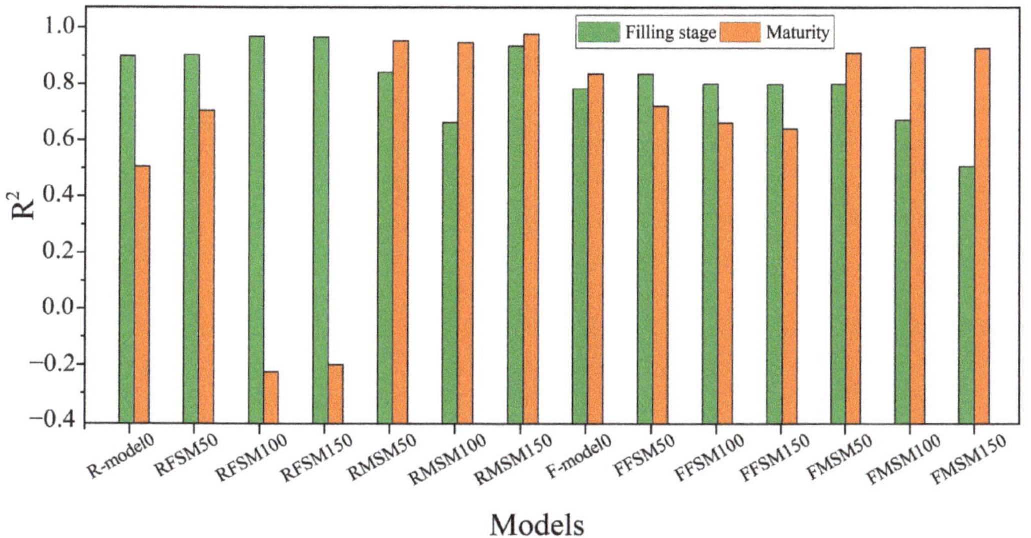

Figure 11. The accuracy of wheat ear recognition by different models in different growth stages.

R-model0 and F-model0 in Figure 11 are two initial models trained on the Global WHEAT dataset. It can be seen from Figure 11 that the Faster R-CNN and RetinaNet models trained on images of wheat ears in a specific growth stage obtain good results for identifying wheat ears in the same growth stage. For identifying wheat ears in other growth stages, the recognition performance of the two models decreases; this is mainly due to the change in the color and awns of wheat ears, which result in changes to wheat features extracted by Faster R-CNN and RetinaNet and the corresponding decline in the performance of wheat ear recognition.

Meanwhile, it was found that the RMSM150 model achieves high recognition accuracy for identifying wheat ears at different growth stages ($R^2 = 0.9434$ for filling stage and $R^2 = 0.9865$ for mature stage), indicating that the wheat ears in the mature stage exhibit better characteristics for recognition. Thus, the RetinaNet model achieves more robustness in recognition performance. Although RFSM achieves a higher accuracy for recognizing wheat ears, it does not perform as stable as the FFSM model on the test data set with wheat ears in different growth stages. In order to have the RetinaNet model achieve higher recognition accuracy for wheat ears in different growth stages, the combination of images collected for wheat ears in different growth stages can be used in the future.

3.3. The Recognition Results and Analysis of RetinaNet

A comprehensive analysis of the experiments presented above indicates that the recognition accuracy of the RetinaNet model is better than that of the Faster R-CNN model. However, these experiments only focus on the wheat ears in a single growth stage and do not consider the wheat ears of multiple growth stages. Therefore, the images of wheat ears in different growth stages can be used as the training data set to study recognition performance of RetinaNet and Faster R-CNN models. Firstly, 150 images of wheat ears in the filling stage and 150 images of wheat ears in the mature stage were used for training. The 300 images contain 39,407 wheat ears in total. The images of the same wheat variety and similar shooting environment are used as test samples. As listed in Table 3, there are 180 images in total, including 13,513 wheat ears.

As for the number of wheat ears in 180 images, the relationship between the true value and the recognition value obtained by RetinaNet and Faster R-CNN models is illustrated in Figure 12. For RetinaNet, the slope of the linear equation between the true value and the identification value is 0.9206. The intercept of the linear equation is 3.3608, and it is approximate to 1, indicating that the use of the RetinaNet to identify wheat ears can obtain a result in good agreement with the ground truth value (only 3.36 wheat ear errors, $R^2 = 0.9722$). Thus, the model can be used for wheat ear identification. Compared with the RetinaNet model, the slope of the linear equation between the recognition value and the true value of the Faster R-CNN model is 0.7106, and the intercept is 16.7735, with $R^2 = 0.8702$. It can be concluded from the above results that the RetinaNet method is more suitable to identify wheat ears.

It can be seen from Table 4 that the F1-score of RetinaNet improved by 8.92% compared with Faster R-CNN. In addition, based on the same Keras framework and operating environment, the running time of different recognition methods was measured and used to calculate the time needed to recognize the wheat ears in 180 images and calculate the average time needed to recognize wheat ears in a single image. The results indicate that the average time of the Faster-RCNN and RetinaNet is 9.19 and 6.51 s, respectively. The RetinaNet method proposed in this paper can meet the requirements of high recognition accuracy and recognition speed.

Table 4. Using different methods for wheat ear recognition.

Methods	F1-Score (%)	Times (s)
Faster R-CNN	82.25	9.19
RetinaNet	91.17	6.51

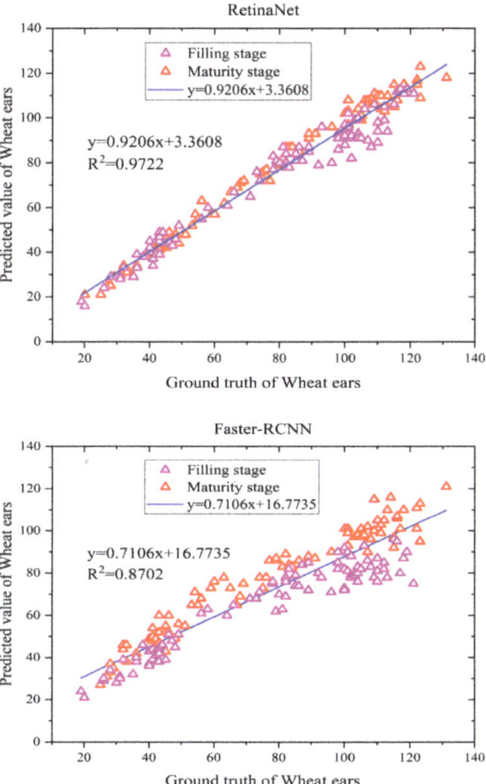

Figure 12. Scatter plot of recognition value and real value of wheat ear in RetinaNet and Faster R-CNN models.

4. Discussion

Earlier studies [27,28] on wheat recognition are limited by the wheat dataset, which causes the studied wheat recognition models not to be well-migrated for application to other regions. Transfer learning [54] transfers the characteristic information of the target in the source domain to the target recognition, which can greatly solve the problem of lack of data in the target domain. The deep learning method needs a large amount of data to train a model and achieve excellent performance, but data in the field for wheat ear recognition is difficult to obtain and costly to label. To our knowledge, there are no other systematic, quantitative assessments of how training sample size and sample selection methods affect the results of wheat identification models. This paper combines deep learning and transfer learning to study a wheat crop detection model with high accuracy and capable of migration. This technique can be extended to any crop identification.

The method proposed in this paper for recognizing wheat ears based on RetinaNet is compared with the method based on Faster R-CNN proposed by Madec et al. [15]. The RetinaNet method achieves the best recognition performance (AP exceeds 82%) by using the Global WHEAT database [43] as the training set; this shows that the RetinaNet method can be applied to most countries in the world, and appliesto the identification of wheat ears under different wheat distribution densities, different wheat varieties, and different growth environments.

Meanwhile, the RetinaNet and Faster R-CNN models trained on the Global WHEAT data set are applied to the field-collected wheat data based on transfer learning. Compared with the Faster R-CNN method, the recognition performance of the RetinaNet is greatly improved, and the RetinaNet performs the best for recognizing wheat ears in different growth stages; this also indicates that the RetinaNet model has stronger transfer learning ability and better wheat ear recognition performance. These abilities are due to how RetinaNet exploits FPN to extract low-level high-resolution and high-level low-resolution semantics and then uses horizontal connections to combine the corresponding feature maps with reconstructed layers, helping the recognizer better predict the position. Therefore, RetinaNet is more sensitive to the target and achieves excellent recognition performance with transfer learning. Meanwhile, ResNet50 acts as a feature extraction network for both RetinaNet and Faster R-CNN. By adopting a feature pyramid with multi-size feature extraction and output, RetinaNet has an advantage in small target detection such as wheat ears, whereas Faster R-CNN only exploits the last layer of features of the underlying network. Therefore, compared with Faster R-CNN, RetinaNet performs better and is more suitable for wheat ear detection. Since ResNet50 is used as a feature extraction network for RetinaNet, the feature extraction network can also be optimized further to improve the accuracy of wheat recognition in the future.

Additionally, the recognition effect of RetinaNet and Faster R-CNN methods for wheat ears in different growth stages is analyzed. The analysis results show that the R^2 of recognizing wheat ears in the filling stage by RetinaNet and Faster R-CNN methods are 0.978 and 0.844, respectively. Meanwhile, the R^2 of recognizing the wheat ears in the mature stage by RetinaNet and Faster R-CNN methods are 0.986 and 0.941, respectively. It can be seen that RetinaNet achieves the best recognition effect in the mature stage; this is mainly because the characteristics of wheat change with the growth stage. When the wheat grows to maturity, the shape and awns of wheat ears tend to be stable, and the contrast between the wheat ears and leaf background is enhanced, causing difficulty in identifying wheat ears; this is consistent with the conclusions of Hasan et al. [8], Madec et al. [15], and Zhu et al. [28]. Simultaneously, using the RetinaNet method, the accuracy for recognizing wheat ears in the filling stage is only 0.8% lower than that in the mature stage, which also indicates that RetinaNet has a better recognition effect. However, the causes of performance differences between the RetinaNet method and Faster R-CNN for recognizing wheat ears in different growth stages are currently unknown, which will be explored in our future work.

The recognition performance of RetinaNet and Faster R-CNN models trained with different fertility data was analyzed. It can be found in Figure 11 that the accuracy of RetinaNet and Faster R-CNN increased with the number of training data samples. It can also be found that RFSM50, RFSM100, and RFSM150 obtain good recognition results for wheat in the filling stage, but the recognition results for wheat ears in the maturity stage are degraded. The first reason for this is that the color, shape, and awn of the wheat ears change significantly as the wheat grows, and the RetinaNet model trained only with the data of the filling stage fails to detect the mature wheat ears well. It demonstrates the relevance of training samples to the performance of the model, consistent with the current studies [55,56]. Furthermore, it is found that the RMSM50, RMSM100, and RMSM150 models trained only with the data of wheat maturity stage perform the best in detecting wheat ears in the maturity stage; they also achieve better recognition accuracy for the wheat ears in the filling stage. Thus, adding the training of wheat ears at maturity can cause the RetinaNet model to obtain higher recognition performance for both filling and maturity stages. This finding can help transfer the RetinaNet method to wheat ears recognition in other growth regions and different growth stages of wheat and can be useful for studying ear detection models for small samples.

Additionally, the recognition speed of the RetinaNet and Faster R-CNN methods is analyzed in this paper. Under the same framework and operating environment, the average time consumption of the two methods is 9.19 and 6.51 s, respectively. It can be seen that the recognition speed of the RetinaNet method is relatively fast, which is mainly due to the time-consuming extraction of candidate frames in the second stage of the Faster R-CNN network.

The comprehensive analysis of the recognition effect and recognition speed of the two methods indicates that the RetinaNet method is more suitable for wheat ear recognition.

5. Conclusions

This paper studies the application of deep learning technology to wheat ear recognition and chooses a better recognition model for recognizing what ears are in different growth stages and the number of wheat ears in a single image. Additionally, the Global WHEAT data set containing images of wheat from different growing environments and varieties are used to generalize training data to create a more robust model. Moreover, the model is integrated with transfer learning to study the transfer ability and recognition performance of the Faster R-CNN and RetinaNet. The comprehensive analysis of the experimental results indicates that the proposed RetinaNet achieves both high recognition performance and recognition speed, which can better meet the requirements of real applications. In our future work we aim to investigate wheat ear recognition on images obtained by unmanned aerial vehicles, which provides a new approach for wheat ear recognition and yield estimation. In order to help researchers reproduce the proposed method, the program file used in our study is provided (https://github.com/lijignbo1024/Program.git, accessed on 8 June 2021).

Author Contributions: Conceptualization, J.L. and C.L.; methodology, J.L.; software, J.L. and Z.X.; validation, J.L., S.F. and C.M.; formal analysis, J.L. and W.C.; investigation, J.L. and J.S.; resources, J.L., W.C. and F.D.; data curation, J.L., Y.W. and Y.L.; writing—original draft preparation, J.L.; writing—review and editing, C.L.; visualization, J.L. and S.F.; supervision, J.L.; project administration, C.L. and C.M.; funding acquisition, C.L. and C.M. All authors have read and agreed to the published version of the manuscript.

Funding: This study was supported by the Natural Science Foundation of China (41871333) and the Important Project of Science and Technology of the Henan Province (212102110238).

Institutional Review Board Statement: Not applicable.

Informed Consent Statement: Not applicable.

Data Availability Statement: Global WHEAT data source: https://www.kaggle.com/c/global-wheat-detection/data, accessed on 12 June 2020. Ground measurement data is available upon request due to privacy.

Acknowledgments: We thank all the authors for their support. The authors would like to thank all the reviewers who participated in this review.

Conflicts of Interest: The authors declare no conflict of interest.

References

1. FAOSTAT. Available online: http://faostat3.fao.org/faostat-gateway/go/to/browse/Q/QC/E (accessed on 2 June 2021).
2. Chen, C.; Frank, K.; Wang, T.; Wu, F. Global wheat trade and Codex Alimentarius guidelines for deoxynivalenol: A mycotoxin common in wheat. *Glob. Food Secur.* **2021**, *29*, 100538. [CrossRef]
3. Powell, J.P.; Reinhard, S. Measuring the effects of extreme weather events on yields. *Weather Clim. Extrem.* **2016**, *12*, 69–79. [CrossRef]
4. Gómez-Plana, A.G.; Devadoss, S. A spatial equilibrium analysis of trade policy reforms on the world wheat market. *Appl. Econ.* **2004**, *36*, 1643–1648. [CrossRef]
5. Zhang, H.; Turner, N.C.; Poole, M.L.; Asseng, S. High ear number is key to achieving high wheat yields in the high-rainfall zone of south-western Australia. *Aust. J. Agric. Res.* **2007**, *58*, 21–27. [CrossRef]
6. Gou, F.; van Ittersum, M.K.; Wang, G.; van der Putten, P.E.; van der Werf, W. Yield and yield components of wheat and maize in wheat–maize intercropping in the Netherlands. *Eur. J. Agron.* **2016**, *76*, 17–27. [CrossRef]
7. Zhou, H.; Riche, A.B.; Hawkesford, M.J.; Whalley, W.R.; Atkinson, B.S.; Sturrock, C.J.; Mooney, S.J. Determination of wheat spike and spikelet architecture and grain traits using X-ray Computed Tomography imaging. *Plant Methods* **2021**, *17*, 26. [CrossRef] [PubMed]
8. Hasan, M.M.; Chopin, J.P.; Laga, H.; Miklavcic, S.J. Detection and analysis of wheat spikes using Convolutional Neural Networks. *Plant Methods* **2018**, *14*, 100. [CrossRef] [PubMed]

9. Fernandez-Gallego, J.A.; Kefauver, S.C.; Gutiérrez, N.A.; Nieto-Taladriz, M.T.; Araus, J.L. Wheat ear counting in-field conditions: High throughput and low-cost approach using RGB images. *Plant Methods* **2018**, *14*, 22. [CrossRef]

10. Dong, S.; Wang, P.; Abbas, K. A survey on deep learning and its applications. *Comput. Sci. Rev.* **2021**, *40*, 100379. [CrossRef]

11. Jin, X.; Zarco-Tejada, P.; Schmidhalter, U.; Reynolds, M.P.; Hawkesford, M.J.; Varshney, R.K.; Yang, T.; Nie, C.H.; Li, Z.; Ming, B.; et al. High-throughput estimation of crop traits: A review of ground and aerial phenotyping platforms. *IEEE Geosci. Remote Sens. Mag.* **2021**, *9*, 200–231. [CrossRef]

12. Lippitt, C.D.; Zhang, S. The impact of small unmanned airborne platforms on passive optical remote sensing: A conceptual perspective. *Int. J. Remote Sens.* **2018**, *39*, 4852–4868. [CrossRef]

13. Mickinney, S.M.; Karthikesalingam, A.; Tse, D.; Kelly, C.J.; Shetty, S. Reply to: Transparency and reproducibility in artificial intelligence. *Nature* **2020**, *586*, E17–E18. [CrossRef]

14. Ren, S.; He, K.; Girshick, R.; Sun, J. Faster R-CNN: Towards real-time object detection with region proposal networks. *IEEE Trans. Pattern Anal. Mach. Intell.* **2017**, *39*, 1137–1149. [CrossRef]

15. Madec, S.; Jin, X.; Lu, H.; De Solan, B.; Liu, S.; Duyme, F.; Heritier, E.; Baret, F. Ear density estimation from high resolution RGB imagery using deep learning technique. *Agric. For. Meteorol.* **2019**, *264*, 225–234. [CrossRef]

16. Girshick, R. Fast R-CNN. In Proceedings of the 2015 IEEE International Conference on Computer Vision (ICCV), Santiago, Chile, 7–13 December 2015; pp. 1440–1448.

17. Redmon, J.; Divvala, S.; Girshick, R.; Farhadi, A. You Only Look Once: Unified, Real-Time Object Detection. In Proceedings of the IEEE Conference on Computer Vision and Pattern Recognition (CVPR), Las Vegas, NV, USA, 27–30 June 2016.

18. Redmon, J.; Farhadi, A. Yolov3: An incremental improvement. *arXiv* **2018**, arXiv:1804.02767.

19. Lin, T.Y.; Goyal, P.; Girshick, R.; He, K.; Dollár, P. Focal Loss for Dense Object Detection. *IEEE Trans. Pattern Anal. Mach. Intell.* **2017**, *99*, 2999–3007.

20. Liu, W.; Anguelov, D.; Erhan, D.; Szegedy, C.; Reed, S.; Fu, C.Y.; Berg, A.C. SSD: Single shot multibox detector. In Proceedings of the European Conference on Computer Vision, Amsterdam, The Netherlands, 8–16 October 2016; pp. 21–37.

21. Lin, T.Y.; Dollár, P.; Girshick, R.; He, K.; Hariharan, B.; Belongie, S. Feature pyramid networks for object detection. In Proceedings of the IEEE Conference on Computer Vision and Pattern Recognition (CVPR), Honolulu, HI, USA, 21–26 July 2017; pp. 2117–2125.

22. Wang, Y.; Wang, C.; Zhang, H.; Dong, Y.; Wei, S. Automatic Ship Detection Based on RetinaNet Using Multi-Resolution Gaofen-3 Imagery. *Remote Sens.* **2019**, *11*, 531. [CrossRef]

23. Santos, A.; Marcato Junior, J.; de Andrade Silva, J.; Pereira, R.; Matos, D.; Menezes, G.; Higa, L.; Eltner, A.; Ramos, A.P.; Osco, L.; et al. Storm-Drain and Manhole Detection Using the RetinaNet Method. *Sensors* **2020**, *20*, 4450. [CrossRef] [PubMed]

24. Chen, Y.; Zhang, X.; Chen, W.; Li, Y.; Wang, J. *Research on Recognition of Fly Species Based on Improved RetinaNet and CBAM*; IEEE Access: Piscataway, NJ, USA, 2020; Volume 8. [CrossRef]

25. Zheng, Z.; Qi, H.; Zhuang, L.; Zhang, Z. Automated rail surface crack analytics using deep data-driven models and transfer learning. *Sustain. Cities Soc.* **2021**, *70*. [CrossRef]

26. Liu, P.Z.; Guo, J.M.; Chamnongthai, K.; Prasetyo, H. Fusion of color histogram and LBP-based features for texture image retrieval and classification. *Inf. Sci.* **2017**, *390*, 95–111. [CrossRef]

27. Zhou, C.; Liang, D.; Yang, X.; Yang, H.; Yue, J.; Yang, G. Wheat ears counting in field conditions based on multi-feature optimization and TWSVM. *Front. Plant Sci.* **2018**, *9*, 1024. [CrossRef] [PubMed]

28. Zhu, Y.; Cao, Z.; Lu, H.; Li, Y.; Xiao, Y. In-field automatic observation of wheat heading stage using computer vision. *Biosyst. Eng.* **2016**, *143*, 28–41. [CrossRef]

29. Schmidhuber, J. Deep learning in neural networks: An overview. *Neural Netw.* **2015**, *61*, 85–117. [CrossRef]

30. Gong, B.; Ergu, D.; Cai, Y.; Ma, B. Real-Time Detection for Wheat Head Applying Deep Neural Network. *Sensors* **2020**, *21*, 191. [CrossRef] [PubMed]

31. Wang, D.; Fu, Y.; Yang, G.; Yang, X.; Liang, D.; Zhou, C.; Zhang, N.; Wu, H.; Zhang, D. Combined use of FCN and harris corner detection for counting wheat ears in field conditions. *IEEE Access* **2019**, *7*, 178930–178941. [CrossRef]

32. He, M.X.; Hao, P.; Xin, Y.Z. A robust method for wheatear detection using UAV in natural scenes. *IEEE Access* **2020**, *8*, 189043–189053. [CrossRef]

33. Ma, J.; Li, Y.; Liu, H.; Du, K.; Zheng, F.; Wu, Y.; Zhang, L. Improving segmentation accuracy for ears of winter wheat at flowering stage by semantic segmentation. *Comput. Electron. Agric.* **2020**, *176*, 105662. [CrossRef]

34. Xu, X.; Li, H.; Yin, F.; Xi, L.; Qiao, H.; Ma, Z.; Shen, S.; Jiang, B.; Ma, X. Wheat ear counting using K-means clustering segmentation and convolutional neural network. *Plant Methods* **2020**, *16*, 106. [CrossRef]

35. Zou, H.; Lu, H.; Li, Y.; Liu, L.; Cao, Z. Maize tassels detection: A benchmark of the state of the art. *Plant Methods* **2020**, *16*, 108. [CrossRef]

36. Lu, H.; Cao, Z.G. TasselNetV2+: A fast implementation for high-throughput plant counting from high-resolution RGB imagery. *Front. Plant Sci.* **2020**, *11*, 1929. [CrossRef]

37. Sadeghi-Tehran, P.; Virlet, N.; Ampe, E.M.; Reyns, P.; Hawkesford, M.J. DeepCount: In-field automatic quantification of wheat spikes using simple linear iterative clustering and deep convolutional neural networks. *Front. Plant Sci.* **2019**, *10*, 1176. [CrossRef] [PubMed]

38. Ampatzidis, Y.; Partel, V. UAV-based high throughput phenotyping in citrus utilizing multispectral imaging and artificial intelligence. *Remote Sens.* **2019**, *11*, 410. [CrossRef]

39. Vit, A.; Shani, G.; Bar-Hillel, A. Length phenotyping with interest point detection. *Comput. Electron. Agric.* **2020**, *176*, 105629. [CrossRef]
40. Nagasubramanian, K.; Jones, S.; Singh, A.K.; Sarkar, S.; Singh, A.; Ganapathysubramanian, B. Plant disease identification using explainable 3D deep learning on hyperspectral images. *Plant Methods* **2019**, *15*, 98. [CrossRef] [PubMed]
41. Mohanty, S.P.; Hughes, D.P.; Salathé, M. Using deep learning for image-based plant disease detection. *Front. Plant Sci.* **2016**, *7*, 1419. [CrossRef]
42. Khaki, S.; Safaei, N.; Pham, H.; Wang, L. WheatNet: A Lightweight Convolutional Neural Network for High-throughput Image-based Wheat Head Detection and Counting. *arXiv* **2021**, arXiv:2103.09408.
43. David, E.; Madec, S.; Sadeghi-Tehran, P.; Aasen, H.; Zheng, B.; Liu, S.; Kirchgessner, N.; Ishikawa, G.; Nagasawa, K.; Badhon, M.A.; et al. Global Wheat Head Detection (GWHD) dataset: A large and diverse dataset of high-resolution RGB-labelled images to develop and benchmark wheat head detection methods. *Plant Phenomics* **2020**, *2020*, 3521852. [CrossRef]
44. Virlet, N.; Sabermanesh, K.; Sadeghi-Tehran, P.; Hawkesford, M.J. Field Scanalyzer: An automated robotic field phenotyping platform for detailed crop monitoring. *Funct. Plant Biol.* **2016**, *44*, 143–153. [CrossRef]
45. Labelimg. Available online: https://github.com/tzutalin/labelImg (accessed on 6 May 2018).
46. Smith, A.R. Color gamut transform pairs. *ACM Siggraph Comput. Graph.* **1978**, *12*, 12–19. [CrossRef]
47. Fredrik Lundh. Python Image Library (PIL). Available online: https://python-pillow.org/ (accessed on 10 September 2020).
48. Uijlings, J.R.R.; van de Sande, K.E.A.; Gevers, T.; Smeulders, A.W.M. Selective Search for Object Recognition. *Int. J. Comput. Vis.* **2013**, *104*, 154–171. [CrossRef]
49. Hosang, J.; Benenson, R.; Schiele, B. Learning non-maximum suppression. *arXiv* **2017**, arXiv:1705.02950.
50. He, K.; Zhang, X.; Ren, S.; Sun, J. Deep residual learning for image recognition. In Proceedings of the 2016 IEEE Conference on Computer Vision and Pattern Recognition (CVPR), Las Vegas, NV, USA, 27–30 June 2016; pp. 770–778.
51. Xiong, X.; Duan, L.; Liu, L.; Tu, H.; Yang, P.; Wu, D.; Chen, G.; Xiong, L.; Yang, W.; Liu, Q. Panicle-SEG: A robust image segmentation methodfor rice panicles in the field based on deep learning and superpixel optimization. *Plant Methods* **2017**, *13*, 104. [CrossRef]
52. Ma, B.; Liu, Z.; Jiang, F.; Yan, Y.; Yuan, J.; Bu, S. Vehicle detectionin aerial images using rotation-invariant cascaded forest. *IEEE Access* **2019**, *7*, 59613–59623. [CrossRef]
53. Salton, G.; McGill, M.J. *Introduction to Modern Information Retrieval*; McGraw-Hill: New York, NY, USA, 1983. Available online: https://trove.nla.gov.au/work/19430022 (accessed on 23 May 2021).
54. Pan, S.J.; Yang, Q. A survey on transfer learning. *IEEE Trans. Knowl. Data Eng.* **2010**, *22*, 1345–1359. [CrossRef]
55. Zhu, W.; Braun, B.; Chiang, L.H.; Romagnoli, J.A. Investigation of Transfer Learning for Image Classification and Impact on Training Sample Size. *Chemom. Intell. Lab. Syst.* **2021**, *211*, 104269. [CrossRef]
56. Millard, K.; Richardson, M. On the Importance of Training Data Sample Selection in Random Forest Image Classification: A Case Study in Peatland Ecosystem Mapping. *Remote Sens.* **2015**, *7*, 8489–8515. [CrossRef]

Article

MSF-Net: Multi-Scale Feature Learning Network for Classification of Surface Defects of Multifarious Sizes

Pengcheng Xu [1,2], Zhongyuan Guo [3], Lei Liang [1] and Xiaohang Xu [3,*,†]

1 College of Computer Science and Technology, Wuhan University of Technology, Wuhan 430070, China; xuguduxia@163.com (P.X.); L30L30@126.com (L.L.)
2 Radar Non Commissioned Officer School, Air Force Early Warning Academy, Wuhan 430019, China
3 School of Electronic Information, Wuhan University, Wuhan 430072, China; guozhongyuan@whu.edu.cn
* Correspondence: xuxiaohang@whu.edu.cn
† Current Address: Shenzhen Mindray Bio-Medical Electronics Co., Ltd., Shenzhen 518057, China.

Abstract: In the field of surface defect detection, the scale difference of product surface defects is often huge. The existing defect detection methods based on Convolutional Neural Networks (CNNs) are more inclined to express macro and abstract features, and the ability to express local and small defects is insufficient, resulting in an imbalance of feature expression capabilities. In this paper, a Multi-Scale Feature Learning Network (MSF-Net) based on Dual Module Feature (DMF) extractor is proposed. DMF extractor is mainly composed of optimized Concatenated Rectified Linear Units (CReLUs) and optimized Inception feature extraction modules, which increases the diversity of feature receptive fields while reducing the amount of calculation; the feature maps of the middle layer with different sizes of receptive fields are merged to increase the richness of the receptive fields of the last layer of feature maps; the residual shortcut connections, batch normalization layer and average pooling layer are used to replace the fully connected layer to improve training efficiency, and make the multi-scale feature learning ability more balanced at the same time. Two representative multi-scale defect data sets are used for experiments, and the experimental results verify the advancement and effectiveness of the proposed MSF-Net in the detection of surface defects with multi-scale features.

Keywords: surface defect classification; deep learning; convolutional neural network; multi-scale features; multi-size defects

Citation: Xu, P.; Guo, Z.; Liang, L.; Xu, X. MSF-Net: Multi-Scale Feature Learning Network for Classification of Surface Defects of Multifarious Sizes. *Sensors* **2021**, *21*, 5125. https://doi.org/10.3390/s21155125

Academic Editors: KWONG Tak Wu Sam, Yun Zhang, Xu Long and Tiesong Zhao

Received: 4 June 2021
Accepted: 13 July 2021
Published: 29 July 2021

Publisher's Note: MDPI stays neutral with regard to jurisdictional claims in published maps and institutional affiliations.

Copyright: © 2021 by the authors. Licensee MDPI, Basel, Switzerland. This article is an open access article distributed under the terms and conditions of the Creative Commons Attribution (CC BY) license (https://creativecommons.org/licenses/by/4.0/).

1. Introduction

With the rapid development of the manufacturing industry, people are paying more and more attention to the surface quality of various industrial products. The surface quality of the product will not only affect the appearance and visual effect of the product, but also affect the internal quality and performance of the product. In order to reduce production costs, improve production efficiency and product quality, it is very necessary to effectively detect surface defects in the product manufacturing process.

At present, the commonly used surface defect detection methods are as follows [1]:

1. Artificial visual inspection, which has the disadvantages of low detection efficiency, high false detection rate and high missed detection rate, high labor intensity, and low speed.

2. The non-contact detection method based on machine vision [2,3] usually adopts image processing algorithms or manual design feature extractors to combine the classifier. Liu T.I. [4] proposed a fuzzy logic expert system for roller bearing defect detection, the system combines frequency response and fuzzy reasoning and has achieved good results. Baygin et al. [5] used Otsu thresholding and Hough transform to extract features from the reference image for the problem of printed circuit board with defects and matched the image to be inspected with the reference image to accurately detect the missing holes on the circuit board. Zhang Lei et al. [6] proposed a fabric defect

classification algorithm combining Local Binary Pattern (LBP) and Gray Level Co-occurrence Matrix (GLCM). The algorithm first uses the LBP algorithm to extract the local feature information of the image and then uses the GLCM to describe the overall texture information, and finally, the feature information of the two parts as a whole constructed as the input of the BP neural network, and a higher classification accuracy is obtained. Denis Sidorov et al. [7] proposed an automatic defect classification method based on the p-median clustering technique, the proposed method uses the p-median combinatorial optimization problem to complete the clustering problem, which can be sued in semiconductor and other manufacturing industries. In general, compared with artificial visual inspection methods, the above methods have the advantages of safety and reliability, high detection accuracy, and long-term operation in complex production environments, which effectively improves production efficiency and quality inspection efficiency. However, in a real and complex industrial environment, there are generally small differences between surface defects and background, low contrast, large differences in defect scales, and various types of defects. The design of image processing algorithm schemes and artificially designed feature extraction schemes typically requires rich expert experience and a large number of experiments, resulting in high cost and time consumption, and the effectiveness and generalization cannot be guaranteed, and it is difficult to obtain better detection results.

In recent years, with the successful application of deep learning models represented by convolutional neural networks (CNNs) [8] in computer vision fields such as face recognition [9], scene text detection [10], target tracking, and autonomous driving [11], the surface defect detection methods based on deep learning have also been widely used in various industrial scenarios and have become the mainstream method in the field of defect detection. Weimer [12] explored the influence of the design of CNN and different hyper-parameters on the accuracy of defect detection results. Ren [13] built a classifier based on the features of image patches, transferred the features from the pre-trained deep learning model, and convolved the trained classifier on the input image to obtain pixel-level predictions, compared with multi-layer perception and support vector machine, its error rate is lower. Masci [14] proposed a max-pooling CNN method for steel defect classification, experiments were performed on seven types of defects, the accuracy rate reached 93%, and its performance is far better than SVM classification trained on feature descriptors. Aiming at the problem of jujube surface defect detection, Guo [15] has done a series of work on data preprocessing, data augmentation, and composite convolutional neural network design, and achieved good results. Deitsch [16] used a modified VGG 19 network to identify solar panel image defects with a resolution of 300 × 300, with an accuracy rate of 88.42%, which exceeds a variety of manual design features and supported vector machine methods. Xu [17] presented a small data-driven convolutional neural network (SDD-CNN) to detect the subtle defects of rollers, the method first used label dilation to solve the problem of the imbalance of the number of classes, then a semi-supervised data augmentation method is proposed, and finally, CNNs were trained, experimental results show that compared with the original CNNs, SDD-CNNs has significantly improved the convergence time and classification accuracy. In addition, some advanced CNN structures have also achieved good detection results, including but not limited to references [18–23].

CNNs are currently used by domestic and foreign researchers and engineers as the preferred architecture for product surface defect classification. However, difficulties and challenges still exist.

For different types of products, the surface defects often have the characteristics of different sizes, uncertain positions, and different shapes; even for the same type of product, the color, texture, shape, and size between different types of defects is also very different. The general CNN often contains a few specific scales of receptive fields and is more inclined to express macroscopic and abstract features. It is not strong in expressing local and small defects, which leads to an imbalance in feature expression capabilities. Therefore, how to design a deep CNN that can simultaneously take into account multi-scale feature extraction has become the focus of research.

Since the receptive fields of the convolution kernels in the CNN are closely related to the sizes of the target features, the CNN's ability to express features at different scales directly determines its ability to detect defects of different sizes [24]. This paper starts with the analysis of the appearance and size characteristics of product surface defects, analyzes the processing mechanism of mainstream CNNs for different scale features, and tries to improve the expression and classification capabilities of deep CNNs for different scale features. On the basis of the above research, a Multi-Scale Feature Learning Network (MSF-Net) based on Dual Module Feature (DMF) extractor is proposed, experiments are carried out on two public multi-scale defect data sets, the experimental results verify the effectiveness and superiority of the MSF-Net proposed in this paper.

The structure of this paper is as follows. Section 2 introduces the related work. Section 3 is the research method. Section 4 is the experimental results and discussion, and Section 5 summarizes the paper.

2. Related Work

The receptive field refers to the input area that neurons can "see" in the CNN [25], as shown in Figure 1, the calculation of an element on the feature map in a CNN corresponds to a certain area on the input image, so the corresponding area is the receptive field of the element. It can be seen from Figure 1 that the receptive field is a relative concept, the elements on the feature map of a certain layer can see different areas on the previous layers.

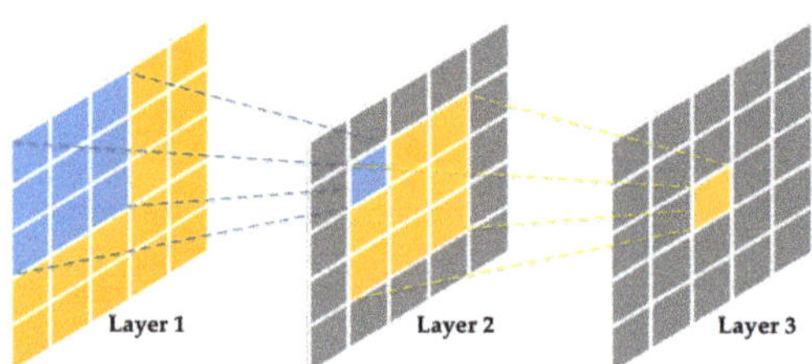

Figure 1. Schematic diagram of the receptive field in CNNs.

The receptive field RF_i of the feature map of the i-th layer is shown by Formula (1):

$$RF_i = RF_{i-1} + (K_i - 1) * \prod_{k=1}^{i-1} S_k \; i \geq 1 \tag{1}$$

where K_i and S_i respectively represent the size of the convolution kernel and the stride of the i-th convolution layer. In addition, for the input layer, $RF_0 = 1$, $S_0 = 1$.

In recent years, some researchers have tried to improve the classification performance of CNNs on multi-scale target data sets by optimizing the CNNs' structure, such as the works of Tang [26] and Kim [27]. In essence, the optimization ideas for these works are derived from the classic CNN architecture, so the distribution of the receptive field in the classic CNNs are analyzed.

2.1. AlexNet and VGGNet

As we all know, there is no branch structure in the two CNNs of AlexNet [28] and VGGNet [29]. Therefore, the receptive field size of the last convolution layer before the fully connected layer is uniform, and the receptive field size is shown in Table 1.

Table 1. The receptive fields of the last convolution layer of feature maps of AlexNet and VGG-16.

CNNs	Feature Maps of the Last Convolution Layer	Receptive Field
AlexNet	pool5	195×195
VGG-16	pool5	212×212

It can be seen from Table 1 that the size of the feature map output by the last convolutional layer of AlexNet is 195×195, while the size of the feature map output by the last convolutional layer of VGG-16 is 212×212, and the input image size of the two CNNs is 224×224. In other words, what the two CNNs finally extract are macroscopic and abstract features, and the extraction of tiny and concrete features is insufficient.

2.2. GoogLeNet and ResNet

Unlike AlexNet and VGGNet, GoogLeNet [30] and ResNet [31] have rich branch structures. This is due to the modularization of the two CNNs, the smallest module unit of GoogLeNet is called the Inception module, and the smallest module unit of ResNet is the Residual Block.

Figure 2 shows the structure of four simplified Inception modules connected in series, for the convenience of calculation, the ratio of the number of output feature maps of the three branches of 1×1, 3×3, and 5×5 is set to 2:1:1. Figure 3 shows the structure of two residual modules connected in series, and the output ratio of feature maps of all branches is set to 1:1.

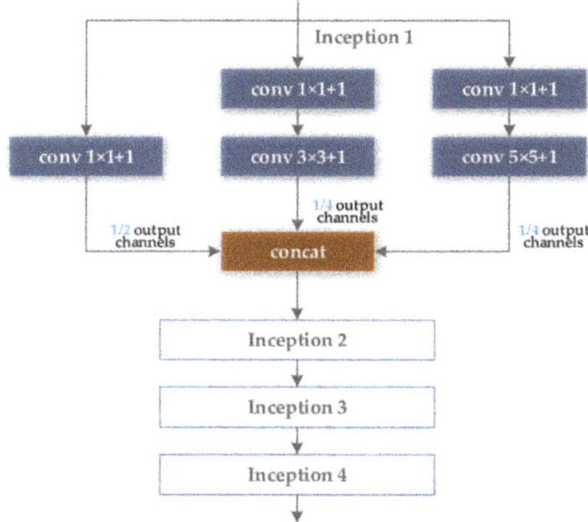

Figure 2. Four Inception modules in series structure.

Figures 4 and 5 respectively reveal the evolution of the receptive fields of the corresponding feature maps of the GoogLeNet and ResNet as the convolutional layer deepens. It can be seen that although the specific number of receptive fields at each scale is not completely the same, both of them reveal a common phenomenon. That is, in the initial stage of feature extraction by CNNs, the size of the receptive field is small, it is sensitive to the micro and local feature information of the image, and the learning ability is strong; in the middle and last stages of feature extraction by CNNs, as the number of convolution operations increases, the feature map becomes more abstract and is more inclined to express macro and global information.

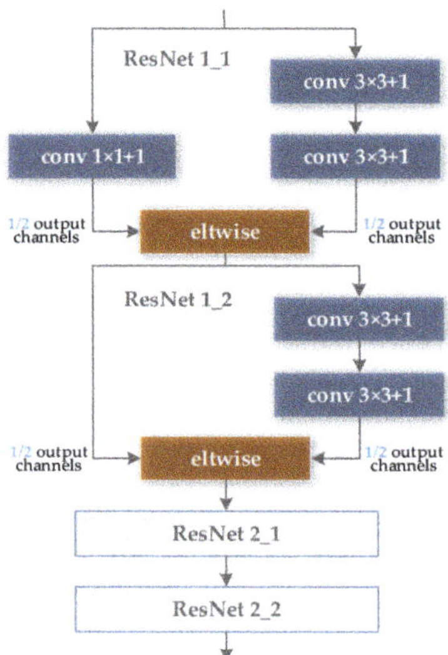

Figure 3. Two residual modules in series structure.

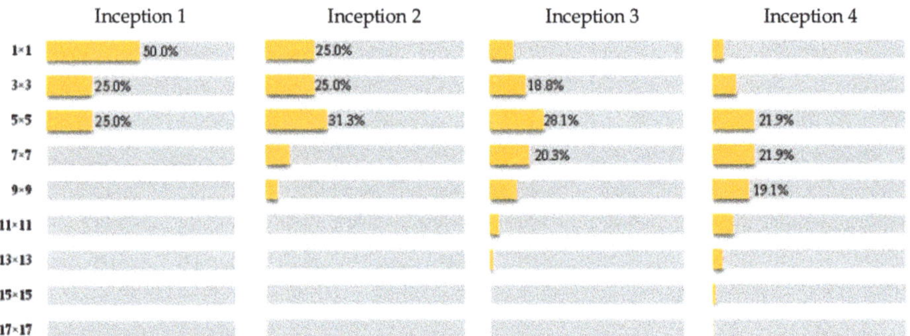

Figure 4. Distribution of the receptive field of the series structure of four Inception modules.

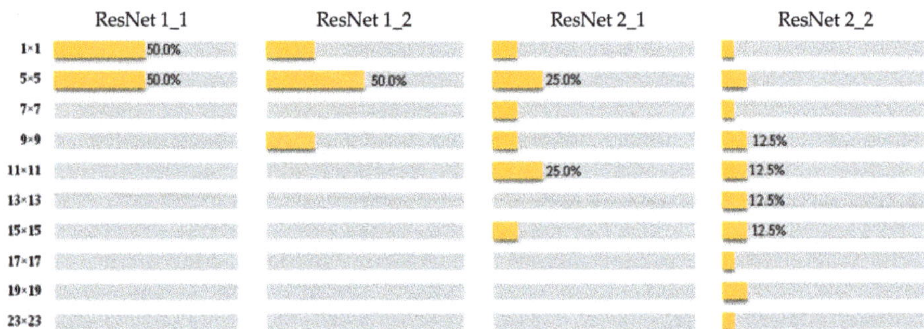

Figure 5. Distribution of the receptive of receptive field of the series structure of two residual blocks.

In addition, although in the distribution map of the fourth module in Figures 4 and 5, three are still small receptive fields such as 1×1, 3×3, and 5×5, but as the convolutional neural network becomes deeper, these local receptive fields will gradually disappear. Table 2 shows the minimum and maximum receptive field scale distributions of the last layer of feature maps in the GoogLeNet and ResNet-18, it can be seen that the local receptive fields have disappeared and been replaced by macroscopic large and even super large receptive fields.

Table 2. The maximum and minimum values of the receptive field of the last convolution layer of feature maps of GoogLeNet and ResNet-18.

CNNs	Feature Map of the Last Convolutional Layer	Minimal Receptive Field	Maximum Receptive Field
GoogLeNet	pool5/7 $\times$ 7_s1	267 $\times$ 267	907 $\times$ 907
ResNet-18	pool5	203 $\times$ 203	627 $\times$ 627

In summary, the existing CNNs have similar characteristics in design, that is, in the early stage of feature extraction of CNNs, they focus on learning the local and concrete information of images, and in the middle and later stages of feature extraction of CNNs, the macro, and abstract feature information are more likely to be learned. Of course, the above law is logical and effective in most classification and detection tasks, but for the surface defect data sets with multi-scale features studied in this paper, the above design ideas of CNNs are obviously not the optimal choice.

3. Methods

3.1. Data Set Preparation

(1) The magnetic tile defect data set [17] was collected by the Institute of Automation of the Chinese Academy of Sciences, the data set has six categories, including non-defective samples, the representative samples of each category are shown in Figure 6. It should be noted that when using a CNN to train the magnetic tile data set, there is a problem with an insufficient number of samples, therefore, in this paper, the Semi-Supervised Data Augmentation (SSDA) method is used, the SSDA method takes into account the shape and location characteristics of the defective target on the basis of the classic data augmentation operation, and maintains the original label attributes of the samples while performing data augmentation, which provides high-quality data support for the training of CNNs [17], and finally get 10,320 sample images, including 6192 images in the training set, 2064 images in the validation set and 2064 image in the test set, the ratio of the training set, validation set and test set is 3:1:1.

(2) Roller surface defect data set [17], which is from the data set collected and published by the Institute of Automation of the Chinese Academy of Sciences (CAS). The roller surface defect data set collects various morphological samples of the roller surface in the air-conditioning compressor. The data set is made from the original image after preprocessing such as ring region expansion, sliding window cutting, image enhancement, etc. The sample examples and numbers of each category are shown in Table 3. Among them, EFQ and CQ are non-destructive surface samples, and the other categories are defective samples. After data augmentation, 22,400 samples are finally obtained, among them, there are 13,440 images in the training set, 4480 images in the validation set, and 4480 sheets in the test set, the ratio between them is also 3:1:1.

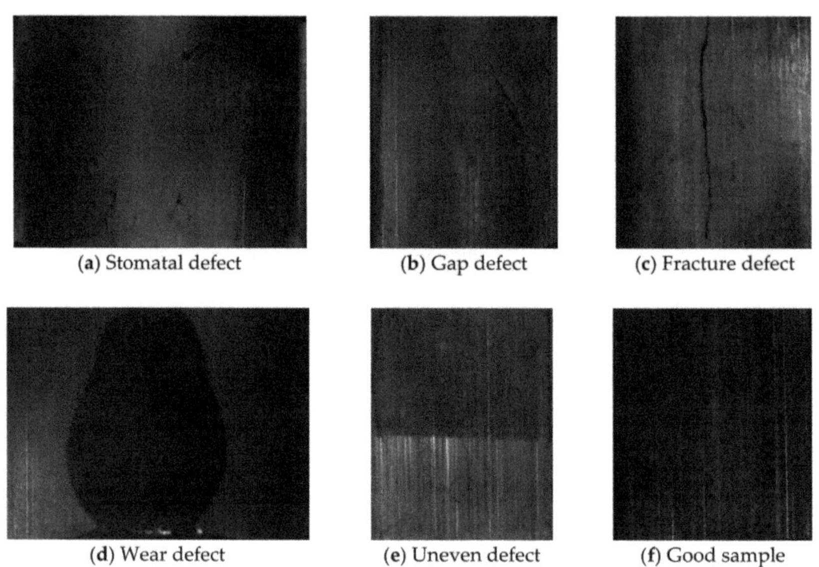

(a) Stomatal defect	**(b)** Gap defect	**(c)** Fracture defect
(d) Wear defect	**(e)** Uneven defect	**(f)** Good sample

Figure 6. Samples of magnetic tile data set.

Table 3. The sample examples and numbers of each category in the roller surface defect data set.

Category Name	EFQ *	EFC	EFI	EFSc	EFSt	EFSF
Number of samples	1500	470	70	160	90	220
Sample example						

Category name	CQ	CC	CI	CSc	CSt
Number of samples	1000	350	155	30	105
Sample example					

* EFQ: end-face qualified; EFC: end-face cracks; EFI: end-face indentations; EFSc: end-face scratches; EFSt: end-face stains EFSF: end-face serious fracture CQ: chamfer qualified CC: chamfer cracks CI: chamfer indentations CSs: chamfer scratches CSt: chamfer stains.

3.2. Sample Defect Size Analysis

Figures 7 and 8 respectively show samples with large differences in defect scales in the two datasets and their corresponding defect labeling positions, it can be seen that the size differences between different defect types are quite huge, which poses a great challenge to classification and recognition tasks. In addition, even for defects of the same type, the size of the defect varies from sample to sample. Figure 9 shows the area statistics of all defect types in the two datasets. It can be seen that in the defect data set of magnetic tile, the average area of wear defects and uneven defects is more than 60 times that of stomatal defects; the maximum defect area of the wear defect sample exceeds the maximum area of the stomatal defect sample by 140 times.

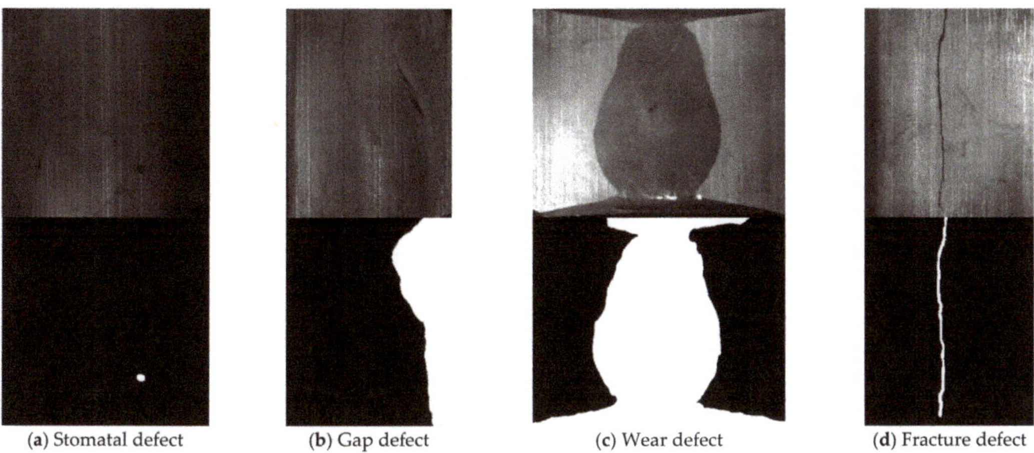

(**a**) Stomatal defect   (**b**) Gap defect   (**c**) Wear defect   (**d**) Fracture defect

Figure 7. The samples with large scale difference of surface defects of magnetic tile and their corresponding defect labeling positions.

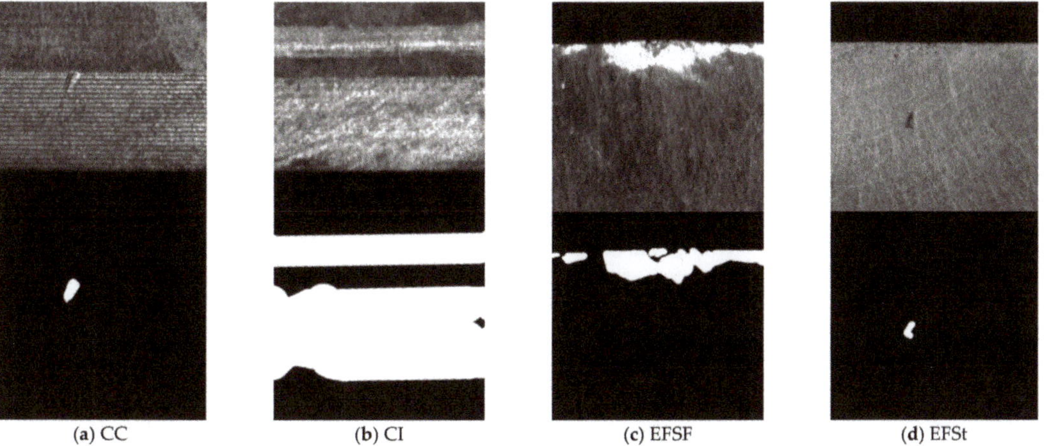

(**a**) CC   (**b**) CI   (**c**) EFSF   (**d**) EFSt

Figure 8. The samples with large scale difference of surface defects of roller and their corresponding defect labeling positions.

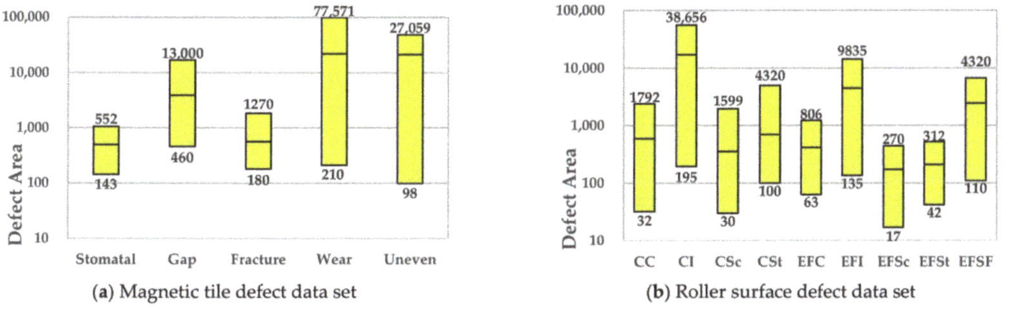

(**a**) Magnetic tile defect data set   (**b**) Roller surface defect data set

Figure 9. Defect area statistics of chart of different types of defect samples in the two data sets (The upper and lower lines of each column in the figure represent the maximum and minimum values respectively, and the middle line of each column represents the average value).

In the roller defect data set, the average defect area of defect samples of CI type is close to 17,000 pixels, while the average defect area of defect sample types such as CC, CSc, CSt, EFC, EFSc, EFSt do not exceed 1000. From the above analysis, it can be seen that these two defect data sets have obvious multi-scale defects problems.

3.3. Multi-Scale Feature Learning Network Based on Dual Module Feature Extractor

In order to solve the problem that the existing CNNs are not strong in learning small-scale defect features, and to improve the classification and recognize the ability of multi-scale surface defects, this paper proposes Multi-Scale Feature Learning Network (MSF-Net) based on Dual Module Feature Extractor (DMF), where DMF is built by the Concatenated Rectified Linear Units (CReLUs) and the Inception modules. The main design ideas of MSF-Net are as follows:

(1) Increase the diversity of the receptive field of a single convolution module in CNNs
From the analysis of the receptive field of CNN in Section 2.1, it can be seen that single branch CNNs, such as AlexNet and VGGNet, have a relatively single receptive field scale. As the number of layers of the CNN deepens, the small-scale receptive field gradually disappears, which is not conducive to the feature expression of subtle defects in the classification task of surface defects. Therefore, this paper chooses the convolution module with branch structure as the basic unit of MSF-Net. Among the many representative modules with branch structures, the Inception module has favored domestic and foreign researchers in the field of target classification and detection, because of its lightweight design ideas and excellent characteristic expression ability and classification accuracy. In this paper, the Inception v3 [32] structure is used as the design prototype, as the feature extraction module in the middle and late stages of MSF-Net; and in the early stage of MSF-Net, in order to reduce the parameter quantity and calculation amount, this paper selects the CReLU module [33] as the prototype design feature extraction module. In this way, the Dual Module Feature (DMF) extractor is formed.

(2) Increase the diversity of the receptive field of the feature map output by the last convolutional layer of CNNs In order to improve the classification accuracy of multi-scale defect samples, it is necessary to ensure that the feature map output by the last convolutional layer before the fully connected layer has sufficient receptive field scales, especially the number of small-scale receptive fields. Therefore, inspired by HyperNet [34], the feature maps of several convolution modules with different scale receptive fields are combined to effectively increase the diversity of the receptive fields of the feature maps of the last convolution layer before the fully connected layer.

(3) Improve training efficiency The improvement of feature expression ability inevitably means the deepening of the number of layers of CNNs. Therefore, it is essential to improve training efficiency. MSF-Net improves training efficiency and avoids over-fitting by using residual shortcuts and batch normalization (BN) layers.

3.3.1. Dual Module Feature Extractor

The Dual Module Feature Extractor (DMF) contains two different convolutional modules: the CReLU modules are mainly used in the early stage of feature extraction of MSF-Net, which aims to reduce the calculation cost and speed up the calculation of forward propagation; in the later stage of feature extraction of MSF-Net, the Inception modules are used to increase the depth and width of MSF-Net and improve feature learning ability.

Optimized CReLU Feature Extraction Modules

The research of Shang et al. [33] revealed that in the early stage of feature extraction of CNNs, the filters in the lower layers form pairs, the phase of each pair of filters is opposite, that is, CNN has a tendency to learn both positive and negative phase information at the same time, however, the Rectified Linear Units (ReLU) [35] will suppress the negative response, making the feature of the lower convolutional layer of the CNN redundant.

CReLU takes these negative responses as the output of the convolutional layer by inverting the feature map and then using the ReLU function, the structure of CReLU is shown in the dashed box in Figure 10. The above operation can convert redundant features into usable features, and extract twice as many feature maps, thereby improving the utilization of features in the lower convolutional layer of CNN. In this paper, the CReLU feature extraction module is designed on the basis of the native CReLU module, as shown in Figure 10. A 1×1 convolutional layer is added to the input and output of the native ReLU module to achieve dimensionality reduction and dimensionality increase of the number of convolution kernel channels, which increases the nonlinearity of CNNs and reduces the amount of calculation. In addition, the shortcut connections in ResNet are also introduced into the CReLU feature extraction module to reduce the loss of information in the transmission process and protect the integrity of the information.

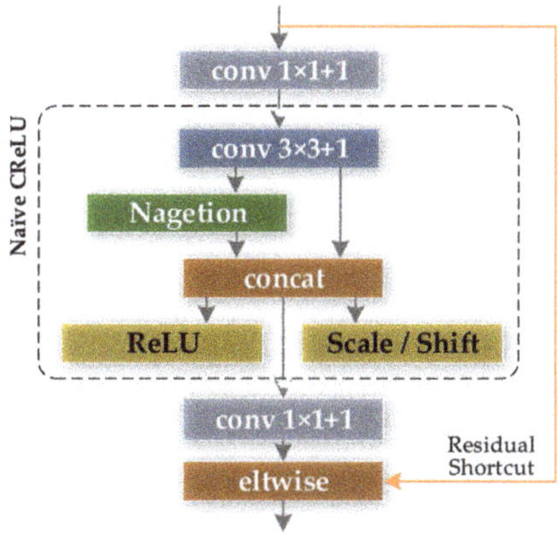

Figure 10. Optimized CReLU feature extraction module.

Optimized Inception V3 Module

The Inception V3 module uses convolution kernels of different sizes such as 1×1, 3×3, and 5×5 to obtain different scales of receptive fields, which improves the diversity of features. In addition, the design of the branch structure saves computational costs while enhancing the width and depth of CNN.

Based on the native Inception V3 module, this paper proposed an optimized version of the Inception feature module, as shown in Figure 11. Similar to the optimized version of the ReLU module, a 1×1 convolutional layer is added to the output of the optimized module to realize the dimension increase of the feature output; in addition, shortcut connections are also introduced into the optimized Inception module.

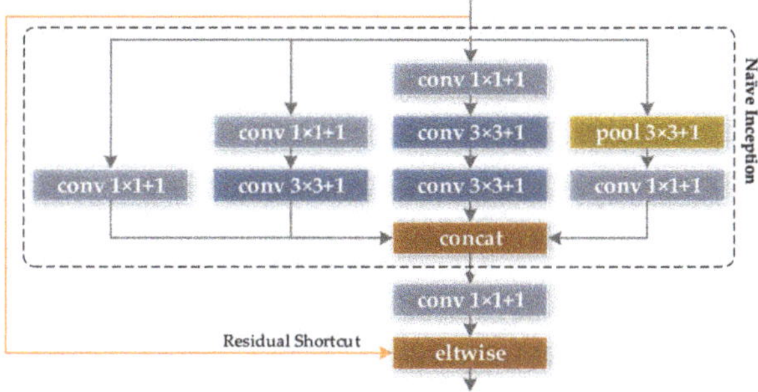

Figure 11. Optimized Inception feature extraction module.

3.3.2. Multi-Scale Feature Learning Network

In order to increase the diversity of the receptive field of the last convolutional layer of CNNs and improve the ability to learn and express features of different scales, especially small and local features, MSF-Net aligns and integrates the feature outputs of several intermediate layers, and then the fully connected layer is used for classification. Figure 12 shows the overall architecture of MSF-Net, and Table 4 lists the specific parameters and indicators in detail. As can be seen from Figure 12, MSF-Net is mainly composed of five convolutional module chains in the feature extraction stage, including two optimized CReLU modules and three optimized Inception V3 modules. In addition, all convolutional layers in MSF-Net, except for the optimized CReLU modules, are designed with batch normalization (BN) layers, scale layers, and ReLU activation layer to better accelerate convergence.

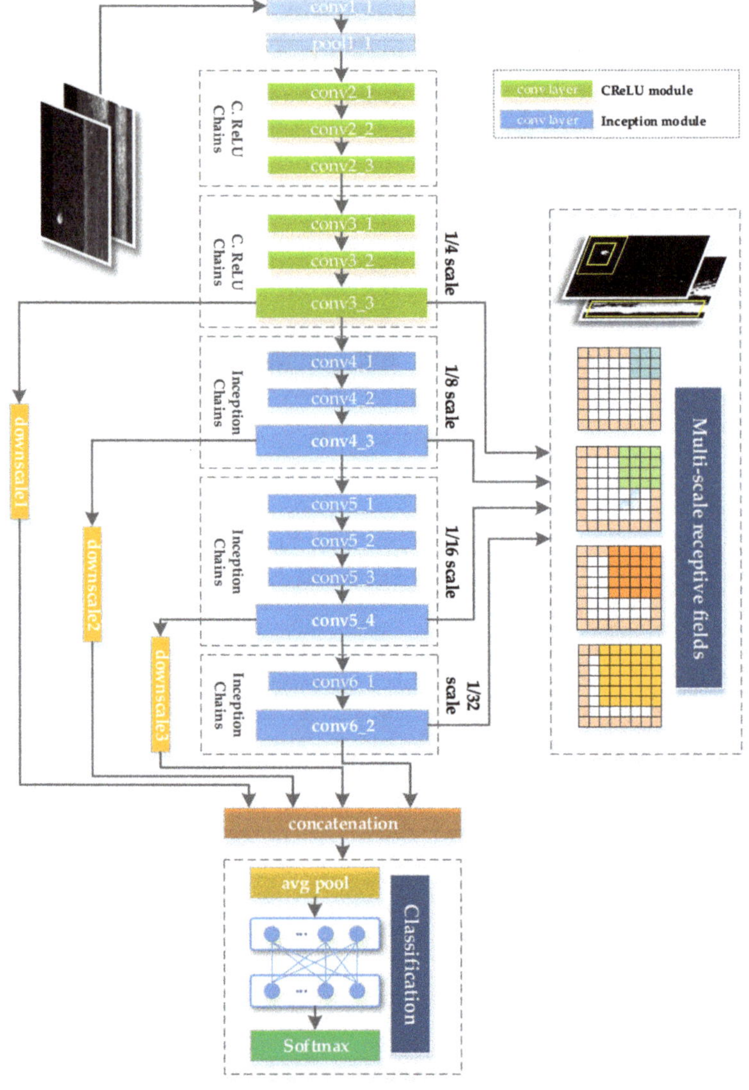

Figure 12. The overall architecture of MSF-Net.

Table 4. The parameters of MSF-Net.

Layername	Type	Output Dimension	Depth	CReLU Output	Inception Output					Parameters
				#1×1-3×3-1×1	Pool Proj	#1×1	#3×3	#5×5	#1×1 out	
conv1_1	3 × 3 CReLU	224 × 224 × 32	1	NA-16-NA			/			480
pool1_1	3 × 3 Max pool.	112 × 112 × 32	0	/						/
conv2_1	3 × 3 CReLU	112 × 112 × 64	3	24-24-64						13,440
conv2_2	3 × 3 CReLU	112 × 112 × 64	3	24-24-64			/			13,440
conv2_3	3 × 3 CReLU	112 × 112 × 64	3	24-24-64						13,440
conv3_1	3 × 3 CReLU	56 × 56 × 128	3	48-48-128						53,760
conv3_2	3 × 3 CReLU	56 × 56 × 128	3	48-48-128			/			53,760
conv3_3	3 × 3 CReLU	56 × 56 × 128	3	48-48-128						53,760
conv4_1	Inception	28 × 28 × 256	4		32	64	96-128	16-32-32	256	322,560
conv4_2	Inception	28 × 28 × 256	4	/	32	64	96-128	16-32-32	256	322,560
conv4_3	Inception	28 × 28 × 256	4		32	64	96-128	16-32-32	256	322,560
conv5_1	Inception	14 × 14 × 512	4		64	128	128-192	32-96-96	512	1,040,384
conv5_2	Inception	14 × 14 × 512	4	/	64	128	128-192	32-96-96	512	1,040,384
conv5_3	Inception	14 × 14 × 512	4		64	128	128-192	32-96-96	512	1,040,384
conv5_4	Inception	14 × 14 × 512	4		64	128	128-192	32-96-96	512	1,040,384
conv6_1	Inception	7 × 7 × 1024	4		128	256	160-320	32-128-128	1024	3,010,560
conv6-2	Inception	7 × 7 × 1024	4	/	128	256	160-320	32-128-128	1024	3,010,560
concat	Concatenation	7 × 7 × 1920	0							/
avg pool	7 × 7 Avg pool.	1 × 1 × 1920	0	/			/			
linear	Inner product	1 × 1 × N_{Class}	1							19,200
Total			47							11,371,616

In order to realize the learning and expression of multi-scale features, the specific design details of MSF-Net are as follows:

(1) At the input of the convolution modules conv3_1, conv4_1, conv5_1, and conv6_1, the convolution kernel with a size of 1 × 1 and a stride of 2 is used instead of the pooling layer to achieve the proportional reduction of the feature map size, so that the feature maps output by conv3, conv4, conv5, conv6 is 1/4, 1/8, 1/16, and 1/32 of the input image, respectively.

$$\begin{cases} Size_{conv3} = \dfrac{Size_{input}}{4} \\ Size_{conv4} = \dfrac{Size_{input}}{8} \\ Size_{conv5} = \dfrac{Size_{input}}{16} \\ Size_{conv6} = \dfrac{Size_{input}}{32} \end{cases} \tag{2}$$

(2) The output feature maps of the convolution modules conv3_3, conv4_3, and conv5_4 are down-sampled, and the average pooling layers with kernel sizes of 8 × 8, 4 × 4, and 2 × 2 are used, so that their respective output feature maps are consistent with the feature map output by conv6_2

$$\begin{cases} f_{c1} = Pool_{Avg}^{8×8}(f_{conv3\_3}) \\ f_{c2} = Pool_{Avg}^{4×4}(f_{conv4\_3}) \\ f_{c3} = Pool_{Avg}^{2×2}(f_{conv5\_4}) \\ f_{c4} = f_{con6\_2} \end{cases} \tag{3}$$

where $f_{conv3\_3}, f_{conv4\_3}, f_{conv5\_4}$ and f_{conv6_2} represent the output feature maps of conv3_3, conv4_3, conv5_4 and conv6_2, respectively, f_{c1}, f_{c2}, f_{c3} and f_{c4} , respectively, represent the four input feature maps of the concatenation layer.

(3) The above four feature maps are combined and connected to form the final output feature map, the formula is as follows:

$$f_{concat} = concat(f_{c1}, f_{c2}, f_{c3}, f_{c4}) \tag{4}$$

where f_{concat} represents the feature map output by the concatenation layer. In this way, the final output features include small-local concrete features and macro-global abstract features.

It is worth mentioning that the architecture design of MSF-Net follows many experiences and guidelines for the design of CNNs, as follows:

(1) Avoiding expression bottlenecks in the early stage of feature extraction. That is, the information flow should avoid highly compressed convolution layers in the forward propagation process, and the width and height of the feature map should be gradually reduced in an orderly manner, especially for surface defect datasets with subtle defect features, it is not wise to compress the feature map too early. Therefore, the convolutional layer conv1_1 (size 3 × 3, stride 1) and pooling layer pool1_1 (size 3 × 3, stride 2) are concatenated to slow down the reduction speed of the feature map.

(2) In the middle and late stages of feature extraction in CNNs, the width and depth of CNNs should be balanced as much as possible. That is, as the CNN deepens, the feature map gradually shrinks, and the output matrix dimension of each convolutional layer should gradually increase. Therefore, the number of modules, the feature map sizes, and the number of channels of the three module chains of conv 4, conv 5, and conv 6 are designed with full reference to Inception V1 [29] and Inception V3 [32] to improve the rationality of CNN's evolution.

(3) Average pooling layer is used to replace the fully connected layer, which can greatly reduce the number of parameters and save calculation costs. Specifically, MSF-Net uses an average pooling layer with a kernel size of 7 × 7 and a strider of 1 to replace the fully connected layer. It can be calculated from Table 4 that this adjustment can reduce 180,635,529 parameters.

(4) The residual shortcut connections are used to effectively accelerate training and promote CNN's convergence. Specially, almost every convolution module in MSF-Net, except conv1_1, uses a shortcut connection, which effectively avoids the problem of gradient disappearance and speeds up training.

4. Experimental Results and Analysis

4.1. Experimental Setup

The performance indicators and parameters of the experimental platform are shown in Table 5.

Table 5. Parameters of the experimental platform.

CPU	Intel E3-1230 V2*2 (3.30 GHz)	RAM	16 GB DDR3	GPU	NVIDIA GTX-1080Ti
OS	Ubuntu 16.04 LTS	Software		Visual Studio Code with Python 2.7	

4.2. CNNs for Comparison

Inception v3 and ResNet-50 are used as the comparison CNNs for the following reasons:

(1) Inception v3 and ResNet-50 are closer to MSF-Net in terms of the number of convolutional layers and parameters, as shown in Table 6, which makes the comparison of experimental results fairer.

(2) MSF-Net is deeply influenced by GoogLeNet v3 in terms of the design of feature extraction modules, the number of modules, and the width and depth of CNN. Therefore, comparing MSF-Net with Inception v3 can more accurately assess the impact of CNN's structure on classification performance.

(3) Except conv1_1, all feature extraction modules of MSF-Net use residual shortcut connections.

Therefore, it is also essential to conduct comparative experiments with ResNet-50, which is similar in size.

Table 6. Comparison of the number of convolutional layers and parameters of three CNNs.

CNNs	Convolution Layers	Parameters
Inception v3	48	24,734,048
ResNet-50	50	~25.5 × 10^6
MSF-Net	56	11,371,616

4.3. Training Efficiency Evaluation of CNNs

Figure 13 shows the comparison of the time and number of iterations required to reach convergence when the three CNN models are trained on two multi-scale surface defect data sets. It can be seen from Figure 13a that MSF-Net achieves convergence in the shortest time on the training performance of the two multi-scale surface defect data sets. Specifically, the convergence time of MSF-Net is shortened by at least 25% compared with that of ResNet-50, and it is also shortened by nearly 10% compared with that of Inception v3. Figure 13b shows the comparison of the number of epochs for the three models to reach convergence. Through the comparative analysis of three CNNs, it can be seen that the efficient training performance of MSF-Net mainly comes from the following two aspects:

(1) Compared with ResNet-50 and Inception v3, MSF-Net's parameters is reduced by more than 54%, which greatly reduces the computational cost, the most intuitive manifestation is that the number of epochs required for MSF-Net to achieve convergence is greatly reduced;

(2) Compared with the Inception modules fully used by Inception v3, MSF-Net uses the optimized CReLU module in the early stage of feature extraction, the optimized CReLU module has outstanding performance in reducing computational costs, which effectively shortens the time consumption of forward back propagation and improves the training performance of MSF-Net.

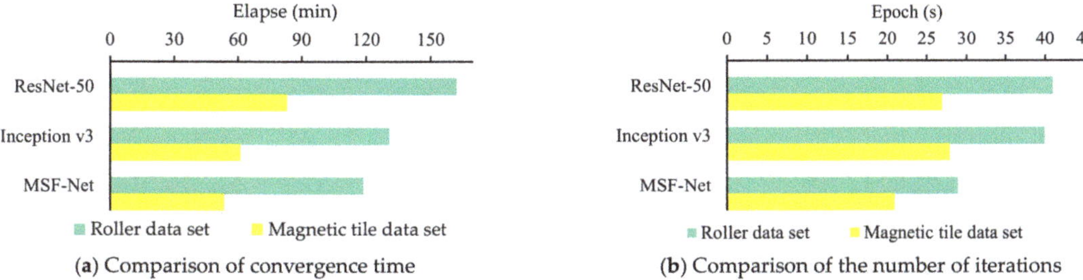

(**a**) Comparison of convergence time   (**b**) Comparison of the number of iterations

Figure 13. Comparison of the time and the number of epochs required for three CNNs to reach convergence.

4.4. Classification Performance Evaluation on Two Multi-Scale Defect Data Sets

Table 7 and Figure 14 respectively show the classification performance of ResNet-50, Inception v3, and MSF-Net on the test of the surface defect data set of roller. It can be seen from Table 7 that the three CNNs have excellent performance in the defect categories of CI, CSs, CSt, EFI, EFSc, and EFSt, with an accuracy rate of 100%. In the remaining five categories, CQ and EFQ are non-defective samples; CC and EFC are small-size defects, and the appearance of the samples is very close to CQ and EFQ respectively, EFSF represents larger defect samples. Therefore, the roller surface defect data set is very suitable for the evaluation and verification of multi-scale feature learning capabilities of CNNs. ResNet-50 and Inception v3 achieved the highest recall rates in the EFC and EFSF defect categories, respectively, while MSF-Net performed better than the other two CNNs in the CC defect category. In addition, MSF-Net has an outstanding performance in the accuracy rate of CQ and EFQ, which has important application value in actual production. Overall, the average recall rate of MSF-Net on the roller defect set is 99.29%, while the recall rates of ResNet-50 and Inception v3 are 98.44% and 99.06%, respectively; at the same time, MSF-Net has the smallest standard deviation in recall rate, showing a more balanced expression and learning ability for defects of different scales. Similarly, compared with ResNet-50 and Inception v3, MSF-Net also achieves the best performance in precision, micro-F1 and macro-F1 indicators, which verified its superiority.

Table 7. Classification performance of three CNNs on the surface defect data set of roller (%).

	CQ	CC	CI	CSc	CSt		Precision	micro-F1
ResNet-50	93.75	96.25	100.00	100.00	100.00		98.46 ± 0.017	98.381
Inception v3	97.50	96.88	100.00	100.00	100.00		99.09 ± 0.008	99.055
MSF-Net	98.13	98.75	100.00	100.00	100.00		99.29 ± 0.006	99.301

	EFQ	EFC	EFI	EFSc	EFSt	EFSF	Recall	macro-F1
ResNet-50	95.42	97.71	100.00	100.00	100.00	98.96	98.44 ± 0.022	98.367
Inception v3	97.92	97.29	100.00	100.00	100.00	99.79	99.06 ± 0.013	99.051
MSF-Net	98.33	97.50	100.00	100.00	100.00	99.58	99.29 ± 0.009	99.298

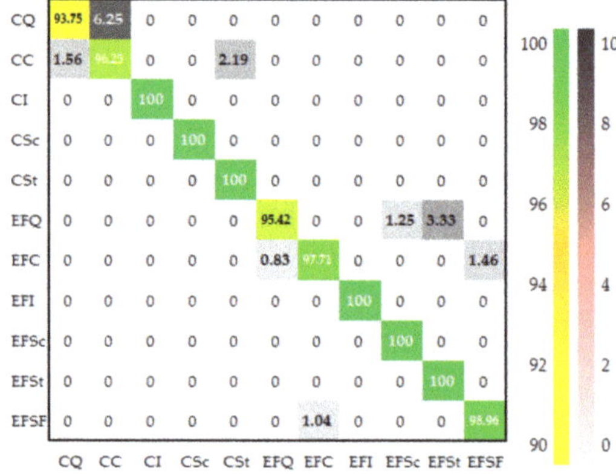

(a) ResNet-50

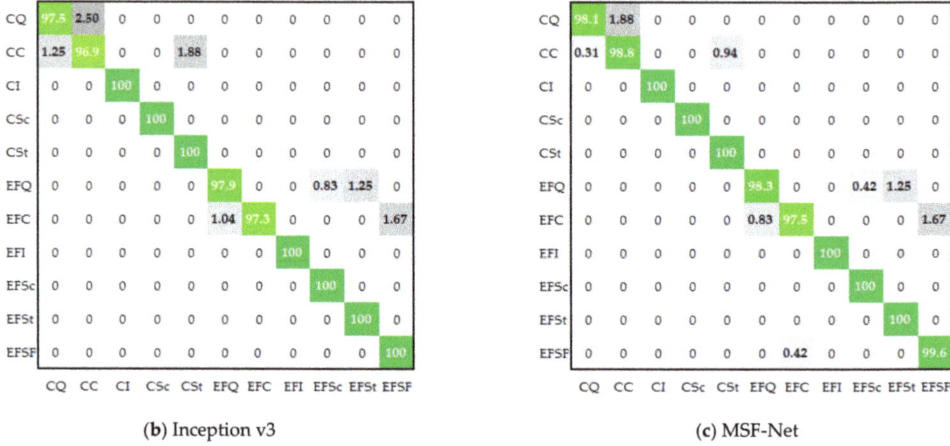

(b) Inception v3

(c) MSF-Net

Figure 14. Confusion matrix of three CNNs on the surface defect data set of roller (unit: %).

Table 8 and Figure 15 show the classification performance of ResNet-50, Inception v3, and MSF-Net on the test set of surface defect data set of magnetic tile. It can be seen from Table 8 that the three CNNs have excellent performance in the categories of fracture defects and good products. MSF-Net has the highest accuracy rate for stomatal and uneven defects, and Inception v3 has the best performance on the gap and wear defect categories. In general, the average recall rate of MSF-Net on the magnetic tile defect set reached 98.93%, while the recall rate of ResNet-50 and Inception v3 were 98.69% and 98.89% respectively; at the same time, among the three CNNs, MSF-Net has the smallest standard deviation in recall rate. Similarly, MSF-Net performs better on precision, micro-F1 and macro-F1 than ResNet-50 and Inception v3.

Table 8. Classification performance of three CNNs on the surface defect data set of magnetic tile (%).

	Stomatal	Gap	Fracture	Precision	Micro-F1
ResNet-50	98.26%	97.67%	100.00%	98.70 ± 0.008	98.70
Inception v3	97.97%	99.13%	99.71%	98.90 ± 0.010	98.90
MSF-Net	98.55%	98.26%	100.00%	98.94 ± 0.007	98.94

	Wear	Uneven	Good	Recall	Macro-F1
ResNet-50	98.55%	97.67%	100.00%	98.69 ± 0.009	98.69
Inception v3	99.42%	97.09%	100.00%	98.89 ± 0.010	98.89
MSF-Net	98.84%	97.97%	100.00%	98.93 ± 0.008	98.93

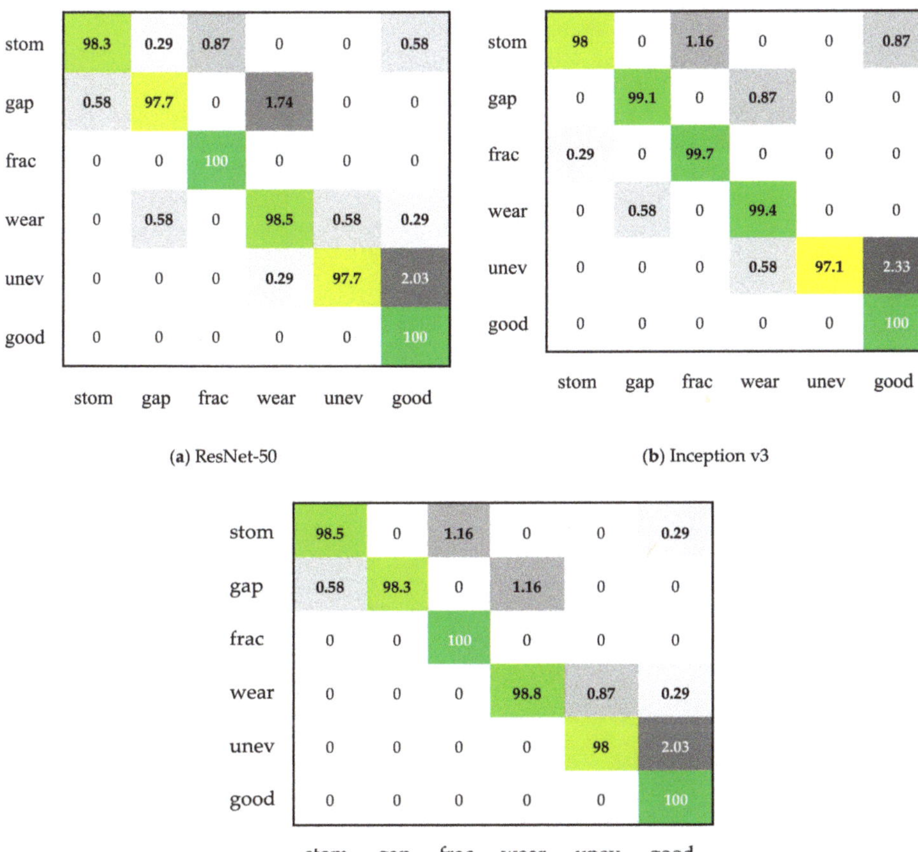

(a) ResNet-50

(b) Inception v3

(c) MSF-Net

Figure 15. Confusion matrix of three CNNs on the surface defect data set of magnetic tile (Abbreviation description—stom: stomatal; frac: fracture; unev: uneven).

5. Conclusions

Aiming at the problem that the commonly used CNNs are not ideal for detecting small and local defects on the products' surface, the multi-scale feature extraction mechanism involved in several mainstream CNNs in the current deep learning field is analyzed, and a multi-scale feature learning network based on dual module feature extractor is proposed, named MSF-Net, the design of the dual module feature extractor, the specific architecture and parameters of MSF-Net, and the optimization to improve training efficiency are introduced in detail. The proposed MSF-Net was trained and tested on two multi-scale surface defect data sets, which verified the advancement and effectiveness in multi-scale defect detection. Future work will focus on exploring the generalization of MSF-Net in more research fields, such as radar image classification and remote sensing image classification.

Author Contributions: Conceptualization, P.X., Z.G., L.L. and X.X.; methodology, P.X., Z.G. and X.X.; software, P.X. and X.X.; validation, P.X., and Z.G.; formal analysis, Z.G. and X.X.; investigation, Z.G. and X.X.; resources, L.L. and X.X.; data curation, Z.G. and X.X.; writing—original draft preparation, P.X., Z.G. and X.X.; writing—review and editing, Z.G. and X.X.; visualization, X.X.; supervision, L.L.; project administration, L.L.; funding acquisition L.L. All authors have read and agreed to the published version of the manuscript.

Funding: This work was supported by the National Key Research and Development Program of China (No. 2020YFF0304902) and the National Natural Science Foundation of China (No. 61771352).

Institutional Review Board Statement: Not applicable.

Informed Consent Statement: Not applicable.

Data Availability Statement: Not applicable.

Conflicts of Interest: The authors declare no conflict of interest.

References

1. Tao, X.; Hou, W.; Xu, D. A Survey of Surface Defect Detection Methods Based on Deep Learning. *Acta Autom. Sin.* **2020**, *46*, 1–18.
2. Jing, J.; Liu, S.; Li, P.; Zhang, L. The fabric defect detection based on CIE L*a*b* color space using 2-D Gabor filter. *J. Text. Inst. Proc. Abstr.* **2016**, *107*, 1305–1313. [CrossRef]
3. Kaewunruen, S.; Sresakoolchai, J.; Thamba, A. Machine learning-aided identification of train weights from railway sleeper vibration. *Insight Non-Destr. Test. Cond. Monit.* **2021**, *63*, 151–159. [CrossRef]
4. Liu, T.I.; Singonahalli, J.H.; Iyer, N.R. Detection of roller bearing defects using expert system and fuzzy logic. *Mech. Syst. Signal Process.* **1996**, *10*, 595–614. [CrossRef]
5. Baygin, M.; Karakose, M.; Sarimaden, A.; Erhan, A.K.I.N. Machine vision based defect detection approach using image processing. In Proceedings of the 2017 International Artificial Intelligence and Data Processing Symposium, Malatya, Turkey, 16–17 September 2017; Institute of Electrical and Electronics Engineers: Malatya, Turkey, 2017; p. 8090292.
6. Zhang, L.; Jing, J.; Zhang, H. Fabric defect classification based on LBP and GLCM. *J. Fiber Bioeng. Inform.* **2015**, *8*, 81–89. [CrossRef]
7. Sidorov, D.; Wei, W.S.; Vasilyev, I.; Salerno, S. Automatic defects classification with p-median clustering technique. In Proceedings of the 2008 10th International Conference on Control, Automation, Robotics and Vision, Hanoi, Vietnam, 17–20 December 2008; pp. 775–780. [CrossRef]
8. LeCun, Y.; Bengio, Y.; Hinton, G. Deep learning. *Nature* **2015**, *521*, 436–444. [CrossRef] [PubMed]
9. He, T.; Huang, W.; Qiao, Y.; Yao, J. Text-Attentional Convolutional Neural Network for Scene Text Detection. *IEEE Trans. Image Process.* **2016**, *25*, 2529–2541. [CrossRef] [PubMed]
10. Prasad, P.S.; Pathak, R.; Gunjan, V.K.; Rao, H.R. Deep learning based representation for face recognition. In *ICCCE 2019*; Springer: Singapore, 2020; pp. 419–424.
11. Grigorescu, S.; Trasnea, B.; Cocias, T.; Macesanu, G. A survey of deep learning techniques for autonomous driving. *J. Field Robot.* **2020**, *37*, 362–386. [CrossRef]
12. Weimer, D.; Scholz-Reiter, B.; Shpitalni, M. Design of deep convolutional neural network architectures for automated feature extraction in industrial inspection. *CIRP Ann. Manuf. Technol.* **2016**, *65*, 417–420. [CrossRef]
13. Ren, R.; Hung, T.; Tan, K.C. A Generic Deep-Learning-Based Approach for Automated Surface Inspection. *IEEE Trans. Cybern.* **2017**, *99*, 1–12. [CrossRef]
14. Masci, J.; Meier, U.; Ciresan, D.; Schmidhuber, J.; Fricout, G. Steel defect classification with Max-Pooling Convolutional Neural Networks. In Proceedings of the International Joint Conference on Neural Networks, Brisbane, Australia, 10–15 June 2012; IEEE: New York, NY, USA, 2012; pp. 1–6.

15. Guo, Z.; Zheng, H.; Xu, X.; Ju, J.; Zheng, Z.; You, C.; Gu, Y. Quality grading of jujubes using composite convolutional neural networks in combination with RGB color space segmentation and deep convolutional generative adversarial networks. *J. Food Process. Eng.* **2021**, *44*, e13620. [CrossRef]
16. Deitsch, S.; Christlein, V.; Berger, S.; Buerhop-Lutz, C.; Maier, A.; Gallwitz, F.; Riess, C. Automatic classification of defective photovoltaic module cells in electroluminescence images. *Sol. Energy* **2019**, *185*, 455–468. [CrossRef]
17. Xu, X.; Zheng, H.; Guo, Z.; Wu, X.; Zheng, Z. SDD-CNN: Small Data-Driven Convolution Neural Networks for Subtle Roller Defect Inspection. *Appl. Sci.* **2019**, *9*, 1364. [CrossRef]
18. Isola, P.; Zhu, J.Y.; Zhou, T.; Efros, A.A. Image-to-image translation with conditional adversarial networks. In Proceedings of the IEEE Conference on Computer Vision and Pattern Recognition, Honolulu, HI, USA, 21–26 July 2017; IEEE: New York, NY, USA, 2017; pp. 1125–1134.
19. Ronneberger, O.; Fischer, P.; Brox, T. U-net: Convolutional networks for biomedical image segmentation. In Proceedings of the International Conference on Medical Image Computing and Computer-Assisted Intervention, Munich, Germany, 5–9 October 2015; Springer: Cham, Switzerland, 2015; pp. 234–241.
20. Milletari, F.; Navab, N.; Ahmadi, S.A. V-net: Fully convolutional neural networks for volumetric medical image segmentation. In Proceedings of the 2016 Fourth International Conference on 3D Vision (3DV), Stanford, CA, USA, 25–28 October 2016; IEEE: New York, NY, USA, 2016; pp. 565–571.
21. Badrinarayanan, V.; Kendall, A.; Cipolla, R. Segnet: A deep convolutional encoder-decoder architecture for image segmentation. *IEEE Trans. Pattern Anal. Mach. Intell.* **2017**, *39*, 2481–2495. [CrossRef] [PubMed]
22. Pelt, D.M.; Sethian, J.A. A mixed-scale dense convolutional neural network for image analysis. *Proc. Natl. Acad. Sci. USA* **2018**, *115*, 254–259. [CrossRef] [PubMed]
23. Rundo, L.; Han, C.; Zhang, J.; Hataya, R.; Nagano, Y.; Militello, C.; Ferretti, C.; Nobile, M.S.; Tangherloni, A.; Gilardi, M.C.; et al. CNN-based prostate zonal segmentation on T2-weighted MR images: A cross-dataset study. In *Neural Approaches to Dynamics of Signal Exchanges*; Springer: Singapore, 2020; pp. 269–280.
24. Luo, W.; Li, Y.; Urtasun, R.; Zemel, R. Understanding the effective receptive field in deep convolutional neural networks. In Proceedings of the 30th International Conference on Neural Information Processing Systems (NIPS), Barcelona, Spain, 5–10 December 2016; pp. 4898–4906.
25. Tang, C.; Sheng, L.; Zhang, Z.; Hu, X. Improving Pedestrian Attribute Recognition with Weakly-Supervised Multi-Scale Attribute-Specific Localization. In Proceedings of the IEEE International Conference on Computer Vision (ICCV), Seoul, Korea, 27 October–2 November 2019; pp. 4997–5006.
26. Kim, Y.; Kang, B.N.; Kim, D. San: Learning relationship between convolutional features for multi-scale object detection. In Proceedings of the European Conference on Computer Vision (ECCV), Munich, Germany, 8–14 September 2018; pp. 316–331.
27. Krizhevsky, A.; Sutskever, I.; Hinton, G.E. ImageNet Classification with Deep Convolutional Neural Networks. *Commun. ACM* **2017**, *60*, 84–90. [CrossRef]
28. Simonyan, K.; Zisserman, A. Very deep convolutional networks for large-scale image recognition. *arXiv* **2014**, arXiv:1409.1556.
29. Szegedy, C. Going deeper with convolutions. In Proceedings of the 2015 IEEE Conference on Computer Vision and Pattern Recognition (CVPR), Boston, MA, USA, 7–12 June 2015; pp. 1–9.
30. He, K.; Zhang, X.; Ren, S.; Sun, J. Deep residual learning for image recognition. In Proceedings of the IEEE Conference on Computer Vision and Pattern Recognition (CVPR), Las Vegas, NV, USA, 27–30 June 2016; pp. 770–778.
31. Huang, Y.; Qiu, C.; Guo, Y.; Wang, X.; Yuan, K. Surface Defect Saliency of Magnetic Tile. In Proceedings of the 2018 IEEE 14th International Conference on Automation Science and Engineering (CASE), Munich, Germany, 20–24 August 2018; pp. 612–617.
32. Szegedy, C.; Vanhoucke, V.; Ioffe, S.; Shlens, J.; Wojna, Z. Rethinking the Inception Architecture for Computer Vision. In Proceedings of the 2016 IEEE Conference on Computer Vision and Pattern Recognition (CVPR), Las Vegas, NV, USA, 27–30 June 2016; IEEE Computer Society: Washington, DC, USA, 2016; pp. 2818–2826.
33. Shang, W.; Sohn, K.; Almeida, D.; Lee, H. Understanding and improving convolutional neural networks via concatenated rectified linear units. In Proceedings of the International Conference on Machine Learning, New York, NY, USA, 20–22 June 2016; pp. 2217–2225.
34. Kong, T.; Yao, A.; Chen, Y.; Sun, F. Hypernet: Towards accurate region proposal generation and joint object detection. In Proceedings of the IEEE Conference on Computer Vision and Pattern Recognition (CVPR), Las Vegas, NV, USA, 27–30 June 2016; pp. 845–853.
35. Nair, V.; Hinton, G.E. Rectified Linear Units Improve Restricted Boltzmann Machines. In Proceedings of the International Conference on Machine Learning (ICML), Haifa, Israel, 21–24 June 2010; pp. 807–814.

Article

Improving the Ability of a Laser Ultrasonic Wave-Based Detection of Damage on the Curved Surface of a Pipe Using a Deep Learning Technique

Byoungjoon Yu [1], Kassahun Demissie Tola [2], Changgil Lee [3] and Seunghee Park [4,5,*

[1] Department of Convergence Engineering for Future City, Sungkyunkwan University, Suwon 16419, Korea; mysinmu123@skku.edu

[2] Department of Civil, Architecture and Environmental System Engineering, Sungkyunkwan University, Suwon 16419, Korea; kastolla@skku.edu

[3] Advanced Infrastructure Convergence Research Department, Korea Railroad Research Institute, Uiwang 16105, Korea; tolck81@krri.re.kr

[4] School of Civil, Architectural Engineering and Landscape Architecture, Sungkyunkwan University, Suwon 16419, Korea

[5] Technical Research Center, Smart Inside Co., Ltd., Suwon 16419, Korea

* Correspondence: shparkpc@skku.edu

Abstract: With the advent of the Fourth Industrial Revolution, the economic, social, and technological demands for pipe maintenance are increasing due to the aging of the infrastructure caused by the increase in industrial development and the expansion of cities. Owing to this, an automatic pipe damage detection system was built using a laser-scanned pipe's ultrasonic wave propagation imaging (UWPI) data and conventional neural network (CNN)-based object detection algorithms. The algorithm used in this study was EfficientDet-d0, a CNN-based object detection algorithm which uses the transfer learning method. As a result, the mean average precision (mAP) was measured to be 0.39. The result found was higher than COCO EfficientDet-d0 mAP, which is expected to enable the efficient maintenance of piping used in construction and many industries.

Keywords: plumbing maintenance; deep learning; ultrasonic wave propagation imaging; CNN; external damage

Citation: Yu, B.; Tola, K.D.; Lee, C.; Park, S. Improving the Ability of a Laser Ultrasonic Wave-Based Detection of Damage on the Curved Surface of a Pipe Using a Deep Learning Technique. *Sensors* **2021**, *21*, 7105. https://doi.org/10.3390/s21217105

Academic Editors: Yun Zhang, Kwong Tak Wu Sam, Xu Long and Tiesong Zhao

Received: 13 October 2021
Accepted: 25 October 2021
Published: 26 October 2021

Publisher's Note: MDPI stays neutral with regard to jurisdictional claims in published maps and institutional affiliations.

Copyright: © 2021 by the authors. Licensee MDPI, Basel, Switzerland. This article is an open access article distributed under the terms and conditions of the Creative Commons Attribution (CC BY) license (https://creativecommons.org/licenses/by/4.0/).

1. Introduction

Piping is widely used as an important material not only in construction but also in many industrial fields such as aviation and machinery, and as a result the economic, social and technical demands for maintenance due to aging are increasing. In many industrial fields, it is required to apply inspection technology that can detect pipe damage at an early stage [1].

Through laser scanning-based research conducted in previous studies, the applicability to steel and bolt loosening was confirmed [2–7]. On account of this, we aim to detect damage to the pipe with ultrasonic wave propagation imaging (UWPI) in the current work.

The UWPI is one of the signal processing methods that excites a test object with a Q-switched pulse laser system, measures it with an acoustic emission (AE) sensor that acquires the wave propagation data, and displays it as waveform data using an image [8].

Many academic research activities are being carried out on piping. Intensity-based optical system [9,10], microwave nondestructive testing [11], pipe NDT inspection using an automated robot [12–14], eddy-current-based crack recognition [15], etc., are among the investigated techniques. However, most damage detection techniques depend on the empirical and subjective judgment of experienced experts. To overcome this problem, a lot of research based on computer vision technology using machine learning [16] is being conducted in the fields of structural health monitoring (SHM) and nondestructive evaluation (NDE) [17].

Recently, research using deep learning technology among various machine learning technologies has been actively conducted. Among the various deep-learning-based technologies, image classification using CNN shows better performance results than existing image classification algorithms and is continuously being researched and developed [18]. To detect pipe damage through CNN-based object detection, a large amount of data is required. In addition, it is difficult to obtain data and a lot of learning time is required. In this regard, we intend to utilize the transfer learning [19] technique that enables efficient learning using a small amount of data. Using the pre-learned COCO 2017 EfficientDet-d0 model, it is proposed to detect a damage in piping by laser scanning utilizing UWPI. The main objectives of this study are as follows. The primary goal of this study is to confirm the possibility of establishing a damage detection system through CNN learning on the ultrasonic wave propagation images found from the laser scanning of a pipe. Next, by applying the transfer learning technique, we want to check whether it is possible to efficiently detect damage with only a small amount of UWPI learning data. The structure of this paper is as follows. Section 2 describes the UWPI system and its theory that utilizes laser scanning technology to create training data. Section 3 describes the CNN algorithm and EfficientDet-d0 model used to detect pipe damage. In Section 4, we present the experiments and experimental results, and in Section 5 we present the conclusion of the study.

2. Ultrasonic Wave Generation Mechanism Using Pulsed Laser

2.1. Ultrasonic Wave Mode Generation Theory

The generation of ultrasonic waves by a pulsed laser and the sensing of the generated waves takes place as shown in Figure 1 [20]. A source of impulsive pressure is applied to the surface and the resulting time records are tracked at different locations on the surface. When a pulsed laser beam collides with a target structure, various physical phenomena can occur. The basic problems of ultrasonic thermoelasticity generation can be divided into three sub-problems: moderate absorption of electromagnetic energy, reflection, transmission of the laser radiation. As a result of these processes, the absorbed laser energy causes local heating of the area, leading to a thermoelastic expansion of the material and the generation of ultrasonic waves [20–22].

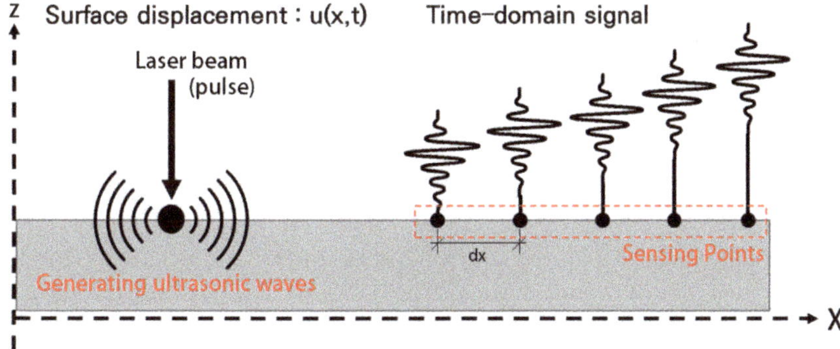

Figure 1. Ultrasonic wave mode generation in a plate [20].

2.2. Ultrasonic Wave Propagation Imaging System Configuration

The UWPI system consists of a Q-switched laser system, a galvanometer (laser mirror scanner), an AE sensor (ultrasonic sensor), a digitizer and an image processor as shown in Figure 2. All devices are synchronized and the ultrasonic response signal is measured by the AE sensor in the digitizer at the same time.

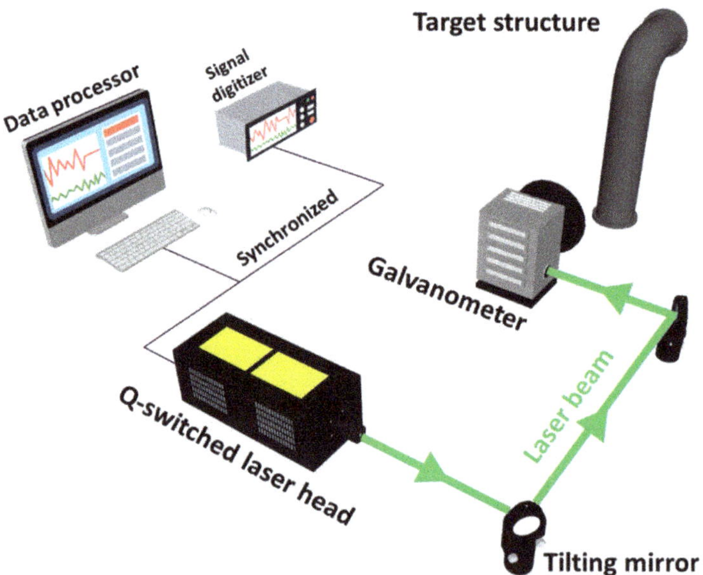

Figure 2. Conceptual diagram of an UWPI system.

A galvanometer is used to target a structure with a laser mirror scanner, to specify a point at the desired location and use it for laser pulse incidence. The laser mirror scanner is driven by two tilting mirrors and is designed to operate at a wavelength of 1064 nm. The maximum angular velocity of the galvanometer is 100 rad/s within the range of ±0.35 rad (±20.05°). The rotation axes of the two tilting mirrors are perpendicular to each other, which allows the laser mirror scanner to scan the 2D scan area at high speed. The scanning takes place as follows: the laser mirror scanner first performs an upward scan on the vertical axis, then moves to the horizontal axis to perform a downward scan after the vertical axis scan is complete. Through these scanning processes, ultrasonic waves are arranged on the target structure in the form of a grid. The details of the laser system are as specified in Table 1.

Table 1. Specifications of the laser system.

Laser Head: Brilliant Ultra GRM100	Galvanometer: Scancube 10
Wavelength: 1064 nm	Wavelength: 1064 nm
Energy per pulse: 100 mJ	Tracking error: 0.16 ms
Pulse repetition rate: 20 Hz	Positioning speed: 10 m/s
Pulse duration: 6.5 ns	Max. angular velocity: 100 rad/s
Beam diameter: 3 mm	(within 0.35 rad)

To drive the UWPI system and acquire the data required for ultrasound images from the acoustic emission sensor (AE sensor) with a built-in amplifier, the UWPI control system was configured as shown in Figure 3 using LabVIEW. The software program consists of a scanning grid configuration for a test object, a parameter setting part (sampling, frequency, number of measured samples, trigger signal level, etc.) necessary for a digitizer, a laser system and laser mirror scanner communication parameter setting part and an ultrasound imaging part.

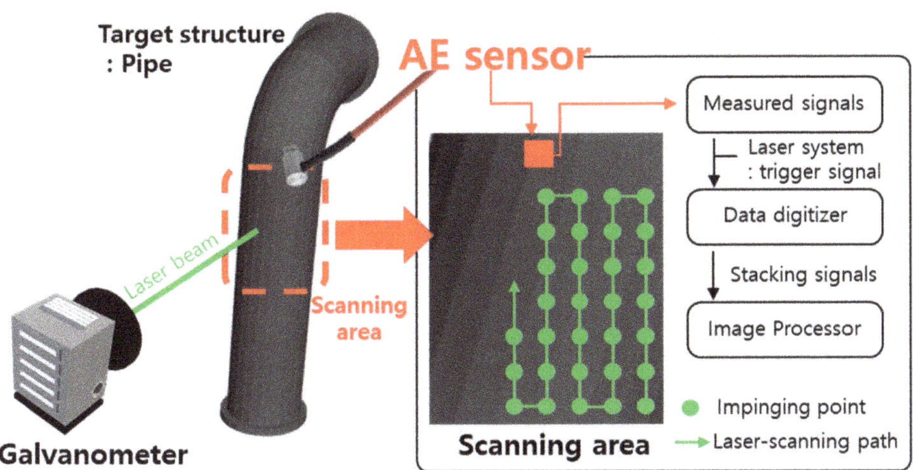

Figure 3. Laser-induced UWPI system.

2.3. Ultrasonic Wave Propagation Imaging Algorithm

The steps to generate an ultrasonic wave image using the ultrasonic signal in the time domain measured by the UWPI system consist of a total of three steps as shown in Figure 4. First, the measured time domain signals are arranged on a vertical plane. At this time, each measurement signal is positioned at the laser beam incident point to construct 3D data of the horizontal axis, the vertical axis, and the time axis. The value at each excitation point on this plane becomes the ultrasonic amplitude value at a specific time instant, and if the image is reproduced repeatedly along the measurement time on the time axis and then played in quick succession, an ultrasonic wave propagation movie can be obtained [5].

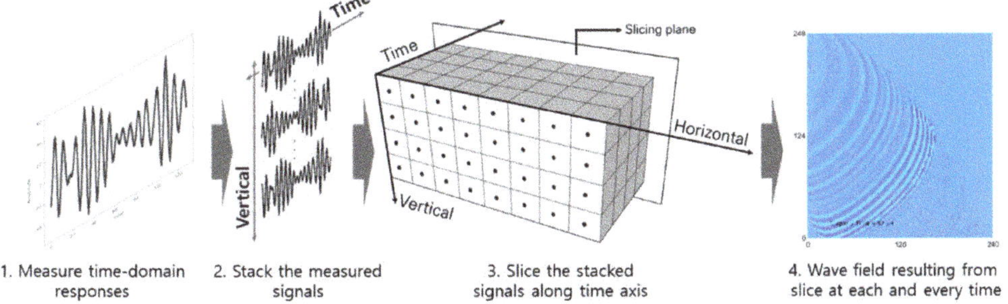

Figure 4. Process of ultrasonic wave propagation imaging (UWPI) system [5].

3. Deep Learning-CNN

Deep learning refers to machine learning techniques that construct a model with a large number of neural layers for pattern recognition problems or feature point learning [23]. Since the publication of the deep belief network paper by Hinton at the University of Toronto in Canada in 2009 [24], deep learning has been developed along with various algorithms in many industries [25]. The neural network structures to which deep learning technology is applied include auto-encoders, restricted Boltzmann machines (RBMs), convolutional neural networks (CNNs), and recurrent neural networks (RNNs) [26–29]. In this study, we intend to utilize a CNN, which has been in the spotlight in image recognition and classification fields, to determine the presence or absence of damage to piping structures through image learning.

3.1. CNN

The conventional neural network (CNN) was devised by LeCun of New York University, USA, and it is one type of deep learning. It is the most popular algorithm in the field of image recognition and classification [30]. CNNs have made great strides in image recognition and classification and has shown tremendous performance in computer vision [31].

The basic structure of the CNN is shown in Figure 5 below. As indicated, it passes the image through the filter of the convolution layer and the pooling layer repeatedly, and classifies the image through the existing fully connected network, multilayer perceptron and softmax algorithm.

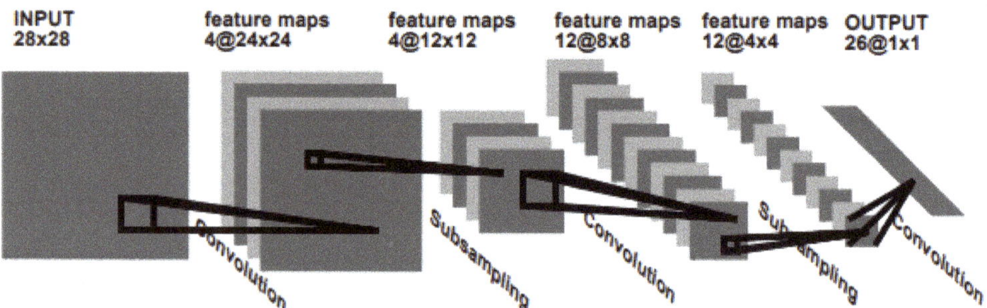

Figure 5. Convolutional neural network (CNN) for image processing [25].

Typically, through TensorFlow [32] and Keras [33], which are open source software provided by Google, people who are not computer developers can use image recognition and classification using deep learning and CNN.

3.2. Object Detection

Object detection (OD) refers to an important computer vision task in digital image processing that can detect instances of visual objects of a specific class (human, animal, car, etc.) [34]. Generally, it is divided into general object detection and detection applications. Detection applications refer to applied detection technologies such as COVID-19 mask detection and automatic vehicle number recognition systems that are commonly seen around. In this study, we intend to perform the learning on laser scanning images of the pipe and detect the damage by using application-specific detection.

3.3. EfficientDet

EfficientDet used in this study ranked first among the models whose performance was measured without extra training data in the 2019 Dataset Object Detection competition on the COCO minival dataset, and it was found that it is an efficient network with good performance, that is, with a low amount of computation (FLOPS) and good accuracy [35]. It is an object detection algorithm that achieved the highest mAP in performance comparison experiments conducted with single-model single-scale and updated SOTA (state-of-the-art, the current highest level of results). Therefore, EfficientDet presents two differences compared with existing models. First, the existing models have developed a cross-scale feature fusion network structure, but EfficientDet pointed out that the contribution to the output feature should be different because each resolution of the input feature is different. To resolve this problem, a weighted bidirectional FPN (BiFPN) [35] structure was proposed as shown in Figure 6. EfficientDet employs EfficientDet [36] as the backbone network, BiFPN as the feature network, and a shared class/box prediction network. Second, the existing models depended on huge backbone networks for large input image size for accuracy, but EfficientDet used compound scaling, a method of increasing the input resolution, depth, and width, which are factors that determine the size and computational amount of the model simultaneously and increase them.

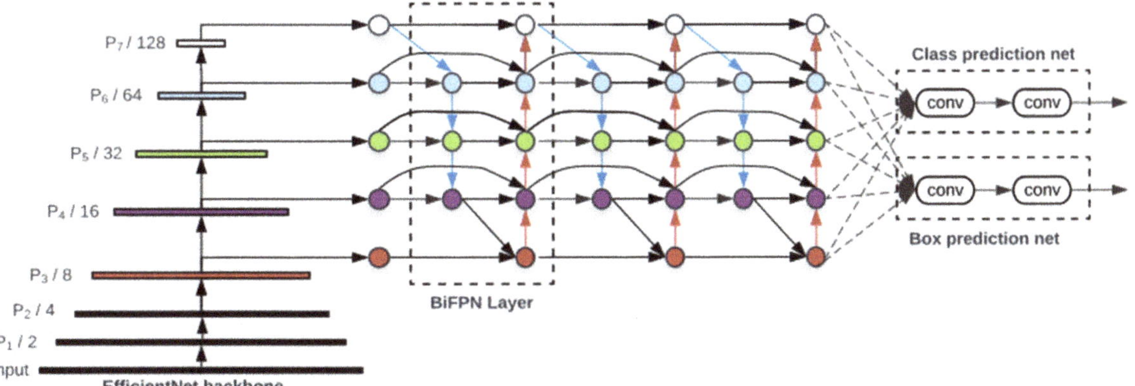

Figure 6. EfficientDet architecture [35].

4. UWPI-System-Based Pipe Damage Detection Experiment and CNN Learning

4.1. Detecting External Damage to Pipe Bends Using UWPI System

To obtain an image of pipe damage to be used in this study, a Nd:YAG pulse laser was used to generate Lamb waves, and an AE sensor was used to measure the waveform. The laser system used in the experiment is shown in Figure 7.

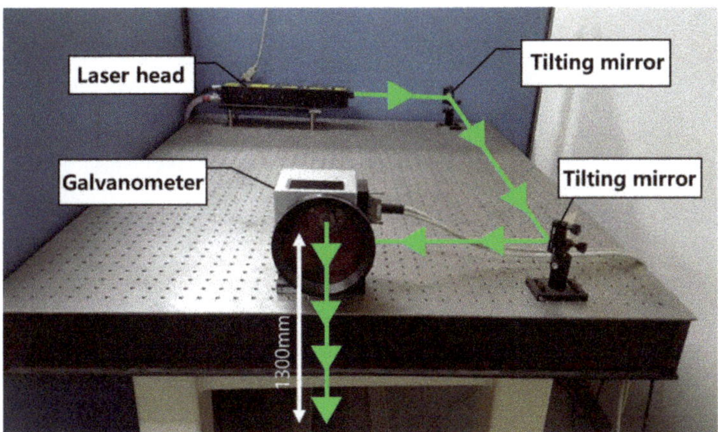

Figure 7. A noncontact laser ultrasonic scanning system composed of a Q-switched Nd:YAG pulsed laser with a galvanometer for ultrasonic excitation scanning [5].

The Q-switched Nd:YAG pulse laser emits a laser beam through a galvanometer after a trigger signal is delivered [5]. Using the mirror inside the galvanometer, the laser beam is emitted to the target point along the scan path, and the measured data are sent to the digitizer through the acoustic emission (AE) sensor. Then, the digitized signal reaches the image processor, where the UWPI process occurs [6].

For the test pipe utilized in this study, a stainless steel 304 specimen was used, and on the curved surface of a pipe a 1 mm deep damage was artificially applied to a diameter of 30 mm, as shown in Figure 8.

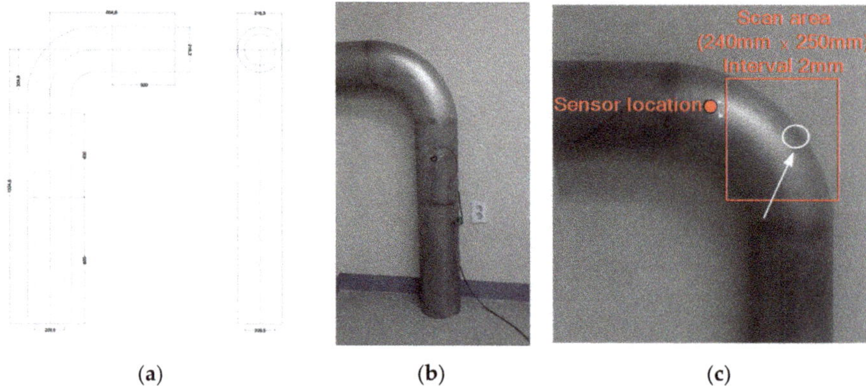

(a)   (b)   (c)

Figure 8. Stainless steel pipe damage diagnosis test specimens. (**a**) Stainless steel 304 specimen drawing; (**b**) stainless steel pipe specimen; (**c**) artificial damage carved on the specimen.

The laser scanning area is 240 mm wide and 250 mm high, and the laser excitation interval is 2 mm. The number of laser excitation points is 15,125, and the scanning time is 12.6 min. The result of the UWPI after scanning the pipe bend is shown in Figure 9.

   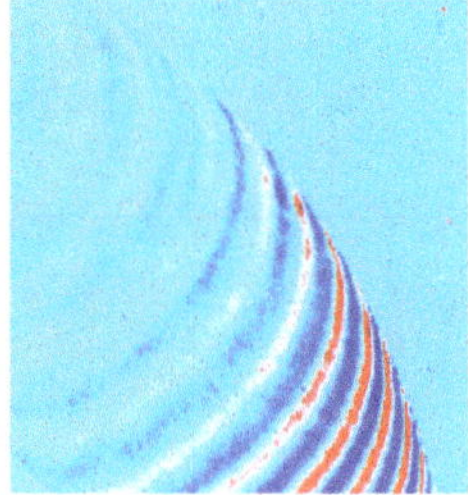

Figure 9. UWPI unfiltered video data to be used for deep learning.

4.2. CNN Learning Using Damage Data

In this study, to find out the applicability of the pipe damage detection model using the laser scanning data of the curved pipe part, dataset construction, data learning and detection, and evaluation were performed in three steps as shown in Figure 10.

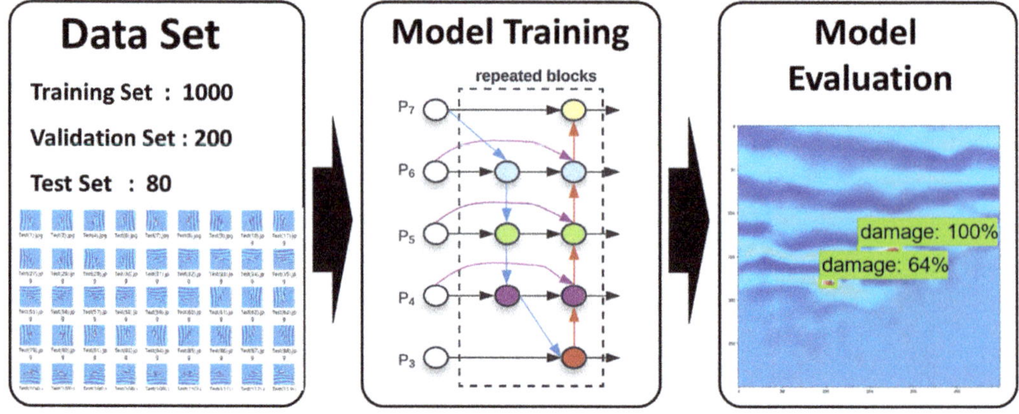

Figure 10. Experimental procedure of the CNN training.

In the first step, an ultrasound image of the pipe was acquired using a laser scanning technique, and an image dataset was constructed using it. In the second step, the CNN (EfficientDet) model was trained using the image dataset. Finally, the learned model was evaluated using the test set.

4.2.1. Transfer Learning

The dataset used in this study comprises about 1280 images, and it is difficult to evaluate it with a general learning method. To this end, a transfer-learning-based EfficientDet model was applied using a COCO dataset [37] that was pretrained with about 330,000 images and 80 categories. The structure of the deep learning network is very complex, and as the amount of training data is small, problems such as overfitting occur and the learning performance deteriorates. As the amount of training data increases, the deep learning network performance improves [38]. In the field of image object detection, when it is difficult to collect specific data, such as an UWPI image used in this study, a transfer learning technique that learns new data using a model pretrained with a lot of data is a widely used technique in various deep learning applications [19,39]. The difference between the existing learning method and the transfer learning is shown in Figure 11. In this study, we train and evaluate the detection and evaluation of pipe bend damage by using the EfficientDet pretrained model using transfer learning.

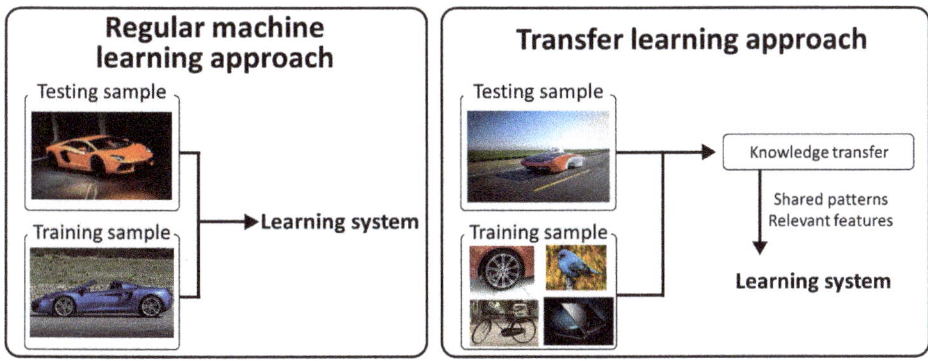

Figure 11. Basic frameworks of traditional machine learning approaches and knowledge transfer approaches [19].

4.2.2. Train Dataset

In general, when developing deep learning algorithms, open image data that are freely available on the Internet such as ImageNet and COCO [37] are used a lot. However, in the case of open image data, the image is object-centered, and the background of the object is often simple and uncomplicated. However, open image data that can be used free of charge on the Internet did not have the UWPI images used in this study. Therefore, the images used in this study were acquired using laser scanning technology, which is detailed in Section 2 of this paper.

The damage to the curved part of the pipe was scanned and the scan data were produced as UWPI image data using MATLAB software. The number of image data produced was 500, with size 1024 × 1024. Of the 500 scanned images, 320 pieces of data that can predict damage information were extracted. A total of 1280 training images were constructed by rotating by 90 degrees, as shown in Figure 12, for accurate deep learning construction.

To increase the resolution consistency and learning precision of the 1280 images, resizing was performed to a size of 512 × 512. The image data were divided into 1000, 200, and 80 images for the training, validation, and test sets, respectively. After dividing the image data set into training, validation, and test sets, the coordinate labeling work of the bounding box (the area of the actual damage location) was performed on each image using

LabelImg software. In the case of the images used in this study, the class name was not designated, as damage was determined based on the laser scanning image of the pipe, and a collective label name "damage" was used.

Original Image   90° Rotation Image   180° Rotation Image   270° Rotation Image

Figure 12. Laser scanning image dataset configuration.

4.2.3. Training Dataset

The hardware specifications used in this study were: Intel Xeon® Silver 4210 CPU, Nvidia GeForce RTX 3060, and 32GB RAM. The main software environment consisted of Anaconda, Python 3.8, TensorFlow 2.5.0, CUDA 11.2, Cudnn 8.1.1. The CNN-based pipe bend damage model was trained using the EfficientDet-d0 model [35].

The model was evaluated using intersection over union (IOU) and mean average precision (mAP), which are evaluation indicators that are often used in object detection. Unlike the existing object classification evaluation method, object detection requires both the evaluation of the class classification and the bounding box to find the position. In this study, since there is only one class (damage), the bounding box was evaluated.

The calculation method of IOU is as shown in Figure 13, and it is an indicator of how well the bounding box is predicted. IOU indicates the ratio of the intersection of the bounding box labeled during the composition of the predicted area and the actual dataset to their union. In general, if the IOU value exceeds 0.5, it is judged as the correct answer [40].

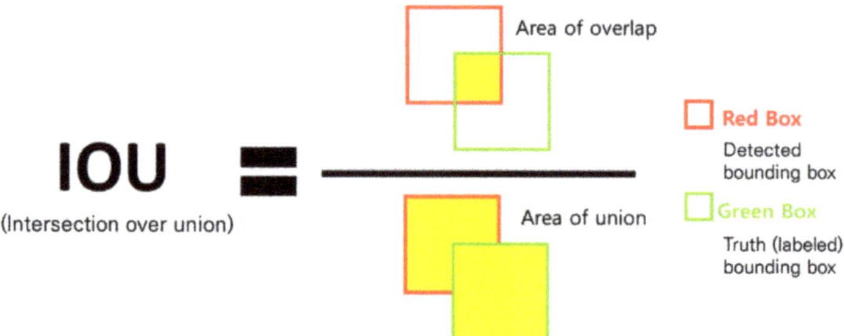

Figure 13. Intersection over union (IOU).

Finally, the mAP is an index used as an evaluation criterion in PASCAL VOC, and it represents the performance of the object detection algorithm as an index, i.e., as an average value of average precision (AP) for each classification class [41]. Precision and recall are commonly used to evaluate the performance of detection models. Precision shows the ratio of detection of the true value to the total detection of data as in Equation (1), and recall refers to the ratio of detection of the true value to the cases of correct detection as in Equation (2). Since the two indicators are correlated with each other, AP, which is the area under the graph, is used in the precision–recall graph. The closer the AP value is to 1, the higher the performance of the object detection algorithm.

$$\text{Precision} = \frac{\text{True positive}}{\text{True positive} + \text{False positive}} \tag{1}$$

$$\text{Recall} = \frac{\text{True positive}}{\text{True positive} + \text{false negative}} \tag{2}$$

4.2.4. UWPI Data Deep Learning Result

Prior to conducting this study, a transfer learning technique using a pretrained model used in object detection was applied to compensate for the lack of training data. Through the learning process, it was possible to know whether the used model was learning the image data well, by looking at the predicted values and the actual values. Learning was carried out in three stages as shown in Table 2. The same hardware specifications as well as the same batch size were applied for accurate comparison. For the batch size, step, and epoch values applied to training, Equation (3), which is widely used in the field of object detection, was used.

$$\text{Batch Size} \times \text{Step} = \text{Epoch} \times \text{No. of samples} \tag{3}$$

Table 2. Pipe damage detection CNN training configuration information.

Batch Size	Steps	Epochs	No. of Samples
8	10,000	80	1000
8	30,000	240	1000
8	50,000	400	1000

Figure 14 shows the learning results after 10,000, 30,000 and 50,000 steps. The sum of damage detection loss and bounding box regression loss for learning according to each step is summarized as total loss. From the results of a total of three learning stages, it was confirmed that the total loss was less than 0.2. Comparing results after 10,000 steps and 50,000 steps, the loss decreases as repeated learning progresses to 0.188 and 0.1441, respectively. In addition, the learning progresses normally.

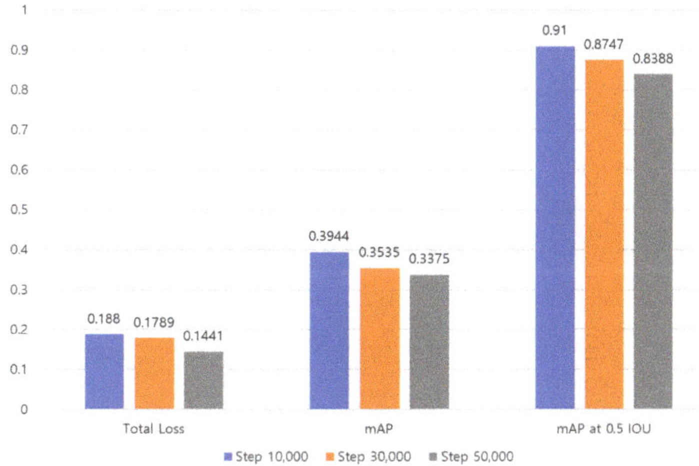

Figure 14. Comparison of deep learning results according to steps (Total loss, mAP, mAP at 0.5 IOU).

As a result of performance evaluation for the trained model, the average mAP values of the pipe damage data learning were calculated as 0.3944, 0.3535, and 0.3375, (as shown in Figure 13) and the average mAP values at 0.5 IOU were calculated as 0.91, 0.8747, and 0.8388, after 10,000, 30,000, and 50,000 steps, respectively. Observing that the average mAP value of the COCO 2017 pretrained CNN (EfficientDet-d0) algorithm used in this study was 0.336 [35], it can be deduced that the learning proceeded normally. The evaluation was conducted using a preclassified test image data set before the learning. As a result of evaluating a total of 80 test images as evaluation data, the results shown in Table 3 below were obtained.

Table 3. Damage detection rate of test images for each step.

Test Image	Step 10,000	Step 30,000	Step 50,000
1	79%	89%	97%
2	79%	88%	96%
3	79%	90%	97%
4	78%	91%	96%
5	77%	91%	98%
6	72%	90%	98%
7	73%	90%	94%
8	77%	89%	94%
9	77%	89%	94%
10	77%	90%	92%
11	78%	87%	84%
12	77%	87%	92%
13	82%	83%	86%
14	85%	88%	90%
15	86%	85%	89%
16	85%	81%	89%
17	85%	80%	91%
18	82%	75%	88%
19	76%	76%	90%
20	66%	76%	94%
21	67%	81%	91%
22	66%	83%	90%
23	62%	78%	95%
24	59%	78%	93%
25	58%	73%	89%
26	67%	77%	88%
27	68%	72%	85%
28	68%	70%	81%
29	58%	72%	87%
30	67%	74%	83%
31	71%	80%	75%
32	73%	83%	68%
33	77%	90%	85%

Table 3. *Cont.*

Test Image	Step 10,000	Step 30,000	Step 50,000
34	80%	92%	94%
35	82%	92%	94%
36	82%	92%	94%
37	83%	91%	94%
38	84%	93%	94%
39	86%	93%	96%
40	87%	94%	99%
41	87%	94%	98%
42	87%	92%	98%
43	88%	93%	97%
44	88%	95%	98%
45	88%	94%	98%
46	88%	94%	97%
47	88%	95%	98%
48	89%	96%	98%
49	86%	95%	95%
50	85%	95%	96%
51	82%	93%	96%
52	85%	93%	96%
53	87%	93%	96%
54	88%	93%	98%
55	89%	93%	98%
56	88%	94%	97%
57	85%	92%	96%
58	82%	93%	94%
59	83%	93%	94%
60	84%	90%	92%
61	84%	92%	96%
62	83%	91%	96%
63	83%	94%	93%
64	83%	92%	92%
65	80%	92%	94%
66	82%	92%	94%
67	82%	89%	96%
68	81%	87%	94%
69	77%	89%	97%
70	77%	90%	97%
71	77%	87%	92%
72	63%	68%	64%
73	51%	57%	0%

Table 3. *Cont.*

Test Image	Step 10,000	Step 30,000	Step 50,000
74	0%	66%	52%
75	58%	69%	0%
76	50%	70%	53%
77	0%	53%	0%
78	0%	56%	58%
79	67%	76%	50%
80	87%	93%	98%
Average detection rate	75%	86%	88%

Following the evaluation at the 10,000, 30,000, and 50,000 step, average detection rates of 75%, 86%, and 88%, respectively, were confirmed. When evaluating the performance of the learning model, the average mAP was lower at steps 30,000 and 50,000 compared to step 10,000. However, because of the direct evaluation, the damage detection rate was higher at step 50,000. At step 10,000, the detection rate ranged from 50% to 89%, resulting in an average detection rate of 75%. At step 30,000 it ranged from 53% to 96%, and at step 50,000 it ranged from 50% to 99%. To see the overall aspect of learning, the undetected data are excluded and are shown in a graph in Figure 15. Taking a close look at the graph, it can be seen that the most accurate result was obtained after 50,000 steps.

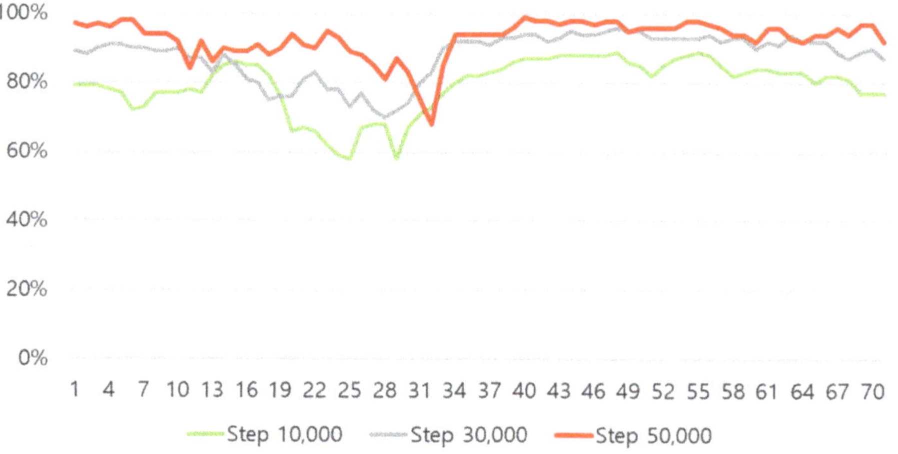

Figure 15. Comparative analysis of detection rate by step using test images.

Figure 16 shows the test result with the highest detection rate compared to the original image data and it can be seen that an average detection rate of 89% or more was achieved compared to the original image data. Observing the overall test, no erroneous detection occurred. However, at steps 10,000 and 50,000, three non-detections occurred as shown in Figure 17.

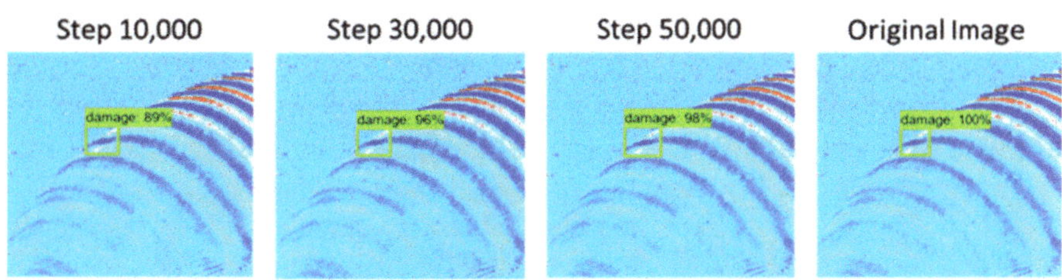

Figure 16. Test result of damage detection with excellent accuracy.

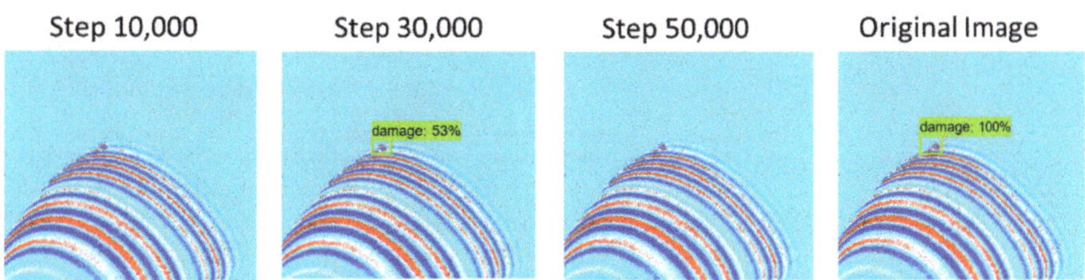

Figure 17. Test result of damage not detected and low accuracy.

In general, the main cause of non-detection in learning results is that there is no difference in color or contrast between an object and the background. This problem is due to a shape that appears depending on the background and physical environment such as color or lighting of the object [42].

From the results of the test, no erroneous detection was found in this study, and three cases of non-detection occurred at steps 10,000 and 50,000. This is thought to be for the following reasons. First, regarding the undetected results, the problem is that there is no difference in contrast between the background color of the image and the color of the damaged part, which is believed to have affected the learning results. Second, it is presumed that some non-detection occurred because there was no experience in learning the UWPI image of this study with the COCO 2017dataset. Therefore, it can be deduced that it will be improved if many pipe UWPI images are acquired and used with deep learning in order to improve detection.

5. Conclusions

In this study, we proposed an automatic damage detection system for pipe bends using a CNN object detection algorithm with laser scanning data to efficiently extend the safety management of pipes used in the construction industry and many industries. Using a Q-switched Nd:YAG pulse laser and an acoustic emission (AE) sensor, UWPI image data were produced for the detection of damage introduced artificially to the pipe bend. A damage detection system was constructed using a total of 1280 training images obtained through post-processing of the UWPI data. Since 1280 images are insufficient to proceed with deep learning, a transfer learning technique using the pretrained COCO 2017 EfficientDet-d0 algorithm was applied.

Examining the learning model using the pipe damage data, it was confirmed that the detection performance index, mAP, was higher than the value of 0.336 from the COCO 2107 Effi-cientDetd-0 model. This indicates that the model training was successful, and it was confirmed that there was no performance difference when comparing the existing methods of learning that use a lot of data with the one implemented through transfer learning with 1280 pieces of data. From the result of the CNN learning using pipe damage data, three cases were not detected after 10,000 steps and 50,000 steps. It was deduced that a small amount of non-detection occurred due to an insufficient quality and quantity of images. Therefore, to supplement the undetected problem, we intend to proceed with the following additional research.

- Through additional experiments and research, we intend to secure UWPI data according to the damage size using laser scanning techniques for the components (curved part, curved pipe part, bolted joint part, welding, etc.) of pipes.
- This study confirmed the possibility of detecting damage to pipes based on laser scanning through the transfer learning technique, and based on this, we intend to propose a better detection technique using new algorithms and large amounts of data.
- To acquire ultrasonic signals in the laser scanning system, this study used the AE sensor installed directly on the pipe. Therefore, we intend to develop a noncontact nondestructive system for efficient pipe damage detection by using laser diameter vibration (LDV) instead of an AE sensor.

In this study, using the UWPI system and CNN, an automatic pipe bend damage detection system was proposed. Therefore, it is expected that efficient maintenance will be possible for piping used in construction and many industries.

Author Contributions: Conceptualization, C.L., K.D.T. and B.Y.; methodology, C.L., K.D.T. and B.Y; formal analysis, B.Y.; software, B.Y.; investigation, B.Y.; resources, S.P.; writing—original draft preparation, B.Y.; writing—review and editing, B.Y.; visualization, K.D.T.; supervision, C.L.; project administration, B.Y.; funding acquisition, S.P. All authors have read and agreed to the published version of the manuscript.

Funding: This work was financially supported by the National Research Foundation of Korea (NRF) grant funded by the Korean government (MSIP) (No. NRF-2017R1A2B3007607) (NRF-2021R1A4A3033128).

Institutional Review Board Statement: Not applicable.

Informed Consent Statement: Not applicable.

Conflicts of Interest: The authors declare no conflict of interest.

References

1. Gunarathna, R.N.P.; Sivahar, V. Challenges in Monitoring Metallic Pipeline Corrosion Using Ultrasonic Waves—A Review Article. *Eng. J. Inst. Eng.* **2021**, *54*, 67–75.
2. Lee, C.; Kang, D.; Park, S. Visualization of Fatigue Cracks at Structural Members Using a Pulsed Laser Scanning System. *Res. Nondestruct. Eval.* **2014**, *26*, 123–132. [CrossRef]
3. Lee, C.; Park, S. Flaw Imaging Technique for Plate-Like Structures Using Scanning Laser Source Actuation. *Shock. Vib.* **2014**, *2014*, 1–14. [CrossRef]
4. Lee, C.; Park, S. Damage visualization of pipeline structures using laser-induced ultrasonic waves. *Struct. Heal. Monit.* **2015**, *14*, 475–488. [CrossRef]
5. Tran, D.Q.; Kim, J.-W.; Tola, K.D.; Kim, W.; Park, S. Artificial Intelligence-Based Bolt Loosening Diagnosis Using Deep Learning Algorithms for Laser Ultrasonic Wave Propagation Data. *Sensors* **2020**, *20*, 5329. [CrossRef]
6. Tola, K.D.; Tran, D.Q.; Yu, B.; Park, S. Determination of Plate Corrosion Dimension Using Nd:YAG Pulsed Laser-generated Wavefield and Experimental Dispersion Curves. *Materials* **2020**, *13*, 1436. [CrossRef]
7. Lee, C.; Zhang, A.; Yu, B.; Park, S. Comparison Study between RMS and Edge Detection Image Processing Algorithms for a Pulsed Laser UWPI (Ultrasonic Wave Propagation Imaging)-Based NDT Technique. *Sensors* **2017**, *17*, 1224. [CrossRef]
8. Michaels, J.E. Ultrasonic wavefield imaging: Research tool or emerging NDE method? *AIP Conf. Proc.* **2017**, *1806*, 020001. [CrossRef]
9. Carmen, V.; David, S.M. Intensity-Based Optical Systems for Fluid Level Detection. *Recent Pat. Electr. Electron. Eng.* **2012**, *5*, 85–95.
10. Safizadeh, M.; Azizzadeh, T. Corrosion detection of internal pipeline using NDT optical inspection system. *NDT E Int.* **2012**, *52*, 144–148. [CrossRef]
11. Chen, G.; Katagiri, T.; Song, H.; Yusa, N.; Hashizume, H. Investigation of the effect of a bend on pipe inspection using microwave NDT. *NDT E Int.* **2020**, *110*, 102208. [CrossRef]
12. Miro, J.V.; Hunt, D.; Ulapane, N.; Behrens, M. Towards Automatic Robotic NDT Dense Mapping for Pipeline Integrity Inspection. In *Field and Service Robotics*; Springer: Manhattan, NY, USA, 2017; Volume 5, pp. 319–333.
13. Kim, S.; Kim, C.H.; Bae, Y.-G.; Na, H.; Jung, S. NDT inspection mobile robot with spiral driven mechanism in pipes. In Proceedings of the IEEE ISR 2013, Seoul, Korea, 24–26 October 2013; pp. 1–2.
14. Krys, D.; Najjaran, H. Development of Visual Simultaneous Localization and Mapping (VSLAM) for a Pipe Inspection Robot. In Proceedings of the 2007 International Symposium on Computational Intelligence in Robotics and Automation, Jacksonville, FL, USA, 20–23 June 2007; pp. 344–349.

15. Dai, L.; Feng, H.; Wang, T.; Xuan, W.; Liang, Z.; Yang, X. Pipe Crack Recognition Based on Eddy Current NDT and 2D Impedance Characteristics. *Appl. Sci.* **2019**, *9*, 689. [CrossRef]
16. Cha, Y.-J.; Choi, W.; Büyüköztürk, O. Deep Learning-Based Crack Damage Detection Using Convolutional Neural Networks. *Comput. Civ. Infrastruct. Eng.* **2017**, *32*, 361–378. [CrossRef]
17. Azimi, M.; Eslamlou, A.D.; Pekcan, G. Data-Driven Structural Health Monitoring and Damage Detection through Deep Learning: State-of-the-Art Review. *Sensors* **2020**, *20*, 2778. [CrossRef] [PubMed]
18. Zhu, Y.; Newsam, S. Dense Net for dense flow. In Proceedings of the 2017 IEEE International Conference on Image Processing (ICIP), Beijing, China, 17–20 September 2017; pp. 790–794.
19. Shao, L.; Zhu, F.; Li, X. Transfer Learning for Visual Categorization: A Survey. *IEEE Trans. Neural Netw. Learn. Syst.* **2015**, *26*, 1019–1034. [CrossRef] [PubMed]
20. Hayward, G.; Hyslop, J. Determination of lamb wave dispersion data in lossy anisotropic plates using time domain finite el-ement analysis. Part I: Theory and experimental verification. *IEEE Trans.* **2006**, *53*, 443–448.
21. Drain, L.E. *Laser Ultrasonics Techniques and Applications*; Routledge: Abington, UK, 2019.
22. White, R.M. Generation of Elastic Waves by Transient Surface Heating. *J. Appl. Phys.* **1963**, *34*, 3559–3567. [CrossRef]
23. Zhu, H.; Ge, W.; Liu, Z. Deep Learning-Based Classification of Weld Surface Defects. *Appl. Sci.* **2019**, *9*, 3312. [CrossRef]
24. Hinton, G.E. Deep belief networks. *Scholarpedia* **2009**, *4*, 5947. [CrossRef]
25. LeCun, Y.; Yoshua, B. Convolutional networks for images, speech, and time series. In *The Handbook of Brain Theory and Neural Networks*; MIT Press: Cambridge, MA, USA, 1997; pp. 255–258.
26. Pascal, V.; Hugo, L.; Isabelle, L.; Yoshua, B.; Manzagol, P.-A. Stacked Denoising Autoencoders: Learning Useful Representations in a Deep Network with a Local Denoising Criterion. *J. Mach. Learn. Res.* **2010**, *11*, 3371–3408.
27. Hinton, G.E.; Osindero, S.; Teh, Y.-W. A Fast Learning Algorithm for Deep Belief Nets. *Neural Comput.* **2006**, *18*, 1527–1554. [CrossRef]
28. LeCun, Y.; Bottou, L.; Bengio, Y.; Haffner, P. Gradient-based learning applied to document recognition. *Proc. IEEE* **1998**, *86*, 2278–2324. [CrossRef]
29. Jozefowicz, R.; Vinyals, O.; Schuster, M.; Shazeer, N.; Wu, Y. Exploring the limits of language modeling. In Proceedings of the 2011 IEEE International Conference on Acoustics, Speech and Signal Processing (ICASSP), Prague, Czech Republic, 22–27 May 2011; pp. 5528–5531.
30. LeCun, Y.; Bengio, Y.; Hinton, G. Deep learning. *Nature* **2015**, *521*, 436–444. [CrossRef]
31. Lee, H.S. A Structure of Convolutional Neural Networks for Image Contents Search. Master's Thesis, Graduate School of Chung-Ang University, Seoul, Korea, 2018.
32. TensorFlow. Available online: https://www.tensorflow.org/ (accessed on 9 November 2015).
33. Keras. Available online: https://github.com/keras-team/keras (accessed on 14 June 2015).
34. Abadi, M.; Barham, P.; Chen, J.; Chen, Z.; Davis, A.; Dean, J.; Devin, M.; Ghemawat, S.; Irving, G.; Isard, M.; et al. TensorFlow: A system for large-scale machine learning. In Proceedings of the 12th USENIX Symposium on Operating Systems Design and Implementation, Savannah, GA, USA, 2–4 November 2016; pp. 256–283.
35. Tan, M.; Pang, R.; Le, Q.V. Efficient Det: Scalable and Efficient Object Detection. In Proceedings of the IEEE/CVF Conference on Computer Vision and Pattern Recognition (CVPR), Seattle, WA, USA, 13–19 June 2020; pp. 10778–10787.
36. Tan, M.; Le, Q.V. Efficient Net: Rethinking Model Scaling for Convolutional Neural Networks. In Proceedings of the 36th International Conference on Machine Learning (ICML 2019), Long Beach, CA, USA, 28 May 2019; pp. 6105–6114.
37. Lin, T.Y.; Maire, M.; Belongie, S.; Hays, J.; Perona, P.; Ramanan, D. 2014.Microsoft COCO: Common objects in context. *Comput. Vis. ECCV* **2014**, *8693*, 740–755.
38. Liu, B.; Wei, Y.; Zhang, Y.; Yang, Q. Deep Neural Networks for High Dimension, Low Sample Size Data. In Proceedings of the Twenty-Sixth International Joint Conference on Artificial Intelligence (IJCAI-17), Melbourne, Australia, 19–25 August 2017; pp. 2287–2293. [CrossRef]
39. Pan, S.J.; Yang, Q. A Survey on Transfer Learning. *IEEE Trans. Knowl. Data Eng.* **2010**, *22*, 1345–1359. [CrossRef]
40. Rezatofighi, H.; Tsoi, N.; Gwak, J.; Sadeghian, A.; Reid, I.; Savarese, S. Generalized Intersection Over Union: A Metric and a Loss for Bounding Box Regression. In Proceedings of the IEEE/CVF Conference on Computer Vision and Pattern Recognition (CVPR), Long Beach, CA, USA, 15–20 June 2019; pp. 658–666.
41. Everingham, M.; Van Gool, L.; Williams, C.K.I.; Winn, J.; Zisserman, A. The Pascal Visual Object Classes (VOC) Challenge. *Int. J. Comput. Vis.* **2010**, *88*, 303–338. [CrossRef]
42. Kim, M.; Shin, S.; Suh, Y. Application of Deep Learning Algorithm for Detecting Construction Workers Wearing Safety Helmet Using Computer Vision. *J. Korean Soc. Saf.* **2019**, *34*, 29–37. [CrossRef]

Article

Lightweight Super-Resolution with Self-Calibrated Convolution for Panoramic Videos

Fanjie Shang [1], Hongying Liu [2,*], Wanhao Ma [1], Yuanyuan Liu [1], Licheng Jiao [1], Fanhua Shang [3], Lijun Wang [4] and Zhenyu Zhou [5]

1 Key Laboratory of Intelligent Perception and Image Understanding of Ministry of Education, School of Artificial Intelligence, Xidian University, Xi'an 710071, China
2 The Medical College, Tianjin University, Tianjin 300072, China
3 College of Intelligence and Computing, Tianjin University, Tianjin 300350, China
4 Hangzhou Institute of Technology, Xidian University, Hangzhou 311231, China
5 Hunan University of Science and Engineering, Yongzhou 425199, China
* Correspondence: hyliu@xidian.edu.cn

Abstract: Panoramic videos are shot by an omnidirectional camera or a collection of cameras, and can display a view in every direction. They can provide viewers with an immersive feeling. The study of super-resolution of panoramic videos has attracted much attention, and many methods have been proposed, especially deep learning-based methods. However, due to complex architectures of all the methods, they always result in a large number of hyperparameters. To address this issue, we propose the first lightweight super-resolution method with self-calibrated convolution for panoramic videos. A new deformable convolution module is designed first, with self-calibration convolution, which can learn more accurate offset and enhance feature alignment. Moreover, we present a new residual dense block for feature reconstruction, which can significantly reduce the parameters while maintaining performance. The performance of the proposed method is compared to those of the state-of-the-art methods, and is verified on the MiG panoramic video dataset.

Keywords: panoramic videos; super-resolution; lightweight network; deformable convolution; self-calibration convolution

Citation: Shang, F.; Liu, H.; Ma, W.; Liu, Y.; Jiao, L.; Shang, F.; Wang, L.; Zhou, Z. Lightweight Super-Resolution with Self-Calibrated Convolution for Panoramic Videos. *Sensors* **2023**, *23*, 392. https://doi.org/10.3390/s23010392

Academic Editors: KWONG Tak Wu Sam, Yun Zhang, Xu Long and Tiesong Zhao

Received: 6 November 2022
Revised: 22 December 2022
Accepted: 26 December 2022
Published: 30 December 2022

Copyright: © 2022 by the authors. Licensee MDPI, Basel, Switzerland. This article is an open access article distributed under the terms and conditions of the Creative Commons Attribution (CC BY) license (https://creativecommons.org/licenses/by/4.0/).

1. Introduction

Video super-resolution (VSR) is a classic problem in computer vision, and aims to recover high-resolution videos from low-resolution ones. VSR technology has been widely used in various areas for high-definition displays, such as network videos, digital TV, and surveillance drones. The panoramic video is one class of videos that are real 360-degree omnidirectional sequences, and its pixels are usually arranged in a spherical shape that can provide an immersive experience for users. The panoramic video is the product of a combination of multiple video technologies. Such a video allows the audience to see a wider field of view and a more realistic field of view experience. Because of its 3D stereoscopic characteristics compared with ordinary videos, it is widely used in entertainment, news, the military, and other fields.

In recent years, due to the emergence of deep learning, various methods for video super-resolution based on deep learning have been proposed. For instance, in SOFVSR [1], an optical flow reconstruction network is presented to infer high-resolution (HR) optical flow from coarse to fine, and the motion-compensated low resolution (LR) is input to a super-resolution network to generate the final super-resolved frames. TDAN [2] proposes a temporal deformable network, which utilizes the features of the reference frame and neighboring frames to dynamically predict the offset of the sampling convolutional kernel, and aligns it adaptively at the feature level. EDVR [3] proposes a pyramid, cascade, and deformable convolution (PCD) module. Unlike TDAN, this module performs alignment in a coarse-to-fine manner and can handle videos with large and complex motions. Moreover, EDVR presents the temporal and spatial attention fusion module, which utilizes temporal attention to concentrate on neighboring frames that are more similar to the reference frame, and uses spatial attention to assign weights to each position in each channel to more effective use of cross-channel and spatial information.

Although the methods mentioned above can achieve good performance for general videos, they may degrade for panoramic videos. Because panoramic videos usually have ultrahigh spatial resolution, they can provide viewers with a strong sense of immersion in the virtual environment. Moreover, the higher the resolution of the camera, the more realistic effect of the panoramic video is. The high resolution requires great hardware performance from camera equipment, and the cost will be largely increased. For this problem, Liu et al. [4] first explored deep learning for super-resolving panoramic videos. Although this method has gained a higher PSNR for panoramic videos, the number of parameters is still high. This issue considerably limits their real-world applications.

In order to balance the performance and the computational cost, we propose a novel lightweight super-resolution framework for panoramic videos. As is known, the alignment between video frames is significant for super-resolution. If more interframe information can be exploited for alignment, it is beneficial to the subsequent reconstruction. Thus, we present a new pooled, self-calibrated convolution (PSCC) for frame alignment, which significantly reduces the complexity of the deformable convolution and achieves accurate alignment in a gradual manner. Moreover, in the reconstruction operation, we design a new lightweight residual dense block to further reduce the complexity of the model. Our method achieves a balance between algorithm performance and complexity.

The main contributions of this work are listed as follows.

- We propose the first lightweight panoramic video super-resolution (LWPVSR) method for panoramic video super-resolution, which can achieve a good balance between performance and complexity. To the best of our knowledge, this is the first proposition of a lightweight panoramic VSR framework.
- Moreover, we present a new pooled, self-calibrated convolution for frame alignment. The self-calibrated convolution is introduced to make the learned offset more accurate in a progressive manner and reduce the complexity of the proposed network.
- Finally, we design a new significantly lighter residual dense block (LWRDB) for feature reconstruction, which achieves the purpose of reducing the complexity of the model while maintaining the performance of our method. Many experimental results verify the advantage of the proposed LWPVSR method against state-of-the-art methods.

The rest of this paper is organized as follows. Some related works on super-resolution of panoramic videos are introduced in Section 2. Section 3 describes the proposed lightweight super-resolution method in detail. In Section 4, we demonstrate the experimental results of our method. Finally, we show the conclusions and future work in Section 5.

2. Related Work

2.1. Super-Resolution Methods for Ordinary Videos

Most of video super-resolution methods (e.g., VESPCN [5], TDAN [2], SOFVSR20 [6], and EDVR [3]) have been proposed to address ordinary videos. They have improved the performance of restored high-resolution videos. For example, Yi et al. [7] proposed a general omniscient framework to leverage the LR framework and estimated hidden states from the past, present, and future frames. Benefiting from the global information feature of OVSR [7], the OVSR method refreshes the metrics on the Vid4 test set.

2.2. Super-Resolution Methods for Panoramic Videos

There are many image super-resolution methods such as [8–13]. For instance, ref. [11] can utilize the plenoptic geometry of the scene to perform alignment between consecutive frames in a video sequence and employ all visual information to generate high-resolution panoramic images. In [10], the spherical Fourier transform (SFT) was calculated based on the nonuniform sampling data on the sphere, which can transform low-resolution panoramic images with arbitrary rotation to reconstruct a high-resolution panoramic image. The joint alignment and super-resolution problem is converted into a least square minimization problem in the SFT domain. In [14], the authors introduced SRCNN [15], which is the earlier work to use deep learning for super-resolution of panoramic images. It fine tuned the SRCNN by optimizing input size and using the panoramic training set to adapt the fine-tuned method to the features of the high-resolution panoramic images. Based on the existing viewport-based panoramic image transmission system, ref. [16] proposed a framework that used the high-resolution content of the viewport to improve the quality of the surrounding low-resolution areas. The adaptive initial viewport of each image was predicted in view of contextual similarity of the sphere, so as to provide more useful information for low-resolution regions.

Only a few works involve the super-resolution of panoramic videos. As we have mentioned in Section 1, Liu et al. [4] designed a single frame and multiframe joint network for the super-resolution of panoramic videos, which explored both the spatial information and the temporal information. In addition, deformable convolutions are employed to eliminate the motion difference between feature maps of the target frame and its neighboring frames. Although it achieves sound results, the network contains a large number of parameters, resulting in low computational efficiency, which is not unfavorable for the promotion of practical applications. Therefore, this paper will propose the first lightweight video super-resolution method for panoramic videos.

3. The Proposed Lightweight Architecture for Panoramic Video Super-Resolution

In this section, we propose the first lightweight panoramic video super-resolution (LWPVSR) method. The proposed LWPVSR method mainly consists of the four main modules: the feature extraction module, the feature alignment module, the reconstruction module, and the dual network module, as shown below.

3.1. Our Network Architecture

As shown in Figure 1, the network structure of our method mainly consists of three main parts, which are the feature extraction module, the feature alignment module, and the reconstruction module. The backbone network learns the residual images of the video frames, then sums them with the direct upsampling results of the target frames to obtain the final super-resolution results. In addition, super-resolution is an ill-posed problem—that is, mapping a low-resolution video to high-resolution is a one-to-many problem. In order to reduce the solution space of the super-resolution, we introduce a dual mechanism to the backbone network, and it learns a dual regression mapping, which can increase the constraints on LR videos—that is, the duality mechanism acts as a subsupervised network to enhance the performance of SR. The whole super-resolution process of the proposed method is expressed as follows,

$$\tilde{I}_t = H_{LWPVSR}(\hat{I}_{t-N:t+N}), \tag{1}$$

where $H_{LWPVSR}(\cdot)$ denotes the proposed algorithm network, N is the temporal radius (e.g., $N = 3$), and $\tilde{I}_t$ and $\hat{I}_t$ are the super-resolution result of the target frame and the low resolution of the target frame, respectively.

3.2. The Feature Extraction Module

The feature extraction module is responsible for extracting the features of the input video frames to prepare for subsequent feature alignment. In the proposed method, the feature extraction module is composed of several residual blocks, which mainly consist of two convolutional layers. The residual block is more conducive to training. Therefore, the number of parameters is very small, and it maintains network performance in combination with other modules. The process of the proposed feature extraction module can be formulated as follows:

$$F = H_{FE}(\hat{I}_{t-N:t+N}). \tag{2}$$

Note that $H_{FE}(\cdot)$ denotes the feature extraction operation, and F denotes the extracted features.

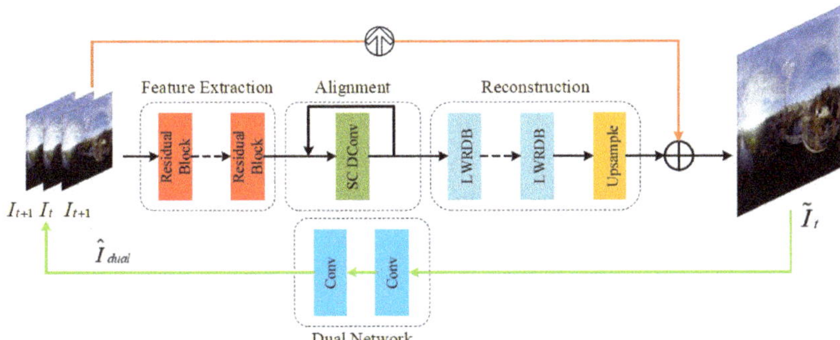

Figure 1. The network architecture of the proposed LWPVSR method. Our LWPVSR method mainly consists of the four modules: the feature extraction module, the feature alignment module, the reconstruction module, and the dual network module.

3.3. The Proposed Feature Alignment Module

In this subsection, we propose a new module for feature alignment between frames based on deformable convolution. As is known, the effectiveness of deformable convolution in video super-resolution has been witnessed and confirmed in EDVR [3]. In EDVR, the deformable convolution was integrated into a pyramid, cascading, and deformable convolution (PCD) module in EDVR, as shown in Figure 2. In fact, PCD has a pyramid-like structure. The top layer is a lower-resolution feature map, and the bottom layer is for the reference frame and neighboring frames. Different layers represent the feature information of different frequencies. PCD first aligns the reference feature map with the smallest resolution to form a rough alignment, and then transfers the offset and aligned feature map to a layer with a larger resolution, so that the offset and continuously aligned feature map are passed to the bottom layer every time. Thus, an implicit motion compensation from coarse to fine is formed from top to bottom. However, PCD introduced a large number of parameters and computational cost. In order to reduce the parameters, here we design a new pooled, self-calibrated convolution (PSCC) to replace the pyramid cascading structure and maintain the multi-scale learning capability, as shown in Figure 3.

Inspired by the self-calibrated convolution in [17], our PSCC module employs a pooling for downsampling to expand the receptive field, so as to learn more contextual information without increasing the network complexity. Specifically, after the target features and the neighboring features are merged, a convolution operation is performed, and then through channel splitting, the channel is divided into two, one channel only performs a simple convolution. In the other channel, we utilize a new upsample operation to make the learned features match with the scale. The learned features via the Sigmoid activation are again multiplied with the features through channel splitting. Finally, the features from the two channels are concatenated for output. Our PSCC module is embedded in the deformable convolution to learn the information about the neighbors of the reference frame, rather than the global information, so as to avoid the pollution information of other irrelevant frames and achieve more accurate frame alignment. From the perspective of the structures of PCD and PSCC, PCD uses multiple deformable convolutional networks (DCNs) and convolutional networks to cascade and form a pyramid structure, and each DCN is based on a feature map of different levels. Although our PSCC only adopts one deformable convolutional network, and combines it with self-correcting convolution, which not only reduces the number of parameters, but also learns more accurate offsets to achieve better alignment. This will be verified by the subsequent experimental results. Compared with the PCD module with 1.38 M parameters in EDVR, the number of the parameters of our PSCC module is only 0.04 M.

The process of the feature alignment in our LWPVSR method is expressed as follows,

$$F_{t\pm i}^{a} = H_{Alignment}(F_t, F_{t\pm i}), \tag{3}$$

where $H_{Alignment}(\cdot)$ denotes our alignment operation and F_t and $F_{t\pm i}$ denote the features of the target frame and the nearest neighboring frame, respectively. $F_{t\pm i}^{a}$ denotes the aligned features of each frame. Here, we use F^a to represent the result of all the aligned frames.

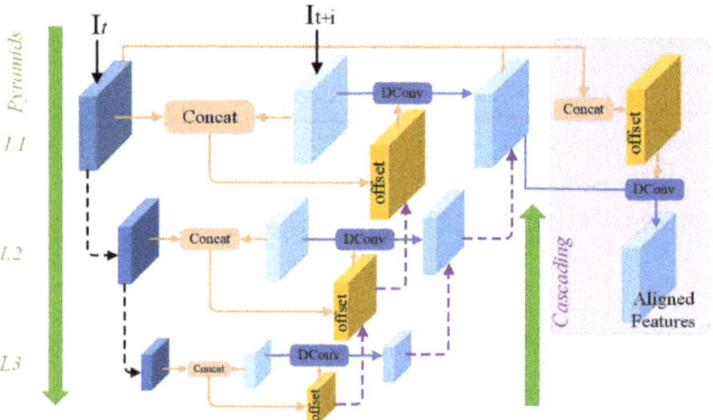

Figure 2. The structure of the PCD module in EDVR [3].

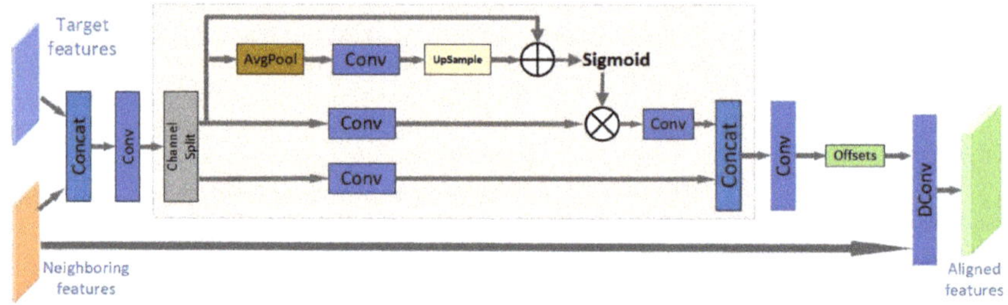

Figure 3. Our proposed pooled self-calibrated convolution (PSCC) module for feature alignment.

3.4. Our Reconstruction Module

In the proposed reconstruction module, inspired by the residual dense blocks (RDB) in [18], as shown in Figure 4a, a lightweight residual dense block (LWRDB) is designed for feature transformation and restoration. Its detailed structure is shown in Figure 4b.

First, the input features F_{in} through a layer of convolution output the corresponding feature maps. The channel is then shuffled by a channel shuffle operation, followed by a channel split operation which divides the number of channels into two proportionally. One part is fed into the convolution, and the other is connected to the output of the convolution in a jump connection. The channel shuffle and channel split operations are executed again, and so on. Finally, the number of channels is reduced through a 1×1 convolution, and the output is added to the initial input of the module to obtain the final result F_{out} of the module. The process is given by

$$F_{out} = H_{LWRDB}(F_{in}), \tag{4}$$

where $H_{LWRDB}(\cdot)$ denotes the operation of the LWRDB module. It is noted that in our design, the introduction of the channel shuffle [19] and the channel split [20] is important compared with that of Figure 4a. The purpose of the channel shuffle operation is to break up the output of the previous layer of convolution in channel dimension, and the purpose of the channel split operation is to split the channel into two according to a presetting ratio, as shown in Figure 4b. The purpose of combining these two operations is to reduce the number of parameters while still making full use of all levels of features like residual dense blocks, so that the number of parameters can be reduced. Meanwhile, it maintains high performance. It should be noted that we have verified in our experiment that the number of parameters of the RDB module is 1.08 M, while that of ours is only 0.81 M. Obviously, the number of parameters of our proposed LWRDB is smaller. In fact, our lightweight residual block in the reconstruction module introduces both channel shuffle and channel split operations. The channel shuffle can strengthen the exchange of information between channels, and channel split can reduce the number of parameters. Compared with the reconstruction module in Figure 4a, our method has fewer parameters and can maintain the performance of the network.

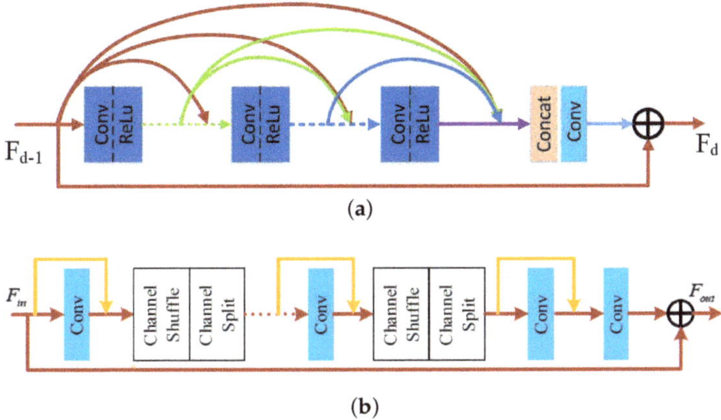

Figure 4. Comparison of the structures of the residual dense block (RDB) used in [19,20] and our lightweight RDB. (**a**) Existing RDB [19,20]. (**b**) Our lightweight RDB.

3.5. Our Dual Network Module and Loss Function

In our proposed method, the final super-resolution result is obtained by adding the output of the reconstruction module to the result of the upsampled target frame. It is expressed as follows,

$$\tilde{I}_t = H_{reconst}(F^a) + \hat{I}_t \uparrow, \tag{5}$$

where $H_{reconst}(\cdot)$ denotes the mapping function of reconstruction module. Here $\uparrow$ denotes upsampling, and $\tilde{I}_t$ denotes the super-resolution result of the target frame.

In order to show the important content in the equatorial region in the panorama video, we introduce a weighted mean square error (WMSE) loss, which is defined as follows,

$$\frac{1}{\sum_{i=0}^{M-1}\sum_{j=0}^{N-1}\omega(i,j)} \sum_{i=0}^{M-1}\sum_{j=0}^{N-1} \omega(i,j) \cdot (\tilde{I}_t(i,j) - I_t(i,j))^2, \tag{6}$$

where M and N denote the width and height of one frame, respectively, (i,j) represents the coordinate position of each pixel in a frame, and $\omega(i,j)$ is the weight at the corresponding pixel position, which is allocated according to the pixel position and is given by

$$\omega(i,j) = \cos\frac{(j + 0.5 - \frac{N}{2})\pi}{N}, \tag{7}$$

where $i = i_0, i_0 + 1, \cdots, i_0 + wd - 1$ and $j = j_0, j_0 + 1, \cdots, j_0 + h - 1$. Here, (i_0, j_0) represents the upper left corner of the patch, wd is the width of the patch, and h is the height of the patch.

In order to explain the weight change in each frame more intuitively, we show it visually in Figure 5. The black and white color represent the distribution of weights. The lighter the color, the greater the weight is, and the darker the color, the smaller the weight is. That is, the weights gradually decrease from the equator to the two polar regions. The weights are assigned on the whole frame during data processing.

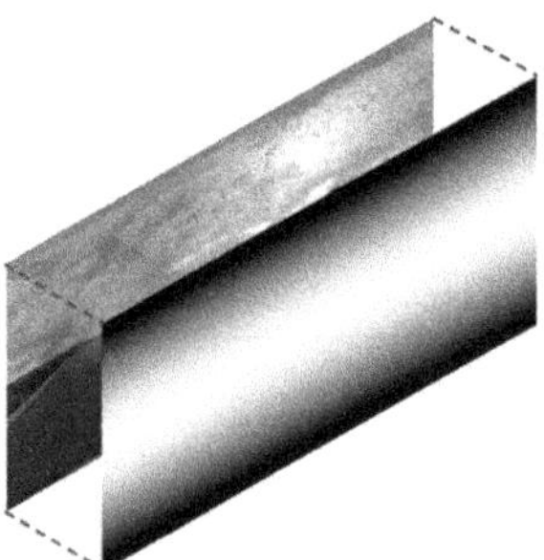

Figure 5. The weight diagram of the loss function.

The loss function in our architecture is composed of two parts. One is from the main branch—$L_{primary}$ (i.e., input, feature extraction, alignment, reconstruction, and output)—and the other is from the dual subnetwork: L_{dual}. The overall loss function of the proposed method is formulated as follows,

$$L_{total} = L_{primary} + \lambda L_{dual},\tag{8}$$

where $L_{primary}$ and L_{dual} are both calculated by Equation (6). The parameter λ is a balance factor between $L_{primary}$ and L_{dual}.

It is noted that compared with ordinary videos, the information of panoramic video is distributed on a sphere instead of a plane. The panoramic video, which is essentially a spherical video, cannot directly use the storage structure and encoding algorithm designed for ordinary videos. The current mainstream solution is to use the mapping relationship to project the spherical video onto the plane and compress the obtained plane video, and the equirectangular projection (ERP) is widely used. In this case, the important content is usually displayed in the equatorial region, and the less content is at the poles. Moreover, because the information of the panoramic video is distributed in a spherical shape, the features in the same dimensionality are more uneven, and the video is more prone to be distorted. Addressing the particularity that more content distributed at the equator and less content at the poles, we used the weighted loss function, as shown in Equation (6). It aims to increase the weight of the equatorial region and reduce the weight of the polar region. Addressing the features distributed on the same dimension are more uneven or the offset is too large, we think that using deformable convolution is not sufficient to solve this issue. Therefore, we propose to adopt self-correcting convolution combined with deformable convolution—that is, our PSCC module to learn these offset features. It is more conducive to achieving better alignment results.

4. Experimental Results

In this section, we compare the proposed LWPVSR method with eight state-of-the-art super-resolution algorithms for panoramic video super-resolution tasks.

4.1. Datasets

The MiG dataset [4] is utilized for evaluating the performance of super-resolution of the proposed LWPVSR method. The data set has 200 videos for training and eight videos for test. We adopt the bicubic interpolation algorithm to $2\times$ downsample each video frame as the ground truth (GT). Then, we further perform $4\times$ downsampling on GT to obtain the corresponding LR video. Moreover, in order to demonstrate the superiority of the proposed LWPVSR method, we also collected another video sequence from the Internet, named Clip_009, and adopted it for performance evaluation.

The peak signal-to-noise ratio (PSNR) and structural similarity (SSIM) are usually used as indicators to measure the performance of all the video super-resolution algorithms. In order to make a fair comparison, similar to other works, all indices are calculated on the Y channel for all the algorithms. Different from ordinary videos, we also use the two video quality metrics (i.e., WS-PSNR and WS-SSIM) in [4] to measure the performance of all the methods.

4.2. Training Setting

We implemented all the models in the PyTorch framework and used two NVIDIA Titan XP GPUs for training. The training schemes and parameters of other methods are listed below.

(1) SR360 [8]: The batch size is set to 16. The weights of all the layers were initialized randomly and the network was trained from the scratch. The network used the Adam solver with a learning rate, 1×10^{-4}.

(2) VSRnet [21]: The batch size is 240, a learning rate of is 1×10^{-4} used for the first two layers, 1×10^{-5} for the last layer and a weight decay rate of 0.0005 are set as in [21].

(3) FRVSR [22]: The Adam is an optimizer. The learning rate is fixed at 1×10^{-4}. Each sample in the batch is a set of 10 consecutive video frames, i.e., 40 video frames are passed through the networks in each iteration.

(4) VESPVN [5]: The initial batch size is 1. Every 10 epochs the batch size is doubled until it reaches a maximum size of 128. The optimizer is Adam with a learning rate, 1×10^{-4}.

(5) TDAN [2]: The batch size is set to 64. The Adam is the optimizer. The learning rate is initialized to 1×10^{-4} for all layers and decreases half for every 100 epochs.

(6) SOFVSR [6]: The batch size is 32. The optimizer is Adam. The initial learning rate is 1×10^{-3} and divided by 10 after every 80 K iterations.

(7) EDVR [3]: The batch size is set to 32. The learning rate is initialized to 4×10^{-4}, and initializes deeper networks by parameters from shallower ones for faster convergence.

(8) OVSR [7]: The batch size is 16. The optimizer is Adam. The initial learning rate is 1×10^{-3} and decays linearly to 1×10^{-4} after 120 K iterations, which keeps the same until 200 K iterations. Then the learning rate is further decayed to 5×10^{-5} and 1×10^{-5} until convergence.

In our method, the feature extraction module is composed of three residual blocks, each residual block consists of two layers of convolution, and the number of channels is set to 64. The reconstruction module includes five LWRDB blocks, each block is composed of six convolutional layers, and the number of channels is 64. In our experiment, we convert the video frames from the RGB space to the YCbCr space and then use the Y channel as the input to our network. Unless stated otherwise, the network takes three consecutive video frames as inputs. The input patch size is 64×64, and the batch size is set to 32. Moreover, we also employ data enhancement techniques as in other methods, including reflection, random cropping, and rotation. Furthermore, we defined the ratio for channel split by experience. If the ratio is larger than 0.5, it means that more features do not participate in the subsequent calculations but are directly cascaded to the subsequent feature maps. Then the following convolutional layers will be meaningless. If the ratio is smaller than 0.5, the model parameters will increase and it results in a higher computational cost. Therefore, the ratio equaling to 0.5 is a balanced choice. During training, we optimize the network by using the Adam optimizer with $\beta_1 = 0.9$ and $\beta_2 = 0.999$. The initial learning rate is set to 2×10^{-4}, and then is reduced by half after every 20 epochs. In our loss function, through experiments and experiences, the value of the parameter λ is set to 0.1. And the performance of each method has been optimized with its hyperparameter tuning to show their best results in our experiments.

4.3. Quantitative Comparison

We also implemented nine other state-of-the-art VSR algorithms for performance comparison. They include bicubic, SR360 [8], VSRnet [21], VESPCN [5], FRVSR [22], TDAN [2], SOFVSR20 [6], EDVR [3], and OVSR [7]. The quantitative results including PSNR/WS-PSNR, SSIM/WS-SSIM, inference time, and floating point operations per second (FLOPs) of all the methods on representative video clips are shown in Tables 1 and 2, respectively.

Table 1. Comparison of all the methods in terms of PSNR (top) and SSIM (bottom).

	Bicubic	SR360 [8]	VSRnet [21]	FRVSR [22]	VESPCN [5]	TDAN [2]	SOFVSR20 [6]	EDVR [3]	OVSR [7]	LWPVSR
Clip_005	26.38	26.58	26.59	25.36	26.71	26.73	26.72	26.73	26.69	**26.81**
	0.6868	0.7101	0.7075	0.7095	0.7203	0.7213	0.7203	0.7217	0.7207	**0.7251**
Clip_006	30.09	30.69	30.39	29.70	31.04	31.11	31.16	31.48	29.72	**31.58**
	0.8494	0.8580	0.8573	0.8700	0.8723	0.8740	0.8775	0.8685	0.8902	**0.8902**
Clip_007	27.65	29.29	28.10	28.90	29.50	29.61	29.54	30.18	29.31	**30.99**
	0.8119	0.8406	0.8245	0.8458	0.8490	0.8534	0.8527	0.8630	0.8622	**0.8700**
Clip_008	31.88	32.22	32.15	31.83	32.63	32.74	32.70	32.91	32.81	**33.02**
	0.9005	0.9001	0.9069	0.9162	0.9134	0.9150	0.9147	0.9186	0.9183	**0.9183**
Average	29.00	29.69	29.30	28.95	29.97	30.05	30.03	30.32	29.97	**30.60**
	0.8121	0.8272	0.8241	0.8353	0.8388	0.8409	0.8413	0.8479	0.8457	**0.8507**
Params. (M)	-	0.58	0.16	5.05	0.86	1.96	1.05	20.60	3.48	2.30
Time (ms)	-	64.30	2.52	71.57	122.59	16.11	76.86	670.80	69.55	92.31
FLOPs (T)	-	0.457	0.018	0.348	0.007	0.558	0.135	0.954	0.201	0.204

Table 2. Comparison of all the methods in terms of WS-PSNR (top) and WS-SSIM (bottom).

	Bicubic	SR360 [8]	VSRnet [21]	FRVSR [22]	VESPCN [5]	TDAN [2]	SOFVSR20 [6]	EDVR [3]	OVSR [7]	LWPVSR
Clip_005	26.39	26.62	26.60	25.37	26.75	26.78	26.77	26.80	26.73	**26.84**
	0.6888	0.7131	0.7118	0.7257	0.7263	0.7267	0.7257	0.7293	0.7260	**0.7298**
Clip_006	28.64	29.37	28.94	28.27	29.63	29.71	29.74	30.04	29.72	**30.15**
	0.8274	0.8422	0.8386	0.8574	0.8569	0.8594	0.8622	0.8744	0.8685	**0.8759**
Clip_007	29.75	30.76	30.15	30.23	31.24	31.42	31.29	31.57	31.55	**31.86**
	0.8009	0.8214	0.8165	0.8374	0.8379	0.8406	0.8392	0.8464	0.8465	**0.8482**
Clip_008	30.46	30.85	30.72	30.34	31.19	31.28	31.24	31.43	31.34	**31.52**
	0.8685	0.8726	0.8779	0.8869	0.8854	0.8885	0.8880	0.8929	0.8924	**0.8938**
Average	28.81	29.40	29.10	28.55	29.70	29.80	29.76	29.96	29.84	**30.09**
	0.7964	0.8123	0.8112	0.8268	0.8266	0.8288	0.8288	0.8358	0.8333	**0.8369**
Params. (M)	-	0.58	0.16	5.05	0.86	1.96	1.05	20.60	3.48	2.30
Time (ms)	-	64.30	2.52	71.57	122.59	16.11	76.86	670.80	69.55	92.31
FLOPs (T)	-	0.457	0.018	0.348	0.007	0.558	0.135	0.954	0.201	0.204

It can be seen that our LWPVSR method obtains the highest PSNR and SSIM results, and the amount of its parameters is relatively small. Our LWPVSR method performs much better than EDVR in terms of PSNR/WS-PSNR and SSIM/WS-SSIM, and the the former has significantly fewer parameters than the latter (i.e., 2.30 M vs. 20.60 M). That is, LWPVSR is nearly 1/10 size of EDVR. It is because the proposed PSCC module in our LWPVSR plays an important role, and decreases the PCD module in EDVR by man parameters but maintains the performance. In addition, compared with FRVSR, our model parameters are 2.7 M smaller, and the PSNR of model is 1.65 dB higher than FRVSR. SR360, VSRnet, VESPCN, TDAN, and SOFVSR20 are relatively lightweight video super-resolution architectures, with model parameters below 2.0 M. However, the performance of all of them is significantly lower than that of the proposed method. Moreover, the PSNR and WS-PSNR results of all the methods on other video clips are shown in Tables 3 and 4. We can see that the results of our LWPVSR method are much better than those of the state-of-the-art methods. All the experimental results show that our LWPVSR method can achieve a good balance between the model complexity and performance.

Table 3. Comparison of all the methods in terms of PSNR (top) and SSIM (bottom) on other video clips.

	Bicubic	SR360 [8]	VSRnet [21]	FRVSR [22]	VESPCN [5]	TDAN [2]	SOFVSR20 [6]	EDVR [3]	OVSR [7]	LWPVSR
Clip_001	27.57	29.06	27.75	28.80	28.56	29.20	29.08	29.44	29.25	**29.68**
	0.8659	0.8833	0.8742	0.8920	0.8909	0.8965	0.9004	0.9108	0.9122	**0.9176**
Clip_002	26.06	27.26	26.54	27.42	27.20	27.43	27.39	27.58	27.87	**27.75**
	0.7426	0.7866	0.7650	0.8045	0.7976	0.8052	0.8073	0.8138	0.8378	**0.8231**
Clip_003	25.68	26.45	25.95	26.39	26.52	26.63	26.55	26.66	26.57	**26.82**
	0.8240	0.8495	0.8359	0.8551	0.8568	0.8623	0.8607	0.8663	0.8737	**0.8700**
Clip_004	30.61	31.46	31.08	32.25	32.17	32.44	32.46	33.03	32.72	**33.66**
	0.8889	0.8931	0.8983	0.9220	0.9196	0.9257	0.9280	0.9379	0.9404	**0.9412**
Clip_009	26.03	27.23	26.50	27.40	27.16	27.40	27.36	27.56	27.79	**29.41**
	0.7515	0.7957	0.7717	0.8123	0.8044	0.8131	0.8136	0.8224	0.8637	**0.8801**
Average	27.19	28.29	27.56	28.45	28.32	28.62	28.57	28.85	28.84	**29.46**
	0.8146	0.8416	0.8290	0.8572	0.8539	0.8606	0.8620	0.8702	0.8856	**0.8864**
Params. (M)	-	0.58	0.16	5.05	0.86	1.96	1.05	20.60	3.48	2.30
Time (ms)	-	64.30	2.52	71.57	122.59	16.11	76.86	670.80	69.55	92.31
FLOPs (T)	-	0.457	0.018	0.348	0.007	0.558	0.135	0.954	0.201	0.204

Table 4. Comparison of all the methods in terms of WS-PSNR and WS-SSIM on other video clips.

	Bicubic	SR360 [8]	VSRnet [21]	FRVSR [22]	VESPCN [5]	TDAN [2]	SOFVSR20 [6]	EDVR [3]	OVSR [7]	LWPVSR
Clip_001	29.84	30.86	30.19	30.95	31.12	31.35	31.35	31.89	31.67	**32.04**
	0.9630	0.8771	0.8731	0.8909	0.8901	0.8948	0.8969	0.9082	0.9053	**0.9071**
Clip_002	25.81	27.02	26.27	27.12	26.89	27.12	27.03	27.32	27.59	**27.45**
	0.7416	0.7792	0.7626	0.7975	0.7916	0.7978	0.8003	0.8082	0.8226	**0.8112**
Clip_003	24.49	25.17	24.75	25.12	25.23	25.33	25.26	25.37	25.26	**25.48**
	0.7807	0.8134	0.7972	0.8197	0.8212	0.8275	0.8252	0.8339	0.8308	**0.8312**
Clip_004	29.88	30.87	30.39	31.66	31.59	31.91	31.90	32.45	32.18	**32.89**
	0.8666	0.8802	0.8796	0.9077	0.9053	0.9122	0.9143	0.9249	0.9242	**0.9263**
Clip_009	25.78	26.99	26.24	27.09	26.87	27.09	27.03	27.29	27.98	**28.11**
	0.7448	0.7815	0.7617	0.7971	0.7907	0.7972	0.7993	0.8076	0.8312	**0.8387**
Average	27.16	28.18	27.57	28.39	28.34	28.56	28.51	28.86	28.94	**29.20**
	0.7993	0.8263	0.8148	0.8426	0.8398	0.8459	0.8472	0.8566	0.8628	**0.8629**
Params. (M)	-	0.58	0.16	5.05	0.86	1.96	1.05	20.60	3.48	2.30
Time (ms)	-	64.30	2.52	71.57	122.59	16.11	76.86	670.80	69.55	92.31
FLOPs (T)	-	0.457	0.018	0.348	0.007	0.558	0.135	0.954	0.201	0.204

In order to demonstrate the relation between performance and parameters more clearly, the visualized diagram is also shown in Figure 6. It can be seen that our method attains a higher performance at the cost of lower numbers of parameters.

4.4. Qualitative Comparison

In this subsection, we qualitatively compare our method with the other methods on video sequences Clip_001, Clip_003, Clip_004 and Clip_009, as shown in Figures 7–10, respectively.

It can be seen that our LWPVSR method has achieved much better performance than other methods, including EDVR with 20.60 M parameters, and they have superior visual results in all these figures. For example, in Figure 7, the image recovered by our LWPVSR method seems more real, which is closer to the original high-resolution image. However, the images recovered by other methods, such as TDAN, FRVSR, and SOFVSR20, are blurry.

Similar results can also be observed from Figures 8–10. In general, compared with other methods, our LWPVSR method achieves a better balance between the model complexity and algorithm performance, resulting in less distortion and more reliable results in the panoramic video super-resolution task.

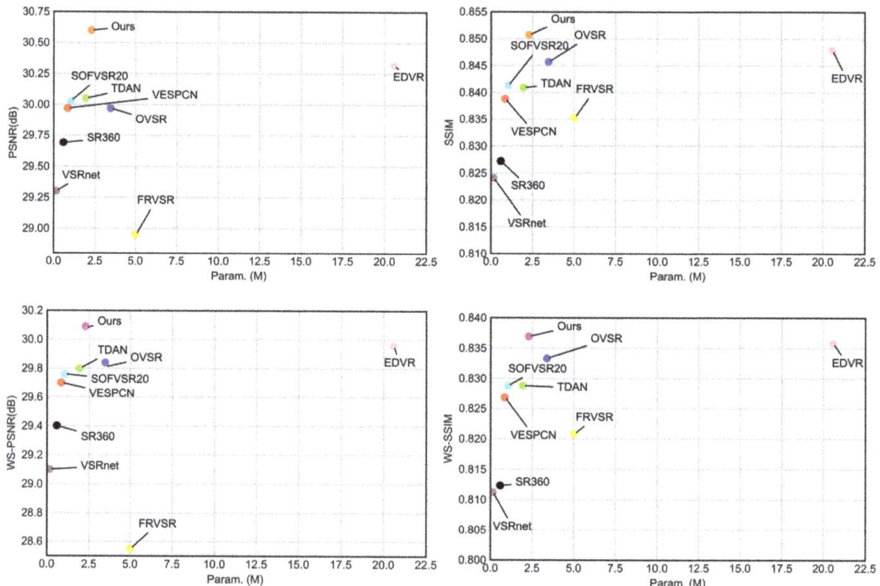

Figure 6. Comparison of all the methods in terms of performance and number of parameters. Note that the *y*-axis represents different performance metrics (including PSNR, SSIM, WS-PSNR, and WS-SSIM), and the *x*-axis corresponds to the number of parameters in different methods.

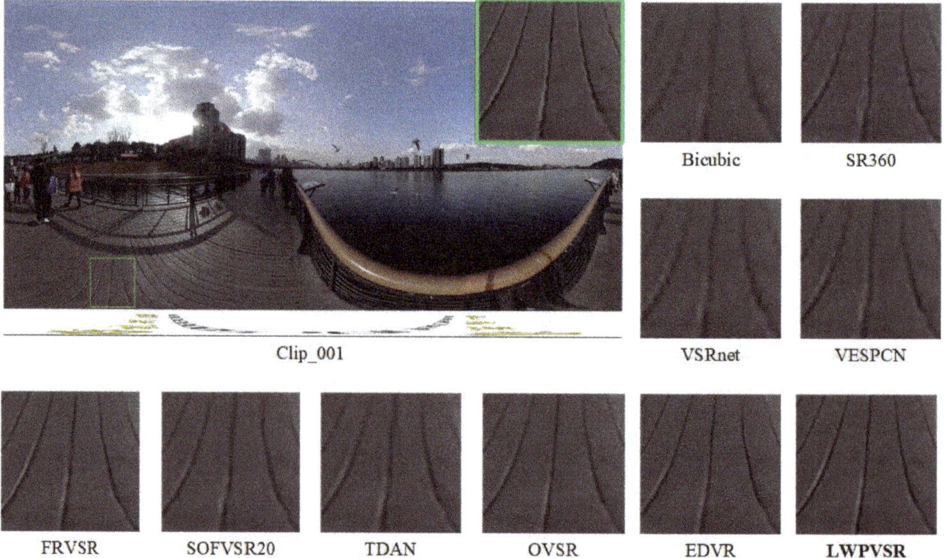

Figure 7. The results of all the algorithms performing 4× super-resolution on Clip_001 of the MiG test set.

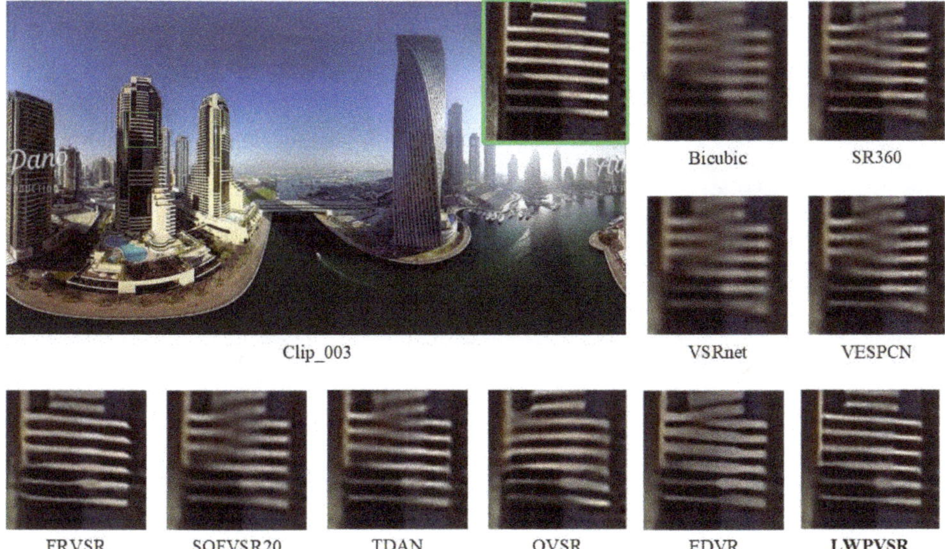

Figure 8. The results of all the algorithms performing 4× super-resolution on Clip_003 of the MiG test set.

Figure 9. The results of all the algorithms performing 4× super-resolution on Clip_004 of the MiG test set.

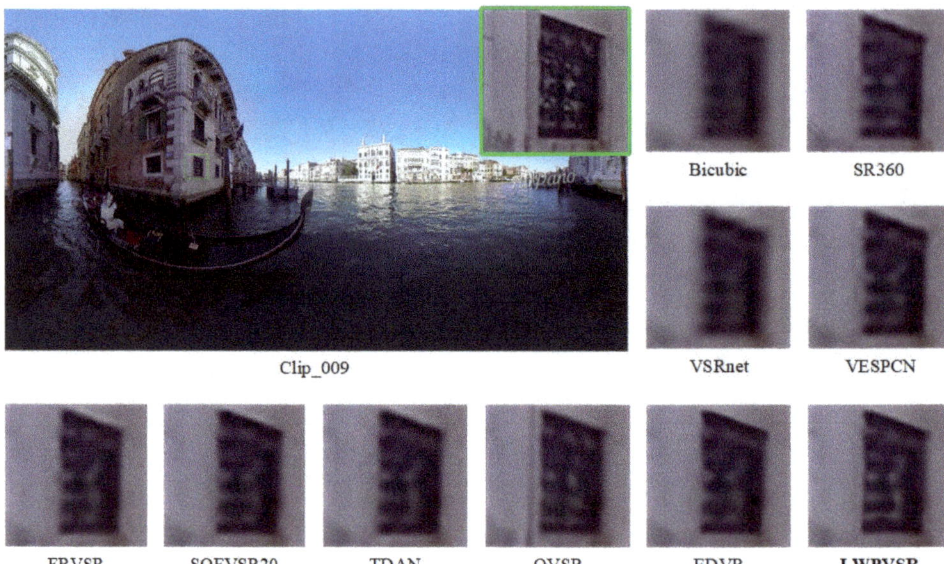

Figure 10. The results of all the algorithms performing 4× super-resolution on Clip_009 of the MiG test set.

4.5. Ablation Studies

In this subsection, we analyze the contribution of each module in our network, mainly including PSCC and LWRDB, as shown in Table 5. The baseline is our architecture, as shown in Figure 1. The PSNR and SSIM results are 30.60 dB and 0.8507, respectively. When the architecture is without the PSCC module, the PSNR drops by 0.30 dB, and the number of parameters decreases 0.04 M. The performance drops by 0.92 dB when the baseline is without the LWRDB module. Moreover, without PSCC and LWRDB, the PSNR result decreases by 0.96 dB. All the results also verify the importance of the proposed modules, including PSCC and LWRDB for the proposed method.

Table 5. Ablation studies for each module in the proposed LWRDB network.

	PSNR	SSIM	Parameters (M)
Ours	30.60	0.8507	2.30
Ours *w/o* PSCC	30.30	0.8471	2.26
Ours *w/o* LWRDB	29.68	0.8290	1.49
Ours *w/o* PSCC and LWRDB	29.64	0.8285	1.44

5. Conclusions and Future Work

In this paper, a lightweight and efficient panoramic video super-resolution method was designed from the perspective of lightweight networks. This method adopts deformable convolution to align the nearest neighbor features with the target feature, in order to further enhance the alignment effect step. In particular, we introduced self-calibrated convolution to gradually implement the alignment operation in a recursive manner. Moreover, we also proposed a lighter and more efficient LWRDB module based on the RDB module. Various experimental results verified the effectiveness of the proposed method. Compared with mainstream video super-resolution algorithms, our proposed method achieves a better balance between performance and algorithm complexity.

In the future, we will design more effective strategies, such as the attention strategy [23] for the lightweight architecture to further enhance the performance while maintaining its cost.

Author Contributions: Methodology, F.S. (Fanjie Shang) and H.L.; Validation, Z.Z.; Formal analysis, Y.L., F.S. (Fanhua Shang) and Z.Z.; Data curation, W.M.; Project administration, L.W.; Funding acquisition, L.J. All authors have read and agreed to the published version of the manuscript.

Funding: This work was supported by the National Natural Science Foundation of China (Nos. 61976164, 62276182 and 61876221), and Natural Science Basic Research Program of Shaanxi (Program No. 2022GY-061).

Acknowledgments: We thank all the reviewers for their valuable comments.

Conflicts of Interest: The authors declare no conflict of interest.

References

1. Wang, L.; Guo, Y.; Lin, Z.; Deng, X.; An, W. Learning for video super-resolution through HR optical flow estimation. In Proceedings of the Asian Conference on Computer Vision (ACCV), Perth, Australia, 2–6 December 2018; pp. 514–529.
2. Tian, Y.; Zhang, Y.; Fu, Y.; Xu, C. TDAN: Temporally-deformable alignment network for video super-resolution. In Proceedings of the IEEE Conference on Computer Vision and Pattern Recognition (CVPR), Seattle, WA, USA, 13–19 June 2020; pp. 3360–3369.
3. Wang, X.; Chan, K.C.K.; Yu, K.; Dong, C.; Loy, C.C. EDVR: Video restoration with enhanced deformable convolutional networks. In Proceedings of the IEEE Conference on Computer Vision and Pattern Recognition (CVPR) Workshops, Long Beach, CA, USA, 15–20 June 2019; pp. 1954–1963.
4. Liu, H.; Ruan, Z.; Fang, C.; Zhao, P.; Shang, F.; Liu, Y.; Wang, L. A single frame and multi-frame joint network for 360-degree panorama video super-resolution. *arXiv* **2020**, arXiv:2008.10320.
5. Caballero, J.; Ledig, C.; Aitken, A.; Acosta, A.; Totz, J.; Wang, Z.; Shi, W. Real-time video super-resolution with spatio-temporal networks and motion compensation. In Proceedings of the IEEE Conference on Computer Vision and Pattern Recognition (CVPR), Honolulu, HI, USA, 21–26 July 2017; pp. 4778–4787.
6. Wang, L.; Guo, Y.; Liu, L.; Lin, Z.; Deng, X.; An, W. Deep video super-resolution using HR optical flow estimation. *IEEE Trans. Image Process.* **2020**, *29*, 4323–4336. [CrossRef] [PubMed]
7. Yi, P.; Wang, Z.; Jiang, K.; Jiang, J.; Lu, T.; Tian, X.; Ma, J. Omniscient video super-resolution. In Proceedings of the IEEE/CVF International Conference on Computer Vision (ICCV), Montreal, BC, Canada, 10–17 October 2021; pp. 4409–4418.
8. Ozcinar, C.; Rana, A.; Smolic, A. Super-resolution of omnidirectional images using adversarial learning. In Proceedings of the 21st International Workshop on Multimedia Signal Processing (MMSP), Kuala Lumpur, Malaysia, 27–29 September 2019; pp. 1–6.
9. Isola, P.; Zhu, J.-Y.; Zhou, T.; Efros, A.A. Image-to-image translation with conditional adversarial networks. In Proceedings of the IEEE Conference on Computer Vision and Pattern Recognition (CVPR), Honolulu, HI, USA, 21–26 July 2017; pp. 5967–5976.
10. Arican, Z.; Frossard, P. Joint registration and super-resolution with omnidirectional images. *IEEE Trans. Image Process.* **2011**, *20*, 3151–3162. [CrossRef] [PubMed]
11. Bagnato, L.; Boursier, Y.; Frossard, P.; Vandergheynst, P. Plenoptic based super-resolution for omnidirectional image sequences. In Proceedings of the IEEE International Conference on Image Processing (ICIP), Hong Kong, China, 26–29 September 2010; pp. 2829–2832.
12. Rivadeneira, R.E.; Sappa, A.D.; Vintimilla, B.X.; Hammoud, R. A Novel Domain Transfer-Based Approach for Unsupervised Thermal Image Super-Resolution. *Sensors* **2022**, *12*, 2254. [CrossRef] [PubMed]
13. Kim, B.; Jin, Y.; Lee, J.; Kim, S. High-Efficiency Super-Resolution FMCW Radar Algorithm Based on FFT Estimation. *Sensors* **2021**, *21*, 4018. [CrossRef] [PubMed]
14. Fakour-Sevom, V.; Guldogan, E.; Kämäräinen, J.-K. 360 panorama super-resolution using deep convolutional networks. In Proceedings of the International Conference on Computer Vision Theory and Applications (VISAPP), Funchal, Portugal, 27–29 January 2018; Volume 1.
15. Dong, C.; Loy, C.C.; He, K.; Tang, X. Learning a deep convolutional network for image super-resolution. In Proceedings of the European Conference on Computer Vision (ECCV), Zurich, Switzerland, 6–12 September 2014; pp. 184–199.
16. Li, S.; Lin, C.; Liao, K.; Zhao, Y.; Zhang, X. Panoramic image quality-enhancement by fusing neural textures of the adaptive initial viewport. In Proceedings of the IEEE Conference on Virtual Reality and 3D User Interfaces Abstracts and Workshops (VRW), Atlanta, GA, USA, 22–26 March 2020; pp. 816–817.
17. Liu, J-J.; Hou, Q.; Cheng, Mi.; Wang, C.; Feng, J. Improving convolutional networks with self-calibrated convolutions. In Proceedings of the IEEE/CVF Conference on Computer Vision and Pattern Recognition (CVPR), Seattle, WA, USA, 13–19 June 2020; pp. 10093–10102.
18. Huang, G.; Liu, Z.; Maaten, L.V.D.; Weinberger, K.Q. Densely connected convolutional networks. In Proceedings of the IEEE Conference on Computer Vision and Pattern Recognition (CVPR), Honolulu, HI, USA, 21–26 July 2017; pp. 4700–4708.
19. Zhang, X.; Zhou, X.; Lin, M.; Sun, J. Shufflenet: An extremely efficient convolutional neural network for mobile devices. In Proceedings of the IEEE/CVF Conference on Computer Vision and Pattern Recognition (CVPR), Salt Lake City, UT, USA, 18–23 June 2018; pp. 6848–6856.
20. Ma, N.; Zhang, X.; Zheng, Ha.; Sun, J. Shufflenet V2: Practical guidelines for efficient CNN architecture design. In Proceedings of the European Conference on Computer Vision (ECCV), Munich, Germany, 8–14 September 2018; pp. 122–138.

21. Kappeler, A.; Yoo, S.; Dai, Q.; Katsaggelos, A.K. Video super-resolution with convolutional neural networks. *IEEE Trans. Comput. Imaging* **2016**, *2*, 109–122. [CrossRef]
22. Sajjadi, M.S.M.; Vemulapalli, R.; Brown, M. Frame-recurrent video super-resolution. In Proceedings of the IEEE Conference on Computer Vision and Pattern Recognition (CVPR), Salt Lake City, UT, USA, 18–23 June 2018; pp. 6626–6634.
23. Du, J.; Cheng, K.; Yu, Y.; Wang, D.; Zhou, H. Panchromatic Image super-resolution via self attention-augmented wasserstein generative adversarial network. *Sensors* **2021**, *21*, 2158. [CrossRef] [PubMed]

Disclaimer/Publisher's Note: The statements, opinions and data contained in all publications are solely those of the individual author(s) and contributor(s) and not of MDPI and/or the editor(s). MDPI and/or the editor(s) disclaim responsibility for any injury to people or property resulting from any ideas, methods, instructions or products referred to in the content.

sensors

MDPI

Article

NRA-Net—Neg-Region Attention Network for Salient Object Detection with Gaze Tracking

Hoijun Kim [1], Soonchul Kwon [2,*] and Seunghyun Lee [3]

[1] Department of Plasma Bio Display, Kwangwoon University, 20 Kwangwoon-ro, Nowon-gu, Seoul 01897, Korea; hoi97@kw.ac.kr
[2] Department of Smart Convergence, Kwangwoon University, 20 Kwangwoon-ro, Nowon-gu, Seoul 01897, Korea
[3] Ingenium College of Liberal Arts, Kwangwoon University, 20 Kwangwoon-ro, Nowon-gu, Seoul 01897, Korea; shlee@kw.ac.kr
* Correspondence: ksc0226@kw.ac.kr; Tel.: +82-2-940-8637

Abstract: In this paper, we propose a detection method for salient objects whose eyes are focused on gaze tracking; this method does not require a device in a single image. A network was constructed using Neg-Region Attention (NRA), which predicts objects with a concentrated line of sight using deep learning techniques. The existing deep learning-based method has an autoencoder structure, which causes feature loss during the encoding process of compressing and extracting features from the image and the decoding process of expanding and restoring. As a result, a feature loss occurs in the area of the object from the detection results, or another area is detected as an object. The proposed method, that is, NRA, can be used for reducing feature loss and emphasizing object areas with encoders. After separating positive and negative regions using the exponential linear unit activation function, converted attention was performed for each region. The attention method provided without using the backbone network emphasized the object area and suppressed the background area. In the experimental results, the proposed method showed higher detection results than the conventional methods.

Keywords: autoencoder; convolutional neural network; deep learning; gaze tracking; image processing; salient object detection

Citation: Kim, H.; Kwon, S.; Lee, S. NRA-Net—Neg-Region Attention Network for Salient Object Detection with Gaze Tracking. *Sensors* **2021**, *21*, 1753. https://doi.org/10.3390/s21051753

Academic Editor: Yun Zhang

Received: 25 January 2021
Accepted: 25 February 2021
Published: 4 March 2021

Publisher's Note: MDPI stays neutral with regard to jurisdictional claims in published maps and institutional affiliations.

Copyright: © 2021 by the authors. Licensee MDPI, Basel, Switzerland. This article is an open access article distributed under the terms and conditions of the Creative Commons Attribution (CC BY) license (https://creativecommons.org/licenses/by/4.0/).

1. Introduction

Video processing technology using deep learning has been studied in various fields, such as monochrome image colorization, object detection and recognition, super-resolution technology, character detection and object recognition. Particularly, in the field of object detection and recognition, the underlying method is salient object detection (SOD). The purpose of this method is to detect objects that are of interest to a person, detect objects that cause the line of sight when a person first sees a video or a single image, and track the line of sight based on the results. SOD predicts where the gaze is focused when gaze tracking without equipment. Dramatic scene changes when predicting gaze can cause tracking to fail. SOD can supplement gaze tracking information by learning the area where the gaze is concentrated and predicting the input image. SOD's training dataset generates an area where the eyes of 20 to 30 people are focused through gaze tracking when displaying images on the screen instantly. This method recognizes objects and is used in various areas, such as scene classification, tracking and detection. Other typical object detection methods include the contour-based, division-based, and deep learning-based methods.

SOD is the creation of a saliency map by detecting objects that are of interest to people or objects that are considered to be the most important in a video or image. The correct saliency map of the SOD input image is called the ground truth. Typical split-based SOD methods include superpixel, contour-based, and deep learning-based methods.

The deep learning-based detection method does not require complicated pretreatments and posttreatment processes and shows a high detection rate. Most of the existing SOD methods have autoencoder structures, and the use of some methods results in a significant deformation of the model structure and loss function. The model structure can be improved to reduce losses during the feature extraction process due to the shallow structure and to maximize the error of the loss function during the learning process. Research is underway to improve the detection rate of salient objects based on deep learning-based methods. However, the detection rate drops due to the high degree of similarity between the object and the background or the existence of several objects.

The existing deep learning-based method typically includes a fully convolutional network (FCN) [1], which uses skip connection to minimize losses during the feature extraction process. However, various values are extracted at the feature extraction stage, and feature values can be expressed as negative in this process. Traditional methods focus on positive region values that do not utilize negative region values, which causes feature loss. Also, since deep learning research is being conducted based on the backbone network, a backbone network is always required. A backbone network is a feature extractor that has learned a lot of data in advance. Although it has the advantage of being able to select various features, the type of backbone is limited and shows poor performance when extracting features that have not been learned.

In this study, the negative region that was not used in the existing method is utilized. A new attention module is created by using the spatial attention technique in the negative region. This module minimizes loss of functionality during feature extraction. We also propose a deep learning model called the Neg-Region Attention (NRA), which aims to minimize the feature loss of salient objects due to complex environmental problems.The proposed method did not use a backbone network to extract desired features and does not require additional pre-trained weights. It aims to construct a relatively light model without a backbone network. In addition, it aims to improve the performance of feature extraction by providing a new module using the negative region.

2. Related Works

Deep learning is a machine learning algorithm that summarizes the core contents and features of complex data, with nonlinear transformation methods composed of multiple layers. In the existing machine learning algorithm, a person directly analyzes and evaluates by extracting the kind of features present in the data to be learned. However, in deep learning, necessary features are extracted and learned from the data that the machine automatically learns. In these studies, deep learning was developed as a convolutional neural network-based method with high data recognition and detection performance. In addition, deep learning technology has been established, and excellent performance methods have been developed in the field of SOD, where FCN is a typical example.

2.1. Hand Crafted-Based Detection Method

Superpixel-based split methods [2,3] split the salient object and background with the internal information of the image, such as brightness, color, contrast and texture. Because salient objects have movements in the video, the method considers the position of the object according to the time using a superpixel partition. Contour-based detection methods [4] include detecting salient objects using a fast Fourier transform and a Gaussian filter. Such a method maintains the contour of the object and shows a high detection rate, but it requires a large amount of calculation due to pre-processing and post-processing process requirements. In addition, the detection rate of salient objects decreases due to problems such as complicated backgrounds, high similarity between backgrounds and objects, or the existence of several objects.

2.2. Deep Learning-Based Detection Method

The deep learning-based SOD method shows high accuracy in trained images without requiring complicated pre-processing and post-processing processes. In the present research, this method is performed with an autoencoder structure, and it can be further classified to methods transforming the network structure and transforming the loss function. When the network structure is deformed, the loss and shallow structure in the process of extracting the features are improved, which consequently improves the results. When the loss function is transformed, the loss function is improved, which consequently minimizes the error in the learning process. Both methods show improved performance, and even when transforming the loss function, it is necessary to improve the network structure in the process of minimizing the error [5–9].

Recently, many deep learning-based detection methods have been studied to improve the performance of SOD. However, the detection rate drops due to problems, such as high similarity between the background and object and complicated background. In addition, feature loss occurs during feature extraction through several convolution layers.

2.3. Autoencoder

An autoencoder [10] is a type of artificial neural network used to compress and restore image data. It is a learning model with a structure similar to that of feed-forward neural networks (FNNs) [11]. Different from FNNs, the sizes of the input and output layers of an autoencoder are always the same.

An autoencoder is largely composed of an encoder and a decoder. The encoder is likely a network that extracts features from the input data or compresses it into an internal representation. The decoder is a generation network that converts extracted features and compressed internal representations into the output. An autoencoder is a deep learning network structure that is often used in the field of SOD and partitioning. The autoencoder copies the input to the output only on the same side as the input layer with the same number of nodes in the hidden layer. Therefore, the number of nodes in the hidden layer is smaller than that in the input layer and the data are compressed, as shown in Figure 1. In this method, control is used to represent data efficiently. The upsampling of the decoding process causes feature loss as it is simply used in the process of increasing the size of the feature map.

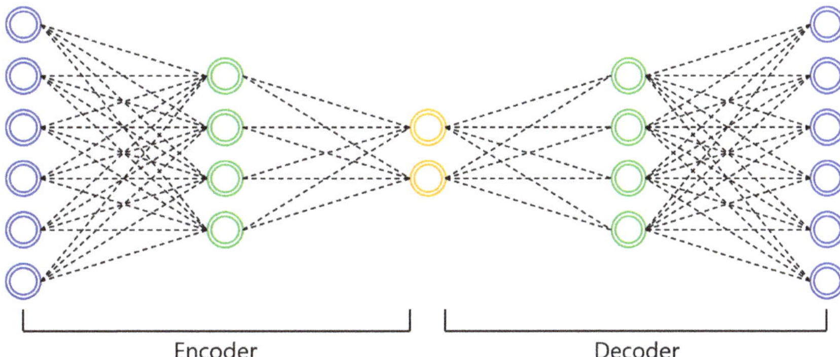

Figure 1. Autoencoder architecture example.

3. Proposed Method

The proposed method, that is, NRA, has an autoencoder structure and research has been conducted to reduce losses that occur in the process of compressing and decompressing features and losses that occur in the process of expanding extracted features. Existing methods use the rectified linear unit (ReLU) [12] as an activation function in the encoder process of extracting features. ReLU treats negative regions as 0, so feature loss occurs, but the learning speed is fast. Instead of using ReLU, the conventional method uses a model in the backbone network where a large dataset is trained to prevent feature loss. The backbone network requires pre-trained weights, and there is a limit to model transformation. Most of the input images are also fixed and the model becomes heavier.

The proposed method aims to improve the heaviness of the model, change the size of the input image according to the user's computer performance, and minimize the feature loss at the encoder stage. ELU is used as an activation function to avoid the feature loss problem that occurs in ReLU. The exponential linear unit (ELU) activation function is used in the encoding process to compress and extract features and utilize them in the negative region. The NRA provided for the extracted features can be used to suppress non-object areas in the negative areas and emphasize the contour and texture information of the objects in the positive areas. The decoding process, which enlarges the extracted features to the size of the input image, utilizes the features in the encoding stage through concatenation. Through this process, an improved saliency map is generated. The detection flowchart using NRA proposed in this paper is shown in Figure 2.

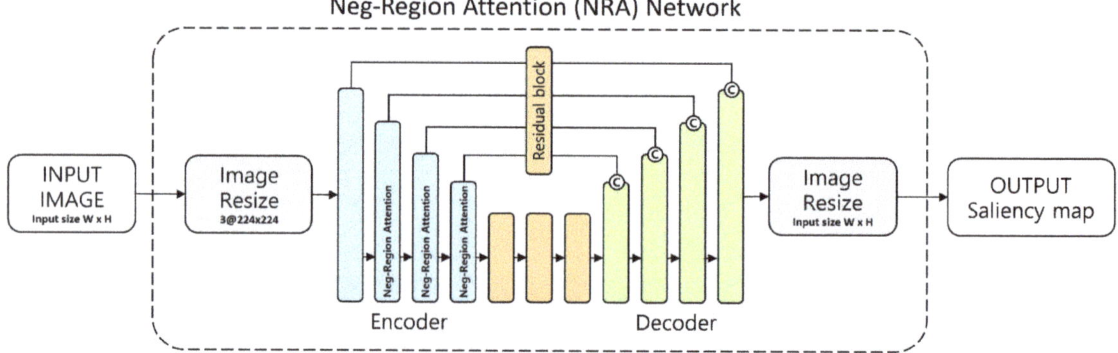

Figure 2. The overall flowchart of our Salient Object detection method. (Neg-Region Attention Network).

3.1. Feature Extraction Using the Proposed Attention

In a deep learning-based method, the convolution operation results involve negative and positive regions. The conventional method uses ReLU activation function to determine which node to pass to the next layer. The ReLU activation function treats the negative region as 0 and causes feature loss in the feature extraction process. Accordingly, the proposed method utilizes the negative region and uses the Exponential Linear Unit (ELU) [13] activation function to prevent feature loss. The positive region of the ELU activation function is processed similarly as the ReLU activation function, and the negative region has a convergence form of (1) and (2). Equations (1) and (2) are the equations of the ReLU and ELU activation functions, respectively, where the $exp()$ function was used in the negative region in ELU. The graphs of the activation functions are presented in Figure 3.

$$ReLU(x) = max(0, x) \tag{1}$$

$$ELU(x) = \begin{cases} x, & \text{if } x \geq 0 \\ a(e^x - 1), & \text{if } x < 0. \end{cases} \tag{2}$$

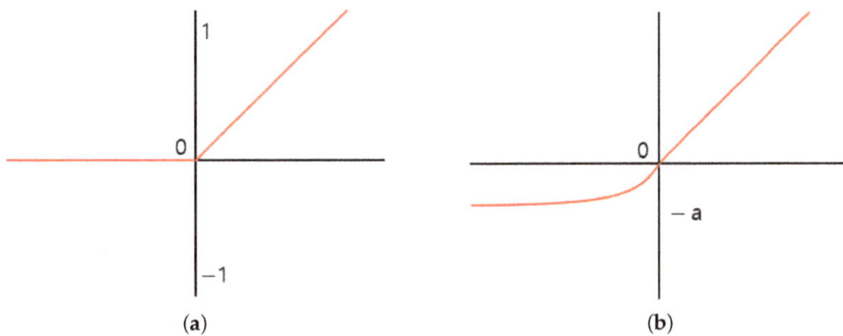

(a)   (b)

Figure 3. The rectified linear unit (ReLU) activation function graph before (**a**) and after (**b**) is an Exponential Linear Unit (ELU) activation function graph.

As shown in Figure 3, the graph of the ELU activation function directly outputs the input in the positive region, but in the negative region, it is normalized so that it is not outputted immediately and converges to $-a$. By setting the values of a, the influence of the negative region can be limited. The proposed method uses all the characteristics of the positive and negative regions by setting a to 1, but the negative region has little effect.

When the value of a is set large, the effect of the negative region becomes large in the complete data representation, and the texture information around the boundary information in which the amount of data change is large is expressed in various ways. The proposed method uses integrated texture information rather than various texture information in the negative area and sets the value of a to 1 to suppress non-object areas.

We propose NRA to suppress the non-object area of the positive region and emphasize the object area using the negative region of the ELU. The proposed method was separately emphasized in the negative and positive regions after applying the ELU activation function. Figure 4 shows the proposed NRA structure.

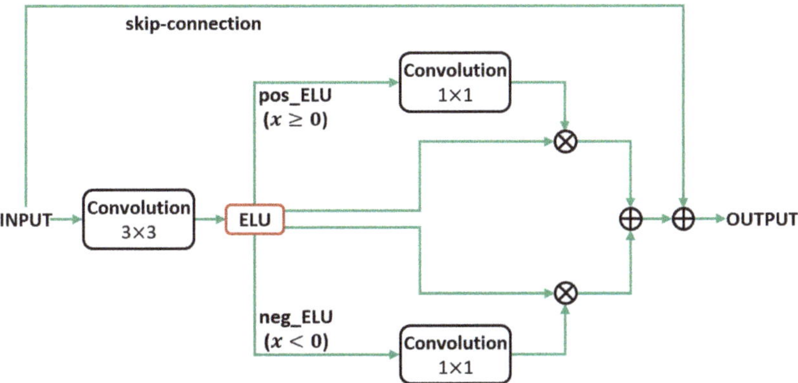

Figure 4. Neg-Region Attention module architecture.

The proposed NRA method is configured as follows after applying ELU, the process of separating the negative and positive areas results in an ELU feature map and an element-wise product spatial attention course using a 1 × 1 convolution in the separated area and an element of the emphasized feature map, which consists of a joint process using an element-wise sum. In Figure 4, neg_ELU represents the negative region of the ELU activation function and pos_ELU represents the positive region.

Equation (2) is the process of separating into a negative region (neg_ELU) and a positive region (pos_ELU), which are respectively shown in (3) and (4), respectively. Figure 5 is a graph showing the separation in the ELU.

$$ELU_n(x) = \begin{cases} 0, & \text{if } x \geq 0 \\ a(e^x - 1), & \text{if } x < 0 \end{cases}, a = 1 \tag{3}$$

$$ELU_p(x) = \begin{cases} x, & \text{if } x \geq 0 \\ 0, & \text{if } x < 0. \end{cases} \tag{4}$$

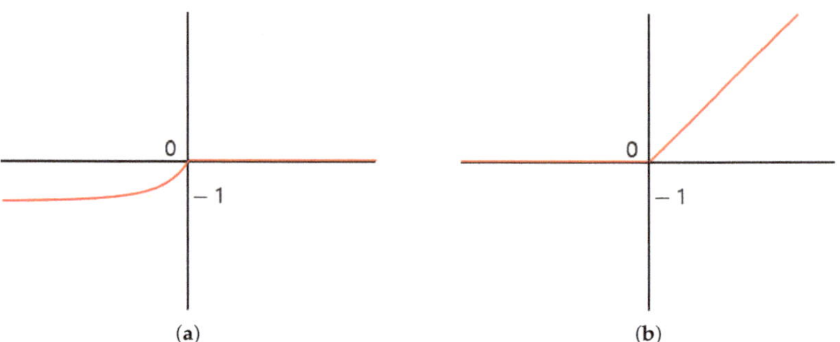

(a)   (b)

Figure 5. The separated ELU activation function graph. (**a**) is the negative region and (**b**) is the positive region.

Using the ELU activation function, the feature map was separated into a negative area and a positive area. In Equation (3), $ELU_n(x)$ is the negative region of the ELU when the input is x, and the positive region is treated as 0. Because a is set to 1 and converges to -1, it can be confirmed that the texture information with a small amount of data change is unified and displayed. In Equation (4), $ELU_p(x)$ is a positive region of the ELU when the input is x and is processed in the same way as the ReLU function. The proposed method separates the positive and negative regions to take advantage of the properties of the ELU.

Figure 6 shows the module of Figure 4, which has a feature map separated into positive and negative regions, as shown in Figure 5. Figure 6c is a feature map showing the separation in the ELU. This map is different from that when the ReLU activation function is applied in the separation of the ELU feature map (b) into a positive region, so the normalization range is different and ReLU shows different results. In the positive region, various textures and boundary information are extracted according to the amount of data change in the image. These features are affected by color and brightness. The texture information of the salient object of lighting is extracted from the features of the non-object area, and the shadow features of the non-object area are extracted in the same way as the salient object. Figure 6d shows the results of dividing the saliency map into negative and positive regions, where a converges to -1.

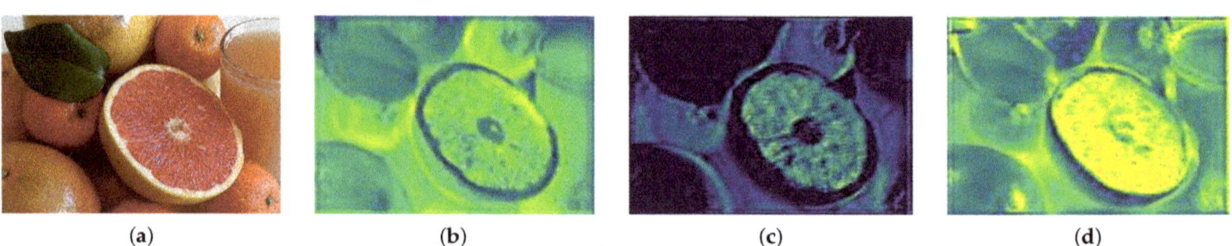

(a)   (b)   (c)   (d)

Figure 6. The ELU activation function and separated results. (**a**) is the input image and (**b**) is the ELU activation function results, (**c**) is the positive region, (**d**) is the negative region.

3.1.1. Extraction of Attention Region from Each Region

Because the SOD needs to detect the area of the salient object, not only the boundary information but also the texture information of the object to be detected is important. Various texture information can be obtained by outputting features through an ELU function with a wide range of feature expressions. However, environmental conditions, such as light reflection, affect color and brightness, and texture information is extracted based on such conditions, resulting in a loss area. Although the loss can be minimized by unifying the various texture information through an emphasis technique, a region with texture information similar to that of a salient object is emphasized, resulting in false detection. Therefore, the proposed method does not use the ELU function as it is and proceeds with the enhancement technique by separating the saliency map into a positive region representing various texture information and a negative region containing mainly unified texture information and boundary information with a large amount of change.

Areas of objects and backgrounds in the image are separated based on the boundary information. Spatial attention can be used for positive areas that have various textures, and boundary information can be used to unify the information of various textures and emphasize the salient object area. The feature separated into the negative region is different from that in the positive region, such that the feature converges to -1 and outputs unified texture information and boundary information with a large amount of data change. If these features are utilized without emphasis, then the area of the salient object can be suppressed as a non-object area. Spatial attention can be used to emphasize only the salient object area based on the boundary information to suppress the non-object area emphasized in the positive area.

The proposed method performs spatial attention in the positive and negative regions using Equations (5) and (6), respectively, and is a transformation of spatial attention. In the case of using the average and maximum pooling, a representative value is outputted in the separate positive region, resulting in a loss of various texture information. Because the sigmoid is normalized to a value between 0 and 1, the amount of data change is altered, so the boundary information is lost. Therefore, in the proposed method, spatial attention through a convolution and element-wise product is used without pooling and sigmoid.

$$A_p(x) = f_{conv}^{1\times1}(ELU_p(f_{conv}^{3\times3}(x))) \bigotimes ELU(f_{conv}^{3\times3}(x)) \tag{5}$$

$$A_n(x) = f_{conv}^{1\times1}(ELU_n(f_{conv}^{3\times3}(x))) \bigotimes ELU(f_{conv}^{3\times3}(x)). \tag{6}$$

In Equations (5) and (6), $A_p(x)$ and $A_n(x)$ represent the positive and negative regions of the input x; $f_{conv}^{1\times1}$ and $f_{conv}^{3\times3}$ are the 1×1 and 3×3 convolutions, respectively; the $\bigotimes$ is an element-wise product; ELU_p and ELU_n are the positive and negative regions separated from the ELU in Equations (3) and (4), respectively; and a was set to 1 in the ELU. The results of such a modified spatial attention equation are shown in Figure 7.

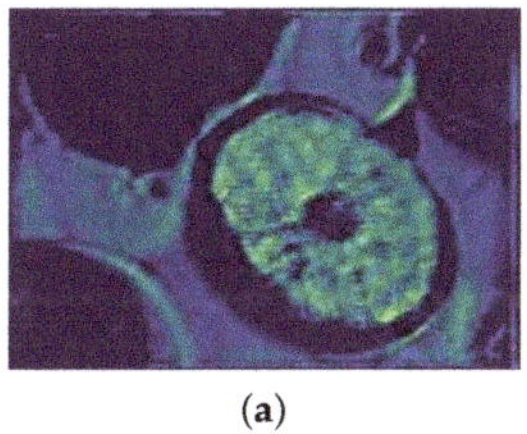 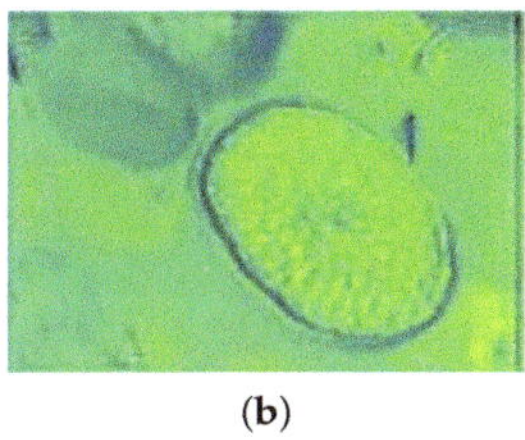 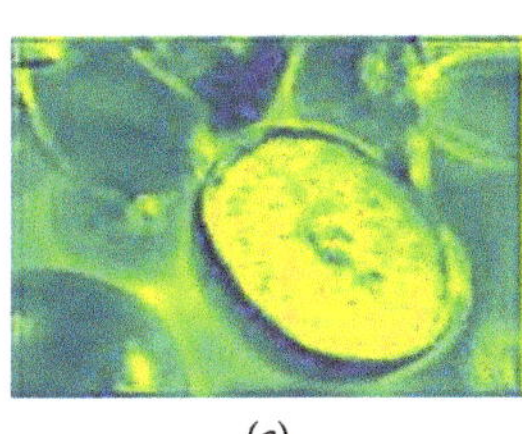 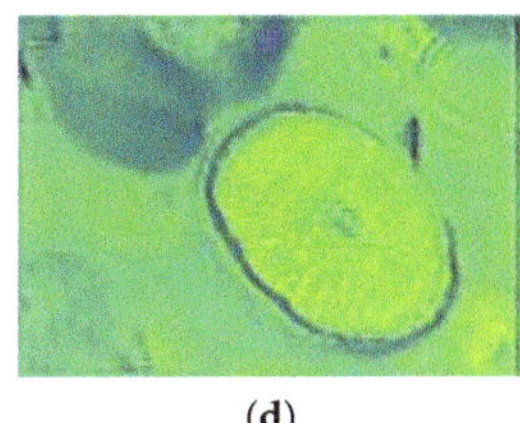

(a)   (b)   (c)   (d)

Figure 7. Results of applying attention in each area. (**a**) is the positive region and (**b**) is the positive attention results, (**c**) is the negative region, (**d**) is the negative attention results.

Figure 7 shows a feature map of each region separated from the ELU and a feature map emphasizing the positive and negative regions. Spatial attention was performed on the salient object to emphasize the detailed information from Figure 7a. Negative areas also emphasized the contour and texture information via spatial attention. Unlike the result of applying the ELU activation function, the area of the salient object was emphasized. The emphasis of the positive area emphasizes the texture information of the entire image, and the negative area suppresses the non-object area and emphasizes the contour and texture information.

3.1.2. Combination of Attention Positive and Negative Region

The texture information of the salient object was emphasized based on the boundary information between the salient object and the background, and the positive area where the shadow area was emphasized and the negative area where the non-object area was suppressed were combined with the element-wise sum. As a result of emphasizing the negative region through the combination, the shadow region, which is a non-object region, is suppressed, and the feature that the region of the salient object is emphasized through the emphasis on the positive and negative regions is obtained. Equation (7) shows the combination of the emphasized feature maps.

$$A_{element}(x) = A_p(x) + A_n(x). \tag{7}$$

The combination of the feature maps emphasized for input x is represented by $A_{element}(x)$. A_p represents a feature map with emphasized positive areas, and A_n represents a feature map with emphasized negative areas.

Figure 8 shows the result of a combination of feature maps and shows an emphasis on each area. Bases on the spatial attention results of the positive region, which contains various detailed information, we have element-wise-summed the positive and negative feature maps emphasized to suppress the shadow features that are non-object regions. By combining the results of emphasizing the unified texture information in (c) and the result of emphasizing various texture information in (b), the non-object area is suppressed, as shown in Figure 8d. The boundary and texture information of the salient objects is also emphasized.

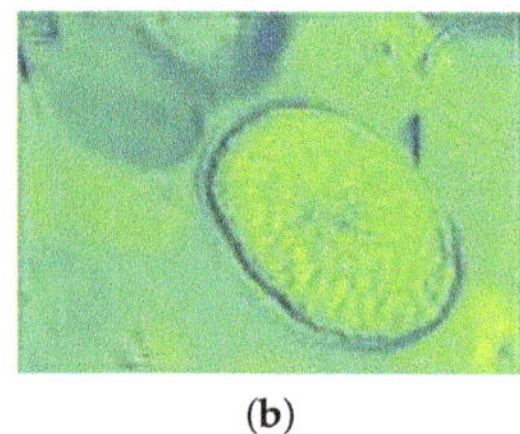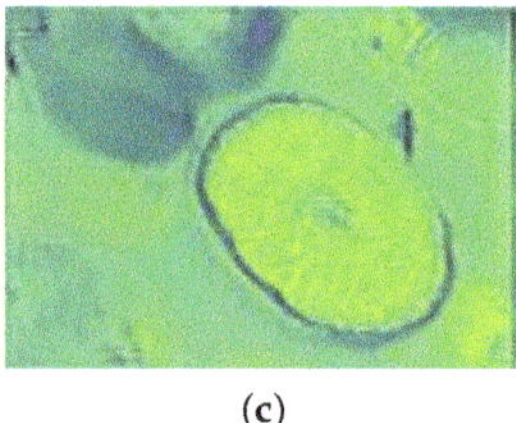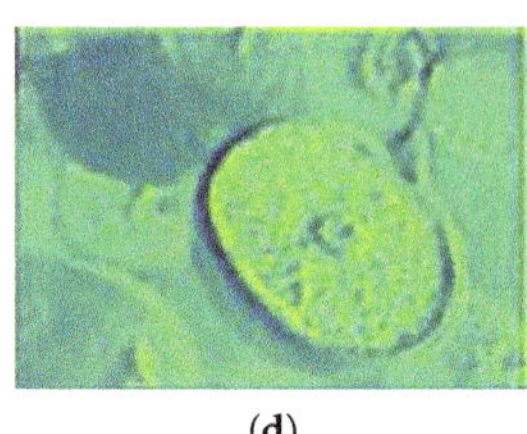

(a)   (b)   (c)   (d)

Figure 8. Sum of the elements of the attention result for each region. (**a**) is hte input image and (**b**) is the positive attention result, (**c**) is the negative attention result, (**d**) is the element-wise sum result.

In the deep learning process, when the distance between the input and output increases as shown above, the slope value is saturated with a large or small value in the backpropagation process in which the weight is transmitted between layers when learning the network, resulting in an ineffective learning, a slope that slows learning, and loss problem. To prevent these problems, a structure that learns the difference between the input value and output value was constructed by applying skip connection after combining the highlighted feature maps.

Figure 9 compares the results of the highlighted feature map combination with the results of applying skip connection to the combined feature map. The problems of weights being propagated directly from the output to the input and the slope disappearing in deep structures when learning the network through a structure that applies skip connection are avoided. The final NRA result is shown in Figure 9c, in which the non-object area is suppressed and the detailed information of the object is emphasized. Equation (8) is the formula for NRA.

$$A_{NA}(x) = A_{element}(x) + x. \tag{8}$$

 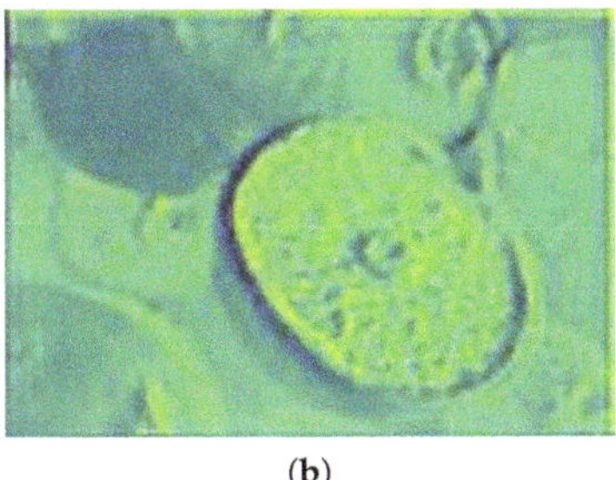 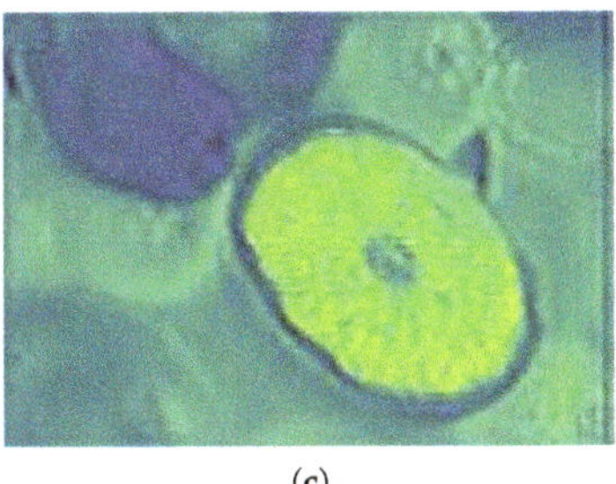

(a) (b) (c)

Figure 9. The skip-connection after element-wise sum. (**a**) is the input image and (**b**) is the element-wise sum result, (**c**) is the skip-connection result.

The result of NRA is represented by $A_{NA}(x)$ on the input x, and $A_{element}(x)$ and the input are element-wise-summed to utilize skip connection. $A_{element}(x)$ is a combination of the positive and negative feature maps emphasized in Equation (7).

Figure 10 is a comparison of the feature map to which the ELU activation function is applied when performing the convolution operation on the input image and the result of applying the proposed NRA to the result emphasized by these feature maps. When spatial attention is applied to the feature map extracted using the ELU activation function, the texture information of the salient object is extracted in the non-object area by lighting, as shown in Figure 10c. The proposed NRA method is separated into positive and negative regions by the ELU activation function, and then spatial attention is applied to each region to suppress non-object regions and emphasize the detailed information of salient objects.

 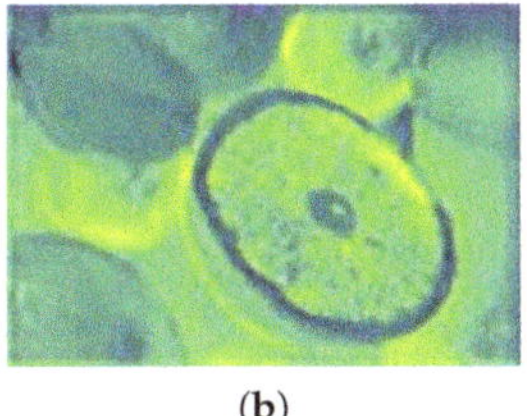 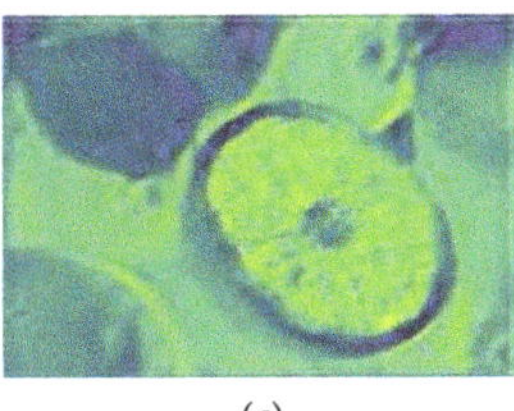 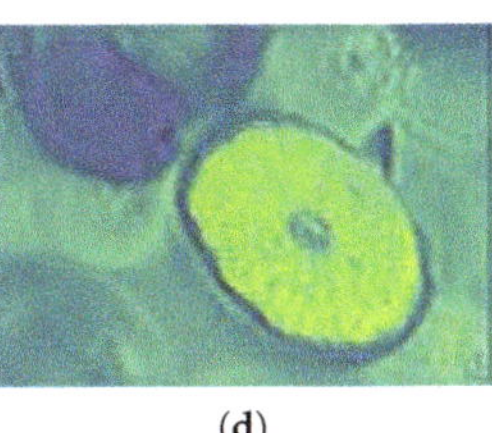

(a) (b) (c) (d)

Figure 10. The Neg-Region Attention module results. (**a**) is the input image and (**b**) is the ELU activation function result, (**c**) is the attention result using the ELU activation function, (**d**) is the Neg-Region Attention module result.

3.2. Decoder for Extending Extracted Features

Because the decoding process is a stage where feature extraction and compressed features are expanded to the size of the input image and restored at the encoder stage, feature loss occurs during the expansion process. In this process, information on the correlation of surrounding pixels is lost. The proposed method uses concatenation to utilize the features of each stage extracted from the encoder at each stage of the decoder to prevent feature loss.

Unlike skip connection, which adds a feature map, concatenation simply follows. The number of feature maps is increasing, following the feature maps of the same size. Equation (9) is the formula for concatenation:

$$Concat(w, h) = [A_{NA}(x)_{w \times h}; D(y)_{w \times h}]. \tag{9}$$

$A_{NA}(x)$ is the result of performing NRA on the input x, and $w \times h$ indicates the size of the resulting feature map. $D(y)$ is the result of the inverse convolution of the input y with the decoder, and $w \times h$ indicates the size of the result feature map. $Concat(w, h)$ is the result of the concatenation of a feature map of the same size as $w \times h$.

3.3. Residual Block in the Process of Concatenation Operation

When the NRA result of the encoder step is directly concatenated to the decoder, only the salient region is expressed because the highlighted feature is not refined. A residual block was used to generate a saliency map close to the ground truth through the feature refinement process of this highlighted region. Figure 11 shows the results of the concatenation of the NRA without feature refinement.

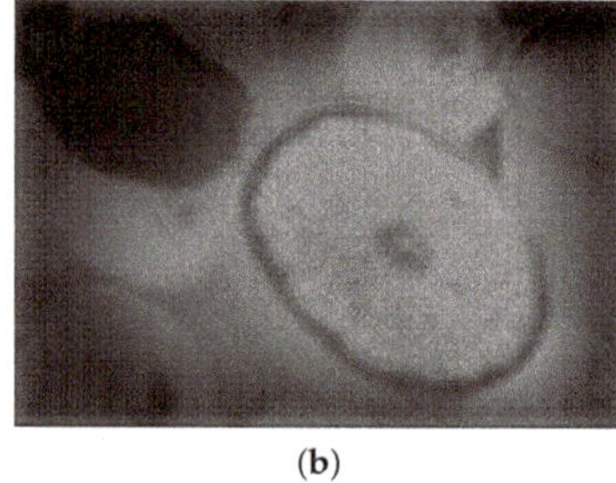
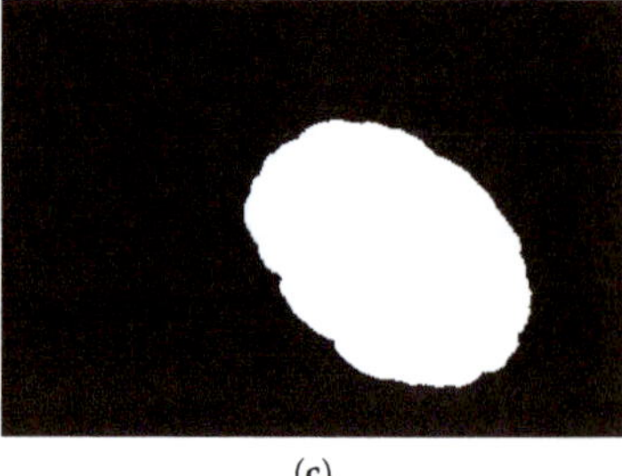

(a)   (b)   (c)

Figure 11. Prediction result of the model using only the attention module. (**a**) is the input image and (**b**) is using only the attention module, (**c**) is groundtruth.

The results show only the emphasized regions where the features have not been purified. The features in the residual block were reconstructed to improve the quality of the saliency maps and generate them closer to the ground truth.

The existing residual block consists of two (3 × 3) convolutions and two ReLU activation functions. Such a structure does not take advantage of the negative region features using the ReLU activation function. Because the proposed method also utilizes features in the negative region, we used the ELU activation function to prevent the loss of features highlighted by the residual block. The information transmitted by skip connection in the proposed method emphasizes the salient area. When the ELU activation function was applied after receiving the emphasized feature information, the features in the emphasized negative region were normalized and feature loss occurred. Therefore, unlike the conventional method, the result was outputted without using the activation function after skip connection. Equation (10) is an equation of the residual block by the proposed method.

$$f_{res}(x) = f_{conv}^{3\times3}(ELU(f_{conv}^{3\times3}(x))) + x. \tag{10}$$

$f_{res}(x)$ is the result of the residual block in the proposed way of the input x, and $f_{conv}^{3\times3}$ is the 3 × 3 convolution. After extracting the features in the 3 × 3 convolution as the input, the features were enabled as an ELU function and the features were extracted again via a 3 × 3 convolution. Then, the input was added to the element-wise sum, and the skip-connection structure was used. Figure 12 shows the structure of the residual block.

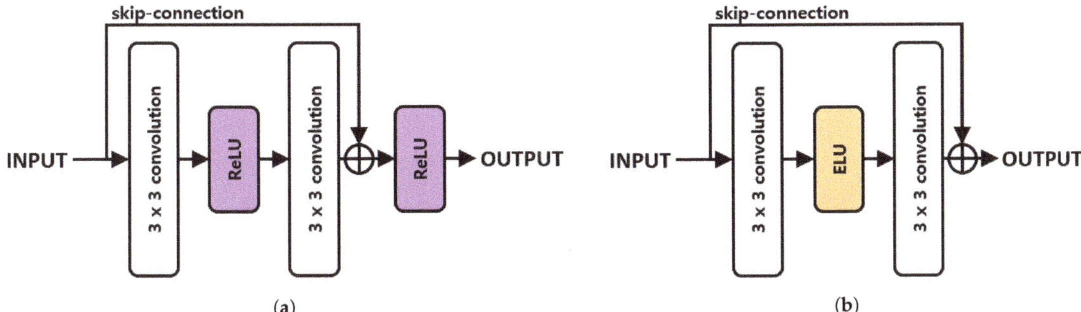

Figure 12. Composition of residual blocks. (**a**) is existing residual block and (**b**) is residual block.

The features of the structure of these proposed residual blocks were purified. This method also avoids the problem of slope disappearance in the skip connection, reduces loss, and generates a saliency map close to the ground truth. Figure 13 shows the final SOD results.

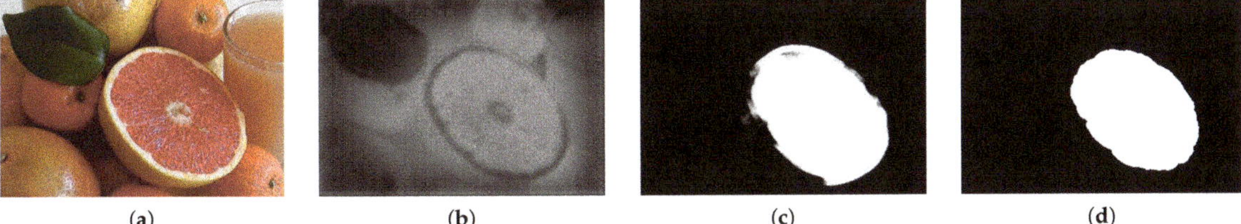

Figure 13. Results of using the proposed residual block. (**a**) is the input image and (**b**) is using only the attention module, (**c**) is using the proposed residual block. (**d**) is ground truth.

4. Experimental Environment

4.1. Environment and Dataset

In this paper, MSRA10K (10,000) Salient Object Database was used as the training dataset, and ECSSD (1000) was used as the validation dataset. The MSRA10K dataset was trained with a total of 80,000 images using rotation ($0°, 90°, 180°, 270°$) and flipping. During the learning process, the verification dataset was used to compare the degree of convergence between the other datasets and to confirm the overfitting phenomenon in which the accuracy of only the training dataset increases. Experimental datasets were compared and analyzed using ECSSD, HKU-IS (4447), and DUT-OMRON (5182). Adaptive Moment Estimation (Adam) optimizer [14] was used as the optimization function. The initial learning rate was set to 0.0001, the batch size was set to 48, and the epoch was set to 80. GPU was trained and experimented using NVIDIA GeForce RTX 3090 24 GB. The learning rate was set through a number of experiments, and if the initial learning rate exceeds 0.0001, the learning convergence speed is fast and learning is not performed. It was adjusted through the learning rate scheduler according to the learning convergence speed. The size of the input image was set to 224×224, which is the most used for comparison with existing methods.

4.2. Loss Function

The proposed method uses the L2 loss function. It is used when there is only one type of object to be detected, such as SOD, or when only correct and incorrect answers are identified. When there are various types of objects to be detected, such as object recognition, loss is calculated for each type of object using cross-entropy. The L2 loss function calculates the error by comparing the saliency map predicted with the mean squared error (MSE) and the squared error of the ground truth. When calculating the error, there are outliers in which the value rapidly changes. MSE is greatly affected by these outliers, and the weights are adjusted accordingly. Equation (11) is the equation of the L2 loss function.

$$f_{Loss}(x) = \sum_{i=1}^{n}(y_i - \hat{y}_i)^2. \tag{11}$$

y represents the ground truth, and $\hat{y}$ represents the saliency map predicted by the proposed method. The result of summing the difference between the ground truth and the predicted saliency map is f_{Loss}, and reducing this value entails the weight adjustment of the learning process.

4.3. Evaluation Index

To compare and analyze the experimental results, the mean absolute error (MAE) [15], precision, recall, and F-measure [16] were used as evaluation indicators. Equation (12) is an expression of the evaluation index MAE.

$$MAE = \frac{1}{W \times H} \sum_{x=1}^{W} \sum_{y=1}^{H} |S(x,y) - G(x,y)|. \tag{12}$$

$S(x,y)$ represents the predicted saliency map, and $G(x,y)$ represents the ground truth. $W \times H$ represents the size of the image. MAE is an error rate that represents the absolute error value between the ground truth and the predicted result, so the lower the value, the better the performance. Equation (13) is the expression of the precision and recall, and Equation (14) is the F-measure.

$$precision = \frac{TP}{TP + FP}, recall = \frac{TP}{TP + FN} \tag{13}$$

$$F_\beta = (1 + \beta^2)\frac{precision \times recall}{(\beta^2 \times precision) + recall}, \beta^2 = 0.3. \tag{14}$$

The precision, recall and F-measure are values that indicate accuracy, and a value higher than 0 indicates better performance. Precision and recall are calculated based on whether the ground truth and the pixel value at the same location in the saliency map are the same.

S-measure (Structure-measure) [17] simultaneously evaluates object-aware structural similarity and region-aware between a predicted saliency map and a ground truth. S-measure is Equation (15).

$$S = \alpha \times S_o + (1 - \alpha) \times S_r, \tag{15}$$

where S_o is the object-aware structural similarity measure and S_r is the region-aware structural similarity measure. S-measure is a combination of two evaluations, and $\alpha = 0.5$ was used.

E-measure(Enhanced-alignment measure) [18] combines the image-level mean value and local pixel values into one. Jointly capture local pixel matching information and image level statistics. E-measure is defined by Equation (16).

$$Q_{FM} = \frac{1}{W \times H} \sum_{x=1}^{W} \sum_{y=1}^{H} \phi_{FM}(x,y), \tag{16}$$

where h and w are height and width of the map.

5. Experimental Results

5.1. Learning Convergence Experiment

Figure 14 is a comparative analysis graph of the loss convergence and precision convergence in the learning process of the conventional FCN method using the ReLU activation function and the FCN method using the ELU activation function. The loss converged faster in the learning process of the FCN (ELU) method using the ELU activation function than the conventional FCN (ReLU). The findings confirmed that using ELU instead of ReLU resulted in a faster convergence.

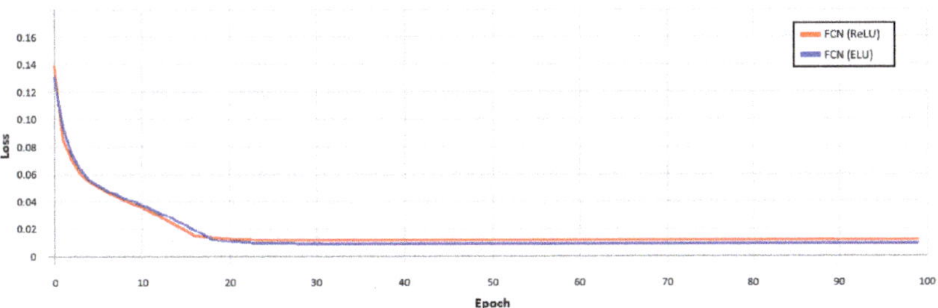

Figure 14. Loss convergence graph for each activation function in the FCN model. Red used the ReLU function, and blue used the ELU function.

5.2. Training MSRA10K Dataset

Figure 15 is a comparison image of the proposed method and other existing methods and experimental results. The experimental results were compared with ECSSD (4 images), HKU-IS (4 images), and DUT-OMRON (4 images) as examples, and the comparison methods were ELD (Encoded Low level Distance map) [19], DS (Deep Saliency) [20], DCL (Deep Contrast Learning) [21], Amulet [22], DGRL (Detect Globally Refine Locally) [23] and AFNet [24], which are all deep learning-based methods. All of these methods use a backbone network, whereas the proposed method was trained without a backbone network. Other detectors use the backbone network for the encoding process, which is the feature extraction step, so various features can be easily extracted. However, the proposed method improves the performance of the feature extraction step by applying the NRA without a backbone to the encoding step and minimizes the loss of texture and contour information. ELD and DS were greatly affected by color and brightness, and detected the surroundings of the target object. DCL mainly detected a single object and detects other objects together. It was vulnerable to multi-object detection and showed a result that is sensitive to contour information. The Amulet detects the area of the target object, but if the input image is complex, the surroundings are also detected. In some of the result images, a background area other than the surrounding area was also detected. DGRL showed a clear detection results, but loss occurred in the detailed part and the surrounding area was detected together. AFNet showed the best performance compared to the previous methods and the area and contour of the object were preserved. If there were multiple small objects, some detection fails, and if the background was complex, the surroundings were detected together. The proposed method showed excellent detection performance for small objects and detected large and multiple objects well. As with the existing methods, when the background is complex, the surroundings were detected together, but false detection was reduced. When compared to existing methods by learning without a backbone network, it showed excellent performance, and the performance of the attention module using ELU was also proven.

Tables 1 and 2 are comparison tables for the evaluation of the proposed method and other deep learning-based algorithms. The number of parameters of the proposed method is less than the average and shows excellent performance without using the backbone network. The MAE, mean F-measure ($\beta^2 = 0.3$), S-measure and E-measure were measured for the datasets ECSSD, HKU-IS, and DUT-OMRON. The best performance numbers are expressed in red, the second is blue, and the third is green. The proposed method showed superior performance in the MAE, mean F-measure, S-measure and E-measure compared to the conventional methods using the backbone network.

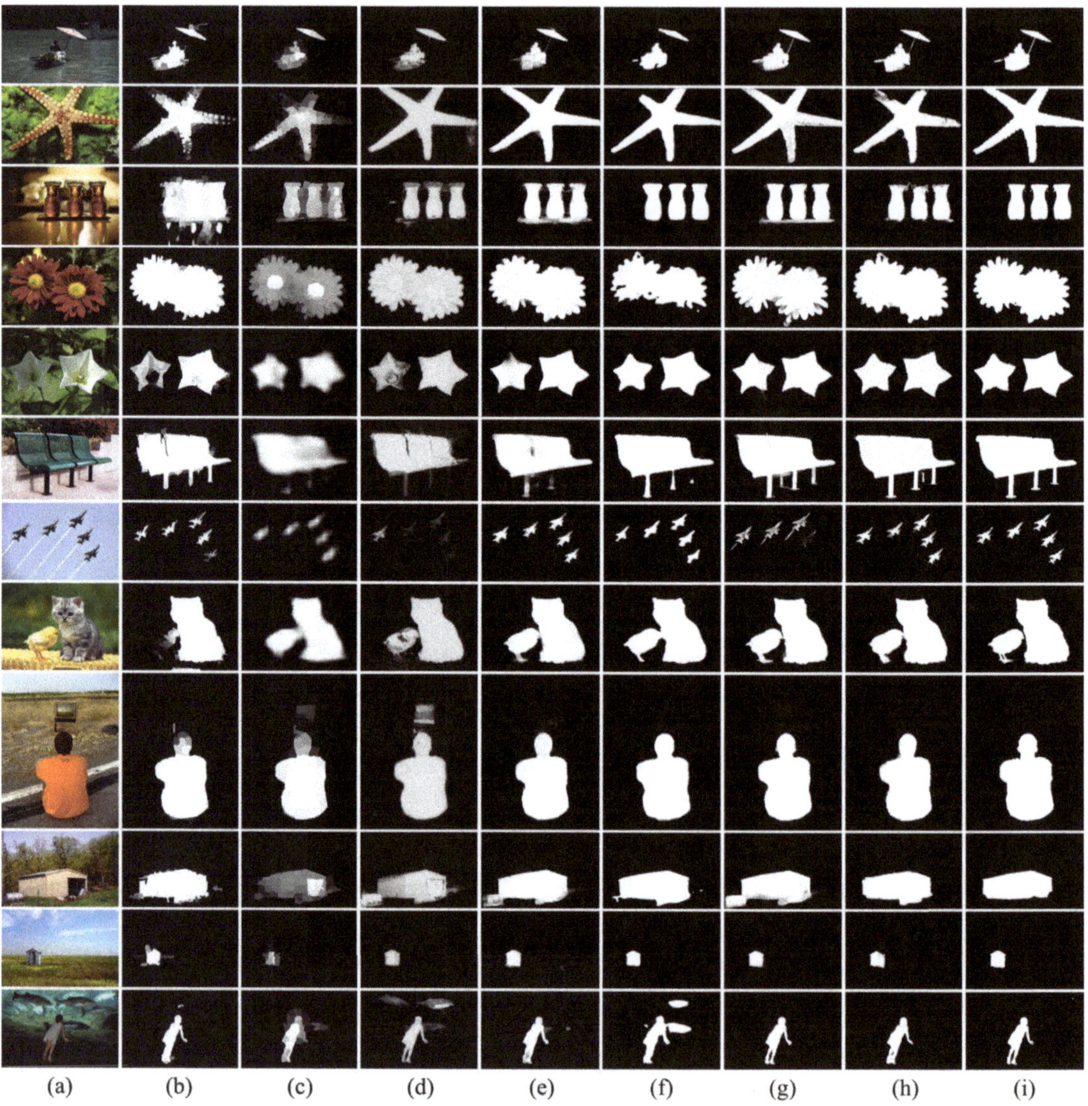

Figure 15. Comparison of experimental results of the proposed method and other deep learning methods. (**a**) Input image, (**b**) ELD, (**c**) DS, (**d**) DCL, (**e**) Amulet, (**f**) DGRL, (**g**) AFNet, (**h**) NRA-Net (proposed method), (**i**) Groundtruth.

Figures 16 and 17 are the comparison diagrams of the precision and recall curves and F-measure curves of the proposed method and other deep learning algorithms. The proposed method shows excellent performance in both indicators. In this curve, the minimum recall value can be used as an indicator of robustness, where the higher the precision value of the minimum recall value, the more accurate the salient object prediction, which means that the background and foreground are well separated.

Table 1. The experimental results of the ECSSD dataset.

Method	Number of Parameters	ECSSD Dataset			
		MAE	F-Measure	S-Measure	E-Measure
ELD	28.37 M	0.0796	0.8102	0.838	0.881
DS	134.27 M	0.1216	0.8255	0.820	0.874
DCL	66.25 M	0.1495	0.8293	0.863	0.885
Amulet	33.16 M	0.0588	0.8684	0.893	0.901
DGRL	126.35 M	0.0419	0.9063	0.903	0.917
AFNet	21.08 M	0.0422	0.9085	0.913	0.918
NRA-Net	56.42 M	0.0489	0.9126	0.898	0.907

Table 2. The experimental results of HKU-IS and DUT-OMRON datasets.

Method	HKU-IS Dataset				DUT-OMRON Dataset			
	MAE	F-Measure	S-Measure	E-Measure	MAE	F-Measure	S-Measure	E-Measure
ELD	0.0741	0.7694	0.820	0.880	0.0923	0.6110	0.750	0.775
DS	0.0780	0.7851	0.852	0.889	0.1204	0.6031	0.750	0.761
DCL	0.1359	0.8533	0.860	0.913	0.0971	0.6837	0.764	0.801
Amulet	0.0521	0.8542	0.883	0.910	0.0977	0.6474	0.780	0.778
DGRL	0.0363	0.8882	0.894	0.943	0.0618	0.7332	0.806	0.848
AFNet	0.0364	0.8904	0.905	0.942	0.0574	0.7382	0.826	0.853
NRA-Net	0.0428	0.8924	0.894	0.919	0.0706	0.7449	0.811	0.836

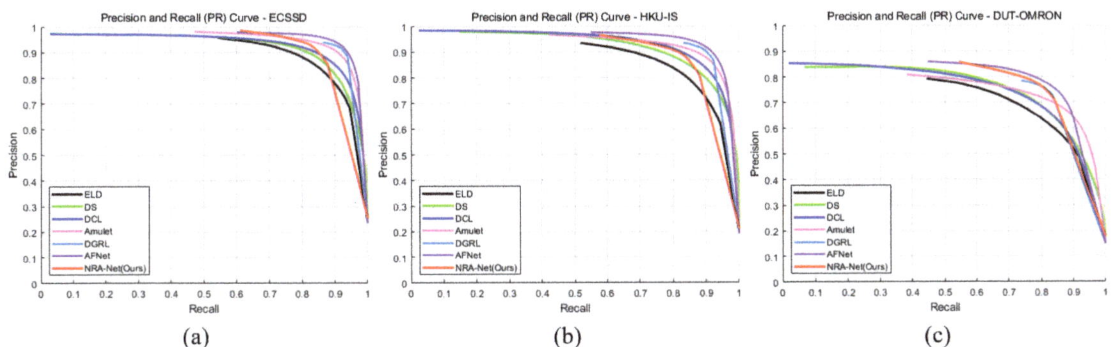

Figure 16. Comparison with other methods of precision and recall curve in each dataset. (**a**) PR curve of ECSSD dataset and (**b**) HKU-IS dataset, (**c**) DUT-OMRON dataset.

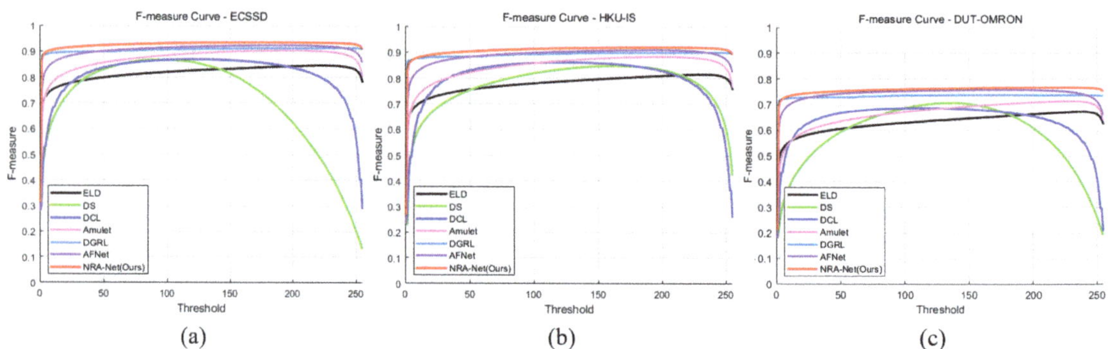

Figure 17. Comparison with other methods of F-measure curve in each dataset. (**a**) F-measure curve of ECSSD dataset and (**b**) HKU-IS dataset, (**c**) DUT-OMRON dataset.

Figure 18 shows the result of detection of protruding objects for the motocross-jump video dataset. When performing gaze tracking, gaze detection may fail if a dramatic scene change occurs as in the video above. When a dramatic scene change occurs, there are factors such as the position of the object or the rotation of the camera. The detection of salient objects compensates for this problem and predicts objects in which human gaze is concentrated even with scene changes. This prediction result can supplement information on which object is mainly focused on gaze detection.

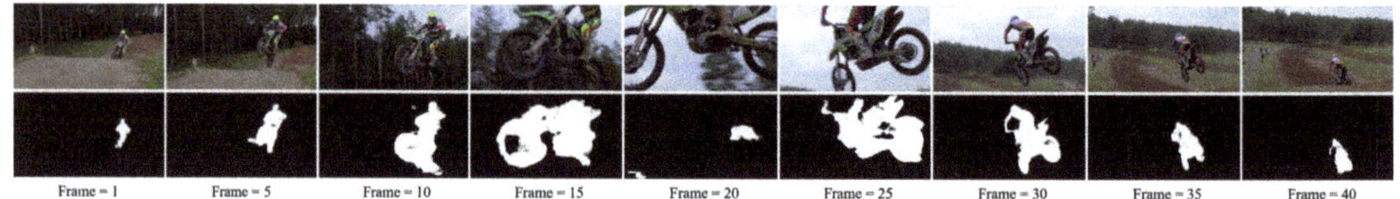

Figure 18. Salient object detection results for motocross-jump video of MIT300 dataset. (**top**) Input video frame and (**bottom**) results of saliency map.

6. Conclusions

In this paper, we propose a deep learning-based method to detect salient objects in images in various environments. Existing deep learning-based methods proceed with an autoencoder structure, and feature loss occurs in the encoding process for extracting and compressing features and the decoding process for expanding and restoring the extracted features. Due to this feature loss, a background other than an object is detected, or an object with complex internal information fails to be detected. Most of the existing methods require a backbone network, and improve the network based on the backbone. However, feature extraction is limited, and it is difficult to extract specialized features for any object. The efficiency of the proposed method to reduce the feature loss in the autoencoder structure was studied. After separating the positive and negative regions through the NRA proposed in the encoding process of the autoencoder structure, the enhancement technique was performed. Positive numbers represent various textures and boundary information, and negative numbers mainly represent boundary information with a large amount of change in data. To utilize this characteristic information, spatial attention technique was performed in each domain. The proposed method prevents feature loss and creates a final saliency map by reconstructing features with a modified residual block. Existing deep learning methods extract features using a backbone network, but the proposed method achieves an excellent performance by extracting features using the attention technique without a backbone network.

Author Contributions: Conceptualization, H.K.; Methodology, H.K.; Software, H.K.; Investigation, H.K., and S.K.; Writing—Original Draft Preparation, H.K.; Writing—Review & Editing, S.L., and S.K.; Supervision, S.L.; Project Administration, S.K. All authors have read and agreed to the published version of the manuscript.

Funding: This research received no external funding.

Institutional Review Board Statement: Not applicable.

Informed Consent Statement: Not applicable.

Data Availability Statement: Code and dataset will be made available on request to the first author's email with appropriate justification. The public site for each dataset is as follows. MSRA10K: https://mmcheng.net/msra10k/; ECSSD: http://www.cse.cuhk.edu.hk/leojia/projects/hsaliency/dataset.html; DUT-OMRON: http://saliencydetection.net/dut-omron/; HKU-IS: https://sites.google.com/site/ligb86/hkuis.

Acknowledgments: This research was supported by Basic Science Research Program through the National Research Foundation of Korea (NRF) funded by the Ministry of Science and ICT (2020R1F1A1069079).

Conflicts of Interest: The authors declare no conflict of interest.

References

1. Long, J.; Shelhamer, E.; Darrell, T. Fully convolutional networks for semantic segmentation. In Proceedings of the IEEE Conference on Computer Vision and Pattern Recognition, Boston, MA, USA, 7–12 June 2015; pp. 3431–3440.
2. Liu, Z.; Zhang, X.; Luo, S.; Le Meur, O. Superpixel-based spatiotemporal saliency detection. *IEEE Trans. Circuits Syst. Video Technol.* **2014**, *24*, 1522–1540. [CrossRef]
3. Li, J.; Liu, Z.; Zhang, X.; Le Meur, O.; Shen, L. Spatiotemporal saliency detection based on superpixel-level trajectory. *Signal Process. Image Commun.* **2015**, *38*, 100–114. [CrossRef]
4. Achanta, R.; Hemami, S.; Estrada, F.; Susstrunk, S. Frequency-tuned salient region detection. In Proceedings of the 2009 IEEE Conference on Computer Vision and Pattern Recognition, Miami, FL, USA, 20–25 June 2009; pp. 1597–1604.
5. Badrinarayanan, V.; Kendall, A.; SegNet, R.C. A deep convolutional encoder-decoder architecture for image segmentation. *IEEE Trans. Pattern Anal. Mach. Intell.* **2017**, *39*, 2481–2495. [CrossRef] [PubMed]
6. Jiang, X.; Gao, Y.; Fang, Z.; Wang, P.; Huang, B. An end-to-end human segmentation by region proposed fully convolutional network. *IEEE Access* **2019**, *7*, 16395–16405. [CrossRef]
7. Meng, F.; Guo, L.; Wu, Q.; Li, H. A new deep segmentation quality assessment network for refining bounding box based segmentation. *IEEE Access* **2019**, *7*, 59514–59523. [CrossRef]
8. Wang, L.; Wang, L.; Lu, H.; Zhang, P.; Ruan, X. Saliency detection with recurrent fully convolutional networks. In Proceedings of the European Conference on Computer Vision, Amsterdam, The Netherlands, 8–16 October 2016; pp. 825–841.
9. Han, L.; Li, X.; Dong, Y. Convolutional edge constraint-based U-net for salient object detection. *IEEE Access* **2019**, *7*, 48890–48900. [CrossRef]
10. Baldi, P. Autoencoders, unsupervised learning, and deep architectures. In Proceedings of the ICML Workshop on Unsupervised and Transfer Learning, Bellevue, WA, USA, 2 July 2012; pp. 37–49.
11. Huang, G.B.; Zhu, Q.Y.; Siew, C.K. Extreme learning machine: A new learning scheme of feedforward neural networks. In Proceedings of the 2004 IEEE International Joint Conference on Neural Networks, Budapest, Hungary, 25–29 July 2004; pp. 985–990.
12. Agarap, A.F. Deep learning using rectified linear units (relu). *arXiv* **2018**, arXiv:1803.08375.
13. Clevert, D.A.; Unterthiner, T.; Hochreiter, S. Fast and accurate deep network learning by exponential linear units (elus). *arXiv* **2015**, arXiv:1511.07289.
14. Kingma, D.P.; Ba, J. Adam: A method for stochastic optimization. *arXiv* **2014**, arXiv:1412.6980.
15. Chai, T.; Draxler, R.R. Root mean square error (RMSE) or mean absolute error (MAE)?—Arguments against avoiding RMSE in the literature. *Geosci. Model Dev.* **2014**, *7*, 1247–1250. [CrossRef]
16. Powers, D.M. Evaluation: From precision, recall and F-measure to ROC, informedness, markedness and correlation. *arXiv* **2011**, arXiv:2010.16061.
17. Fan, D.P.; Gong, C.; Cao, Y.; Ren, B.; Cheng, M.M.; Borji, A. Enhanced-alignment measure for binary foreground map evaluation. *arXiv* **2018**, arXiv:1805.10421.
18. Fan, D.P.; Cheng, M.M.; Liu, Y.; Li, T.; Borji, A. Structure-measure: A new way to evaluate foreground maps. In Proceedings of the IEEE International Conference on Computer Vision, Venice, Italy, 22–29 October 2017; pp. 4548–4557.
19. Lee, G.; Tai, Y.W.; Kim, J. Deep saliency with encoded low level distance map and high level features. In Proceedings of the IEEE Conference on Computer Vision and Pattern Recognition, Las Vegas, NV, USA, 27–30 June 2016; pp. 660–668.

20. Li, X.; Zhao, L.; Wei, L.; Yang, M.H.; Wu, F.; Zhuang, Y.; Wang, J. Deepsaliency: Multi-task deep neural network model for salient object detection. *IEEE Trans. Image Process.* **2016**, *25*, 3919–3930. [CrossRef] [PubMed]
21. Li, G.; Yu, Y. Deep contrast learning for salient object detection. In Proceedings of the IEEE Conference on Computer Vision and Pattern Recognition, Las Vegas, NV, USA, 27–30 June 2016; pp. 478–487.
22. Zhang, P.; Wang, D.; Lu, H.; Wang, H.; Ruan, X. Amulet: Aggregating multi-level convolutional features for salient object detection. In Proceedings of the IEEE International Conference on Computer Vision, Venice, Italy, 22–29 October 2017; pp. 202–211.
23. Wang, T.; Zhang, L.; Wang, S.; Lu, H.; Yang, G.; Ruan, X.; Borji, A. Detect globally, refine locally: A novel approach to saliency detection. In Proceedings of the IEEE Conference on Computer Vision and Pattern Recognition, Salt Lake City, UT, USA, 18–23 June 2018; pp. 3127–3135.
24. Feng, M.; Lu, H.; Ding, E. Attentive feedback network for boundary-aware salient object detection. In Proceedings of the IEEE/CVF Conference on Computer Vision and Pattern Recognition, Long Beach, CA, USA, 16–20 June 2019; pp. 1623–1632.

Article

Small Object Detection in Traffic Scenes Based on YOLO-MXANet

Xiaowei He *, Rao Cheng, Zhonglong Zheng and Zeji Wang

College of Mathematics and Computer Science, Zhejiang Normal University, Jinhua 321004, China; chengrao@zjnu.edu.cn (R.C.); zhonglong@zjnu.edu.cn (Z.Z.); wangzj@zjnu.edu.cn (Z.W.)
* Correspondence: jhhxw@zjnu.edu.cn

Abstract: In terms of small objects in traffic scenes, general object detection algorithms have low detection accuracy, high model complexity, and slow detection speed. To solve the above problems, an improved algorithm (named YOLO-MXANet) is proposed in this paper. Complete-Intersection over Union (CIoU) is utilized to improve loss function for promoting the positioning accuracy of the small object. In order to reduce the complexity of the model, we present a lightweight yet powerful backbone network (named SA-MobileNeXt) that incorporates channel and spatial attention. Our approach can extract expressive features more effectively by applying the Shuffle Channel and Spatial Attention (SCSA) module into the SandGlass Block (SGBlock) module while increasing the parameters by a small number. In addition, the data enhancement method combining Mosaic and Mixup is employed to improve the robustness of the training model. The Multi-scale Feature Enhancement Fusion (MFEF) network is proposed to fuse the extracted features better. In addition, the SiLU activation function is utilized to optimize the Convolution-Batchnorm-Leaky ReLU (CBL) module and the SGBlock module to accelerate the convergence of the model. The ablation experiments on the KITTI dataset show that each improved method is effective. The improved algorithm reduces the complexity and detection speed of the model while improving the object detection accuracy. The comparative experiments on the KITTY dataset and CCTSDB dataset with other algorithms show that our algorithm also has certain advantages.

Keywords: deep learning; computer vision; intelligence transportation; YOLOv3; lightweight

Citation: He, X.; Cheng, R.; Zheng, Z.; Wang, Z. Small Object Detection in Traffic Scenes Based on YOLO-MXANet. *Sensors* **2021**, *21*, 7422. https://doi.org/10.3390/s21217422

Academic Editors: KWONG Tak Wu Sam, Xu Long, Tiesong Zhao and Yun Zhang

Received: 9 October 2021
Accepted: 5 November 2021
Published: 8 November 2021

Publisher's Note: MDPI stays neutral with regard to jurisdictional claims in published maps and institutional affiliations.

Copyright: © 2021 by the authors. Licensee MDPI, Basel, Switzerland. This article is an open access article distributed under the terms and conditions of the Creative Commons Attribution (CC BY) license (https://creativecommons.org/licenses/by/4.0/).

1. Introduction

Object detection is an essential field of computer vision, and its task is to locate and classify objects with the variable number in an image. Object detection in traffic scenes is an essential part of driverless technology, which adopts image processing or deep learning to detect and identify vehicles, pedestrians, and traffic signs in traffic scenes to lay a good foundation for developing intelligent transportation. Object detection algorithms based on convolutional neural networks are mainly divided into two categories: one is the two-stage algorithms represented by RCNN series [1–3], and the other is the one-stage algorithms represented by SSD series [4,5] and YOLO series [6–8]. Object detection algorithms based on anchor-free [9–12] are developing rapidly in the one-stage algorithms. Two-stage algorithms depend on the proposals, and their detection speed is generally slow, in other words, their real-time performance cannot meet the demand of traffic scenes, even though its detection accuracy is constantly improving. The speed of one-stage algorithms based on regression is fast enough to satisfy the requirements of most tasks. However, there is still room for improvement in detection accuracy. At present, many scholars have applied general object detection algorithms to the traffic field. Que Luying et al. [13] proposed a lightweight pedestrian detection engine with a two-stage low-complexity detection network and adaptive region focusing technique, which not only reduced the computational complexity but also maintained sufficient detection accuracy. Yang Xiaoting et al. [14] proposed a novel scale-sensitive feature reassembly network

(SSNet) for pedestrian detection in road scenes. Ma Li et al. [15] studied and solved the problem that YOLOv3-tiny has a high missed detection rate for small-scale objects such as pedestrians in real-time detection; however, the accuracy of their algorithm cannot satisfy the requirements in actual scenes. Guo Fan et al. [16] proposed the traffic sign detection network (YOLOv3-A) based on an attention mechanism to solve the misdetection and omission of small objects. Liu Changyuan et al. [17] proposed the vehicle target detection network (YOLOV3-M2), which promoted the detection efficiency and enhanced the detection ability of small targets; however, it only detected a single class of targets.

In object detection, when the size of an object is small enough relative to the size of the original image, we usually consider the object as a small object. For small objects, some datasets have a clear definition. For example, CityPerson, the pedestrian dataset, defines objects less than 75 px high as small objects in the raw image with the size of 1024×2048. In the MS COCO dataset, small objects are the objects with pixels less than 32×32. In the traffic sign dataset, Zhu et al. [18] defined the objects whose width accounted for less than 20% of the whole image as small objects. The current general object detection algorithms have achieved a good detection effect for large and medium objects. However, because of the smaller coverage area, lower resolution, weaker feature expression ability, and little feature information of small objects, the above general object algorithms are not good at detecting small objects. Recently, many researchers have focused their attention on small object detection. Wang Hongfeng et al. [19] proposed a generative adversarial network (GAN) capable of image super-resolution and two-stage small object detection, which exhibited a better detection performance than mainstream methods. Bosquet Brais et al. [20] introduced STDnet-ST, an end-to-end spatiotemporal convolutional neural network for small object detection in video, which achieved state-of-the-art results for small objects. Lian Jing et al. [21] proposed a small object detection method in traffic scenes based on attention feature fusion, which improved the detection accuracy of small objects in traffic scenes. Zhang Can et al. [22] proposed a neural network for detecting small objects based on original Cascade RCNN, which performed better not only in small object detection but also in industrial applications.

In general, there are multiple objects, small objects, and occluded objects in complex traffic scenes [23], and it is difficult for traditional object detection methods to obtain better detection results. Therefore, it is necessary to study more algorithms of small object detection in traffic scenes. In recent years, the continuous improvement of network performance has led to the increase of model size and computation. With the popularity of mobile embedded devices, the deep neural network can be better applied to mobile devices only when the precision, parameter size, and inference speed are well balanced. Deploying well-performing algorithms on mobile devices is a trend. For example, Chen Rung-Ching et al. [24] developed a real-time monitoring system for home pets using raspberry pie. In our paper, to reduce the amount of calculation and the number of parameters while maintaining a better detection accuracy and speed, YOLO-MXANet is proposed by using the YOLOv3 algorithm for reference. CIoU [25] is adopted to improve the loss function, which makes the bounding box regress better. Although the lightweight network MobileNeXt [26] dramatically reduces the number of parameters and computational effort by using depthwise separable convolution, it has weaker feature extraction capability. To improve the feature extraction capability of MobileNeXt, the Shuffle Channel and Spatial Attention (SCSA) module is embedded into the SGBlock module, which can model long-distance dependency well to highlight the features of small objects. For the dataset, Mosaic [27] and Mixup [28] are used to enhance the robustness of the model. In the process of feature fusion, the Multi-scale Feature Enhancement Fusion (MFEF) network is proposed, in which an additional Down-top path is added, and the four-fold subsampled feature maps are fused to extract the features of small objects effectively. Meanwhile, the idea of CSPNet [29] is utilized to combine the convolution operation to reduce the number of network parameters and amount of calculation. In our work, the SiLU activation function is adopted into the Convolution-Batchnorm-SiLU (CBS) and A-SGBlock module to accelerate the convergence of the model. The experimental results on KITTI and CCTSDB datasets show that YOLO-MXANet in this paper has lower computational complexity and smaller number of parameters while improving the detection accuracy and speed. Compared with the original YOLOv3, the detection performance of the model is greatly enhanced while the speed is promoted, and the complexity of the model is lower. Compared with the latest algorithms, YOLO-MXANet also has certain advantages in detection accuracy and model complexity.

2. Baseline and YOLO-MXANet Algorithm

In this section, firstly, YOLOv3 baseline algorithm will be introduced in Section 2.1. Then, in Section 2.2, our proposed algorithm will be organized through five sub-sections. In Section 2.2.1, the backbone network SA-MobileNeXt will be presented and explained. In Section 2.2.2, Multi-scale Feature Enhancement Fusion Network will be elaborated. In Section 2.2.3, SiLU activation Function will be described in detail. In Section 2.2.4, the data enhancement approach utilized will be explained. In Section 2.2.5, the loss function used will be presented.

2.1. YOLOv3 Baseline Algorithm

YOLOv3 uses the Darknet-53 backbone network to extract features, which integrates the residual idea of ResNet [30]. The advantage of residual structure in the Darknet-53 (named Res_Unit) is that the accuracy can be improved by increasing the depth of network. The Res_Unit block uses the shortcut, which can alleviate the gradient diffusion problem caused by increasing the depth of the network. In addition, YOLOv3 utilizes three different feature layers extracted from the Darknet-53 backbone network to fuse and form three prediction layers for prediction. In the YOLOv3, the feature fusion idea of FPN [31] is adopted. That is, the semantic information and location information of three feature maps with different scales are combined by up-sampling and fusion to obtain feature maps containing rich information for detection. Therefore, YOLOv3 can effectively detect small objects. Specifically, the image with the size of 640×640 is sent into the network, and three feature maps with different scales (e.g., $80 \times 80, 40 \times 40, 20 \times 20$) are obtained. The 32-fold downsampled feature maps from the backbone network pass through five convolution layers. On the one hand, the feature maps generated are directly predicted after passing through one convolution layer. On the other hand, after a convolution layer and an upsampling operation, they are concatenated with the 16-fold downsampled feature maps from the backbone network to obtain the fusion feature maps. The operations of the 16-fold downsampled feature maps from the backbone network are similar to those of the 32-fold downsampled feature maps.

YOLOv3 employs the K-means algorithm to determine the size of the prior box. Although too many prior boxes can guarantee the effect, it greatly affects the detection speed of the model, so it gets nine prior boxes by clustering on the COCO dataset. The feature maps with a single scale utilize three prior boxes, and the corresponding relationship between prior boxes and feature maps with different scales is as follows. In detail, the 32-fold downsampled feature maps use the following three prior boxes: [(116,90); (159,198); (373,326)]; the 16-fold downsampled feature maps apply the following three prior boxes: [(30,61); (62,45); (59,119)]; the 8-fold downsampled feature maps employ the following three prior boxes: [(10,13); (16,30); (33,23)]. Large feature maps with small receptive fields are very sensitive to small-scale objects, so small prior boxes are selected. On the contrary, small feature maps with large receptive fields are suitable for detecting large objects, so large prior boxes are selected.

2.2. YOLO-MXANet Algorithm

2.2.1. SA-MobileNeXt

Although numerous residual modules can extract sufficient feature information, the Darknet-53 has numerous parameters and demands a large amount of computation. The deployment of convolutional neural networks on embedded devices is challenging due to the limited memory and computing resources. In order to balance the complexity, the detection speed, and the detection accuracy of the model, in this paper, we propose the lightweight feature enhancement backbone network called SA-MobileNeXt.

To reduce the number of parameters and computation amount of the network, we chose the lightweight backbone network called MobileNeXt as the basic model for improvement to simplify the network model. In recent years, artificially designed lightweight backbone networks have become popular, such as MobileNet Series (e.g., MobileNetv1 [32], MobileNetv2 [33], and MobileNetv3 [34]), ShuffleNet Families (e.g., ShuffleNetv1 [35] and ShuffleNetv2 [36]), and SqueezeNet [37]. The above manually designed backbone networks are built by stacking basic modules. In our work, firstly, the newly proposed lightweight backbone network-MobileNeXt [26] is utilized, which is made up of stacked SandGlass blocks (SGBlock), and its structure is shown on the left of Figure 1. Many studies have proved that the SGBlock is better than the Inverted Residual (IR) blocks in MobileNetv2 to preserve adequate feature information and promote gradient propagation. The specific structure of SGBlock is shown in the light blue box of Figure 2 (t represents the reduction rate of dimension, and s represents the stride). In detail, two depthwise convolutions are placed at the end of the block, and two pointwise convolutions are placed in the middle of the block. The point convolution can be used to encode the information of internal channels but cannot capture spatial information, and the depthwise convolution can learn more expressive spatial context information. It is worth noting that the first depthwise convolution and the last point convolution utilize the ReLU6 activation function; the first point convolution and the second depthwise convolution directly perform linear output to reduce information loss, and there is no identity mapping in the SGBlock when the input and output channels are different. Mathematically, let $F \in \Re^{D_f \times D_f \times M}$ be the input tensor, and $G \in \Re^{D_f \times D_f \times M}$ be the output tensor of the SGBlock, and the SGBlock can be formulated as follows:

$$\hat{G} = T_{1,p}T_{1,d}(F),$$
$$G = T_{2,d}T_{2,p}(\hat{G}) + F \tag{1}$$

where $T_{i,p}$ and $T_{i,d}$ are the i-th pointwise convolution and depthwise convolution, respectively. The depthwise separable convolution is used in the SGBlock. Compared with the standard convolution, the depthwise separable convolution includes the depthwise convolution and the point convolution. Assume that the size of the input feature maps is $D_f \times D_f \times M$, the size of the output feature maps is $D_f \times D_f \times N$, and the size of standard convolution kernel is $D_k \times D_k \times M$. The computational cost of the standard convolution is $D_k \cdot D_k \cdot M \cdot N \cdot D_f \cdot D_f$, and the computational cost of depthwise separable convolution is $D_k \cdot D_k \cdot M \cdot D_f \cdot D_f + M \cdot N \cdot D_f \cdot D_f$. From the above formulas, we can see that the calculation amount of depthwise separable convolution is much less than the calculation amount of standard convolution [32]. In order to enable the MobileNeXt to be used as the backbone network of YOLOv3, the original MobileNeXt is improved by removing the 7×7 average pooling layer and the fully connected layer to form the backbone network MobileNeXt used in this paper.

Figure 1. The structure of MobileNeXt and SA-MobileNeXt.

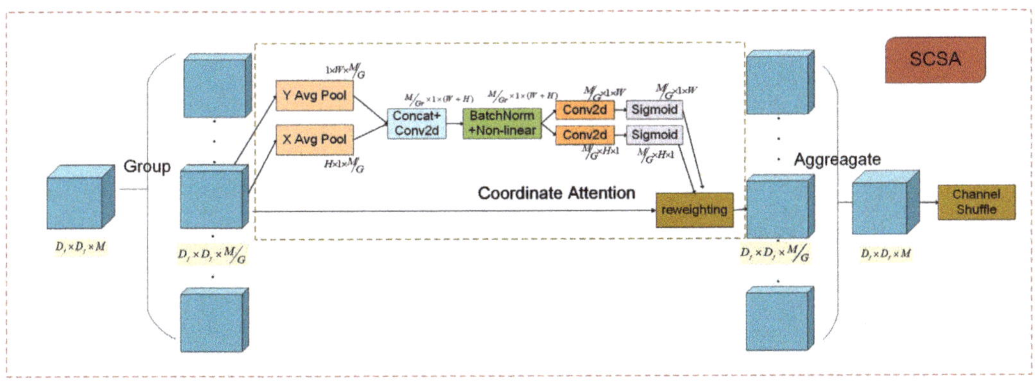

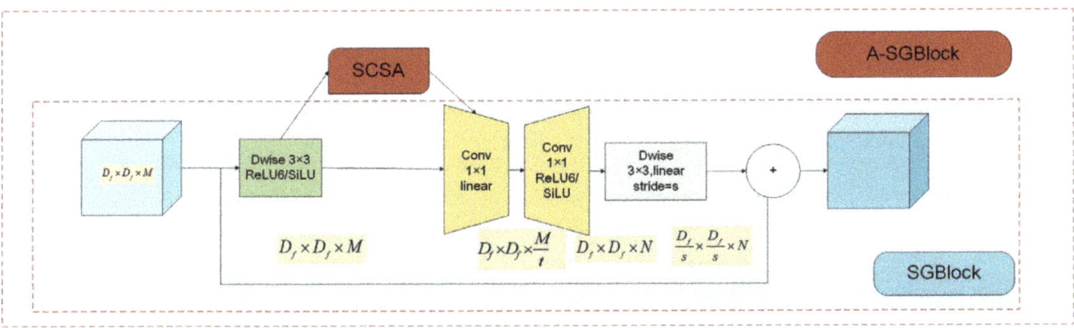

Figure 2. The structure of SGBlock and A-SGBlock. The light blue box represents the specific structure of SGBlock. Based on the SGBlock, A-SGBlock incorporates the SCSA module. SCSA represents the Shuffle Channel and Spatial Attention module. Coordinate Attention includes channel and spatial attention.

Although the lightweight backbone network called MobileNeXt can reduce the amount of computation and the number of parameters in the network, its feature extraction capability is insufficient. The attention module is embedded into the convolutional neural network, which enables the lightweight convolutional neural network to calculate the correlation coefficient of the internal feature points, thus enhancing the internal correlation of the feature maps. Recent studies have found that channel attention (e.g., Squeeze-and-Excitation Attention [38]) is a significant factor in improving the performance of the model. However, they usually ignore the position information. In order to encode more useful position information, the Coordinate Attention (CA) [39] embeds position information into channel attention so that the model could locate the object area more accurately. Although the Coordinate Attention can effectively encode channel and spatial features of small objects, its number of parameters is more than most attention mechanism modules. Therefore, in this paper, the number of parameters of the Coordinate Attention is reduced by grouping features, and the Shuffle Channel and Spatial Attention (SCSA) module is presented, which has fewer parameters than that of the Coordinate Attention module. The embedded position and structure of the SCSA module are shown in Figure 2, and an SGBlock embedded with an SCSA module is called an A-SGBlock, and it can be formulated as follows:

$$\hat{G}' = T_{1,p} SCSA(T_{1,d}(F)),$$
$$G' = T_{2,d} T_{2,p}(\hat{G}') + F \tag{2}$$

where $SCSA$ is the attention module, and whose specific operations are described as follows. Firstly, the input feature maps $X \in R^{D_f \times D_f \times M}$ are divided into G groups along the channel dimension, i.e., $X = [X_1, \ldots, X_G]$, $X_k \in \Re^{D_f \times D_f \times \frac{M}{G}}$. Secondly, the Coordinate Attention (that is, channel and spatial attention) is performed for each group X_k, in which the Coordinate Attention decomposes the channel attention into two one-dimensional feature coding processes that aggregates features along with different directions. The advantage of this process is to capture long-range dependencies along one spatial direction and retain accurate position information along the other spatial direction. Thirdly, each group of feature maps that pass through the Coordinate Attention module are fused. Fourthly, the Shuffle Channel [36] promotes information communication between different groups of features.

Precisely, the Coordinate Attention module consists of two steps: coordinate information embedding and coordinate attention generation. Firstly, each channel is encoded along with the horizontal and vertical coordinates by using pooling kernels with sizes $(H, 1)$ and $(1, W)$, respectively. Mathematically, the output of the m-th channel at height h and the output of the m-th channel at width w can be respectively formulated as follows:

$$Z_m^h(h) = \frac{1}{W} \sum_{0 \leq i < W} x_m(h, i)$$
$$Z_m^w(w) = \frac{1}{H} \sum_{0 \leq i < H} x_m(j, w) \tag{3}$$

A pair of direction-aware and position-sensitive feature maps are obtained. Then generated feature maps are fused in spatial dimensions and fed into a shared 1×1 convolution transformation function T, and this process can be formulated as follows:

$$f = \delta(T_1(z^h, z^w)) \tag{4}$$

where $[.,.]$ represents the concatenation operation along the spatial dimension, δ is a nonlinear activation function, $f \in \Re^{M/Gr \times (H+W)}$ is the intermediate feature map, and r is the reduction rate to control the module size. Next, the feature maps f obtained in the previous step are divided into two separate tensors $f^h \in \Re^{M/Gr \times H}$ and $f^w \in \Re^{M/Gr \times W}$ along the spatial dimension. In the next step, two convolution transformation functions T_h and T_w

are used to transform the channel number of feature maps to make it consistent with the channel number of the input feature maps, and this process can be formulated as follows:

$$
\begin{aligned}
g^h &= \sigma(T_h(f^h)), \\
g^w &= \sigma(T_w(f^w))
\end{aligned}
\tag{5}
$$

where σ is the sigmoid activation function. Finally, the input feature maps are multiplied with a pair of feature maps obtained through the steps of coordinate information embedding and coordinate attention generation, and then, attention feature maps are generated to enhance the representation of the region of interest, and this process can be formulated as follows:

$$
y_m(i,j) = x_m(i,j) \times g_m^h(i) \times g_m^w(j)
\tag{6}
$$

Therefore, in order to enhance the ability of lightweight backbone network, in this paper, we present the feature enhancement backbone network called SA-MobileNeXt, which is based on attention and is shown on the right of Figure 1. "SGBlockn/A-SGBlockn" represents "n SGBlock/A-SGBlock modules are used"; if "/2", it represents "the stride of SGBlock/A-SGBlock is 2", otherwise it represents "the stride of SGBlock/A-SGBlock is 1". The SA-MobileNeXt uses the A-SGBlock module in the front part of the backbone network, which embeds the Shuffle Channel and Spatial Attention (SCSA) module proposed in this paper into the SGBlock. In this work, nine A-SGBlock modules are employed for two reasons. On the one hand, using lots of A-SGBlock modules (especially the A-SGBlock modules with numerous channels located at the back of the backbone network) can increase the number of parameters and computation amount, resulting in a decrease in speed while not improving the accuracy. On the other hand, using A-SGBlock modules in the shallow layer of the backbone network can encode more accurate location information, which is conducive to detecting small objects. In addition, in our SA-MobileNeXt, the ReLU6 activation function used in the original SGBlock modules is replaced with the SiLU activation function, making the model converge faster. The SiLU activation function is described in Section 2.2.3.

2.2.2. Multi-Scale Feature Enhancement Fusion Network

In the process of feature fusion, to better integrate the features extracted from the backbone network, the Multi-scale Feature Enhancement Fusion Network is proposed, which further promotes the performance of small object detection. Its main structure is shown in Figure 3 and is explained as follows.

In the original YOLOv3, the feature fusion method of FPN only integrates 8-fold downsampled feature maps, 16-fold downsampled feature maps, and 32-fold downsampled feature maps. However, the shallow features extracted by the backbone are essential for detecting small objects. As a result, the 4-fold downsampled feature maps from the backbone network are integrated to promote small object detection, and the specific operations of fusion are as follows. Firstly, the 8-fold downsampled fusion feature maps with low-resolution pass through a BottleneckCSP and a CBS module and then are processed by an upsampling operation. Finally, the resulting feature maps are fused with feature maps with the size of $160 \times 160 \times 144$ from the backbone network.

Meanwhile, the feature fusion method in PANet [40] can better preserve the shallow feature information, a Down-top path (Figure 3b) is added by referring to the method in PANet. We take the fusion process of 8-fold downsampled feature maps in the Down-top path as an example, and its operations are detailed as follows. The 4-fold downsampled fusion feature maps pass through a BottleneckCSP module and then are upsampled by a CBS module with stride 2 to become 8-fold downsampled feature maps, and the resulting feature maps are fused with feature maps with the same resolution from the Top-down path. In order to save the number of parameters and make the model converge faster, we still utilize the last three detection for detection.

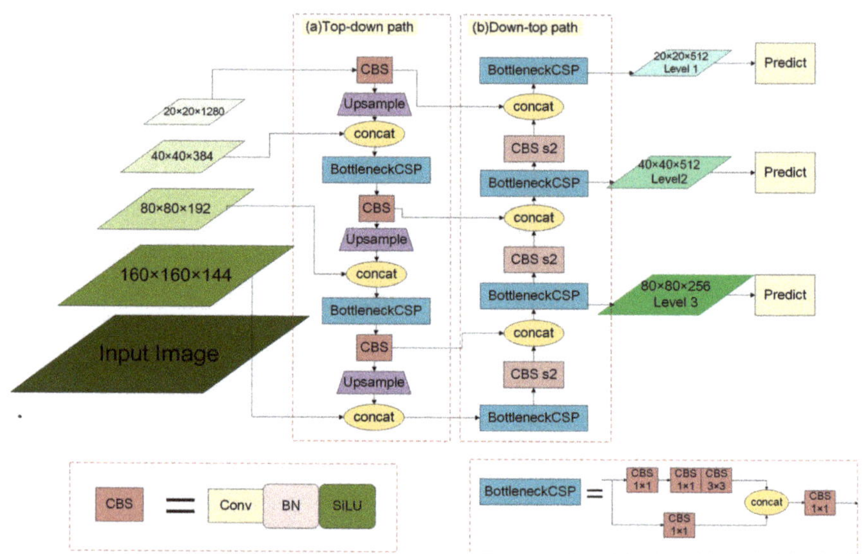

Figure 3. The structure diagram of Multi-scale Feature Enhancement Fusion network. CBS represents the convolution module with stride 1. CBS s2 represents the convolution module with stride 2. BottleneckCSP (BC) represents the combination of convolution module.

The introduction of the 4-fold downsampled feature maps in the backbone network and an additional Down-top path can improve the accuracy of object detection but increase the number of parameters to a certain extent. Therefore, the previous convolution blocks are combined into the BottleneckCSP module by using the idea of CSPNet to further reduce the number of parameters and computation amount in the network without affecting the detection accuracy. The structure of the BottleneckCSP module is shown in the bottom right of Figure 3, and it contains two branches. Firstly, in the first branch, there are three convolution layers. In the second branch, there is a 1×1 convolution layer. Then, the feature maps of the two branches are fused, and finally, the number of channels is transformed by a 1×1 convolution layer. In addition, the CBL modules in the feature fusion network are replaced with the CBS modules, whose structures are shown in the bottom left of Figure 3. As we can see, the optimized CBS modules in this paper use the SiLU (Sigmoid Weighted Linear Unit) to replace the Leaky ReLU.

2.2.3. SiLU Activation Function

In this paper, the optimized SGBlock modules use the SiLU [41] (Sigmoid Weighted Linear Unit) to replace the ReLU6. Meanwhile, the CBS modules utilize the SiLU to replace the Leaky ReLU. The calculation formulas of *SiLU* and its first derivative are as shown in Equation (7) and Figure 4.

$$SiLU = x \cdot sigmoid(x)$$
$$sigmoid(z) = \frac{1}{1+e^{-z}}$$
$$SiLU' = SiLU + sigmoid(1 - SiLU) \tag{7}$$

If the input value is greater than 0, the *SiLU* is approximately the same as the ReLU; and if the input value is less than 0, the value of *SiLU* approaches 0. Compared with the Sigmoid and Tanh, the *SiLU* activation function does not increase monotonously and has a global minimum value of about −0.28. In general, deep convolutional neural networks often encounter the phenomenon of gradient explosion. However, an attractive feature of SiLU is self-stability: when the derivative is zero, the global minimum can play the role of "soft bottom", which can inhibit the update of large weights from avoiding gradient explosion.

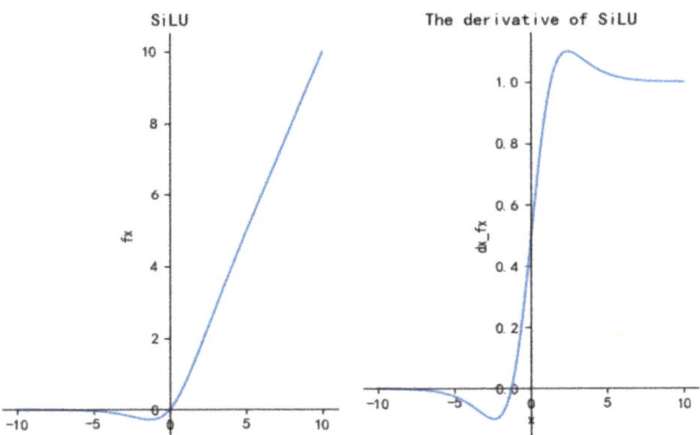

Figure 4. The activation function and derivative curves of SiLU.

2.2.4. Data Enhancement

In deep learning, it is crucial to keep the number of samples be sufficient. Numerous samples will make the trained model have a better effect and generalization ability. However, for the KITTI and CCTSDB datasets used in this paper, their sample quantity and quality are not good enough, which will lead to overfitting. Recently, Dewi, Christine et al. [42] combined synthetic images with original images to enhance datasets and verify the effectiveness of synthetic datasets. Therefore, data enhancement is an effective solution to improve the quality of datasets, which can reduce the overfitting phenomenon of the network. A network with better generalization ability can be obtained by transforming the training images, which can better adapt to the application scenarios. Therefore, two methods of data enhancement, Mosaic and Mixup, are adopted in this paper to improve the quality of the dataset so that the proposed improved algorithm is more suitable for training on a single GPU.

The two types of data enhancement are described in detail below. The Mixup merges the positive and the negative samples into a new group of samples, which doubles the size of the sample. Meanwhile, the objects in each batch after Mixup will be more than the objects in the original batch. The Mosaic combines four training images into one in a certain proportion, enabling the model to learn to recognize smaller objects, which can enrich the background of detecting objects and calculate four kinds of images in Batch Normalization, and the batch size does not need to be large so that a GPU can achieve better results.

In this paper, due to the limitation of GPU and model size, and in order to make a fair comparison between different models, the training batch size is uniformly set as 4. We adopted such a data enhancement strategy that uses only the Mosaic data enhancement strategy in the three batches and uses a combination of the Mosaic and Mixup data enhancement strategy in the one batch. Through the experiments, the model trained by our data enhancement strategy is better than the model trained by the Mosaic only in the four batches.

2.2.5. Loss Function

The total loss function used by the YOLO-MXANet algorithm is shown in Equation (8). *CIoU* regression loss is employed to improve MSE regression loss [43], and the improved loss function is more suitable for detecting small objects in traffic scenes. *CIoU* inherits the advantages of Generalized Intersection Over Union (GIoU) [44] and Distance-IoU (DIoU) [45], which not only considers the distance and overlap ratio but also considers the scale and the aspect ratio between the prediction box and the ground truth box so that it can carry out the bounding box regression better [43]. It consists of three parts: the first is $loss_{CIoU}$, which represents regression loss; The second part is $loss_{obj}$, which represents confidence loss. The third part $loss_{class}$ represents classification loss.

$$
\begin{aligned}
LOSS &= loss_{CIoU} + loss_{obj} + loss_{class} \\
loss_{CIoU} &= 1 - CIoU, \, CIoU = IoU - \frac{\rho^2\left(b, b^{gt}\right)}{c^2} - \alpha v \\
loss_{obj} &= -\sum_{i=0}^{K \times K} \sum_{j=0}^{M} I_{ij}^{obj} \left[\hat{C_i} \log(C_i) + (1 - \hat{C_i}) \log(1 - C_i) \right] \\
&\quad - \lambda_{noobj} \sum_{i=0}^{K \times K} \sum_{j=0}^{M} I_{ij}^{noobj} \left[\hat{C_i} \log(C_i) + (1 - \hat{C_i}) \log(1 - C_i) \right] \\
loss_{class} &= -\sum_{i=0}^{K \times K} I_{ij}^{obj} \sum_{c \in classes} \left[\hat{p_i}(c) \log(p_i(c)) + (1 - \hat{p_i}(c)) \log(1 - p_i(c)) \right]
\end{aligned}
\tag{8}
$$

3. Experimental Results and Analysis

In order to verify the performance of YOLO-MXANet, comparative experiments on the KITTI dataset and CCTSDB dataset are conducted. In this paper, the experimental platforms are Intel® Core™ i7-9700 CPU @ 3.00 GHz processor and NVIDIA GeForce RTX 2080Ti GPU. The algorithms in this paper are programmed in Python 3.8 and implemented in PyCharm Community 2020.2.3 software. To ensure the fairness of test, all models are trained from scratch and trained 200 epochs. To make the training process more stable, the Adam optimizer is used for training. In the training process, the warm-up strategy is adopted in the first three epochs, and the cosine annealing strategy is adopted for training from the fourth epoch to the 200th epoch, which reduces the learning rate from 0.01 to 0.002. The value of Momentum is set to 0.937, and the value of Weight_decay is set to 0.0005.

The following evaluation indexes are used to evaluate the performance of algorithms. The accuracy of detection algorithms is measured by using the Precision, Recall, and F1 score (the harmonic mean value of Precision and Recall). The Average Precision (AP) is used to measure the detection accuracy of each type of object. The mean Average Precision (mAP) is used to measure the average detection accuracy of multi-class objects. The higher the mAP value is, the higher the comprehensive performance of the model in all categories will be. The speed of each image on the GPU is used to measure the detection speed of object detector. The number of parameters and computation amount are used to measure the complexity of the model.

3.1. Ablation Learning on the KITTI Dataset

In order to prove the effectiveness of each improvement method, we conduct ablation experiments on the KITTI dataset, and the results are shown in Table 1. The KITTI dataset is randomly and automatically divided into train set, validation set and test set, and the 8:1:1 ratio is adopted in this study. At the same time, eight classes of objects in the dataset are fused into three types of objects, namely Pedestrian, Car, and Cyclist. In our experiment, we use images with the size of 640 × 640 × 3 for training and testing. We employ original YOLOv3 as our Scheme A. We first established a more robust YOLOv3 baseline, which has a good performance in terms of speed. Meanwhile, YOLOv3 also has a higher detection accuracy, but the number of parameters and computation amount are large. Based on Scheme A, Scheme B adopts CIoU loss function, which can improve the positioning ability of small objects. Based on Scheme B, Scheme C uses MobileNeXt, which causes a slight reduction in detection performance but simplifies the model by reducing the number of network parameters from 61,508,200 to 22,927,784. At the same time, the detection speed of each image is improved from 3.5 ms to 2.4 ms. Based on Scheme C, Scheme D utilizes Mosaic and Mixup to promote the quality of dataset, which makes up for the performance loss caused by the lightweight network while keeping the speed unchanged and improving the detection ability of small object and the generalization of the network, increasing F1 from 0.812 to 0.842 and mAP 0.5 from 0.865 to 0.897. Based on Scheme D, Scheme E introduces the feature fusion method of PANet and integrates the four-fold subsampled feature maps containing small object information to improve the detection ability of small objects. At the same time, the idea of CSPNet is used to combine convolution blocks, which reduces the number of parameters. From the experimental results, Scheme E improves the detection performance of small objects, which reduces the number of parameters from 22,927,784 to 13,870,888 and increases F1 from 0.842 to 0.861 and mAP 0.5 from 0.897 to 0.905. Based on Scheme E, we introduce the SiLU activation function to make the model converge faster and improve the stability of the model, which replaces original activation function of CBL and SGBlock module with SiLU and makes it become our Scheme F, which increases F1 from 0.861 to 0.877, mAP 0.5 from 0.905 to 0.916. Based on Scheme F, Scheme G introduces the Coordinate Attention mechanism to obtain the valuable features of small objects, which increases mAP 0.5 from 0.916 to 0.922. Based on Scheme G, Scheme H proposes the Shuffle Channel and Spatial Attention (SCSA) module to improve the detection accuracy while further simplifying the model, which not only increases F1 from 0.876 to 0.885 and mAP 0.5 from 0.922 to 0.924 but also decreases the number of parameters from 13,987,271 to 13,874,564. In conclusion, compared with YOLOv3 baseline, our final scheme reduces the number of parameters from 61,508,200 to 13,874,564, and the GFLOPS from 154.9 to 37.0, increasing the speed by 0.6 ms. Meanwhile, the detection performance is improved, which increases F1 by 4.8 percentage points and mAP 0.5 by 3.6 percentage points.

Table 1. The ablation experiments on the KITTI dataset.

Scheme	Method	P	R	F1	mAP 0.5	Speed$_{GPU}$/ms	Params	GFLOPS
A	YOLOv3	0.923	0.765	0.837	0.888	3.5	61,508,200	154.9
B	A + CIoU	0.930	0.799	0.860	0.911	3.5	61,508,200	154.9
C	B + MobileNeXt	0.857	0.772	0.812	0.865	2.4	22,927,784	43.4
D	C + DA	0.882	0.806	0.842	0.897	2.4	22,927,784	43.4
E	D + PAN + 4s + BC	0.876	0.846	0.861	0.905	2.5	13,870,888	37.0
F	E + SiLU	0.941	0.822	0.877	0.916	2.5	13,870,888	37.0
G	F + A-MobileNeXt	0.943	0.818	0.876	0.922	2.9	13,987,271	37.1
H	G + SA-MobileNeXt	0.930	0.844	0.885	0.924	2.9	13,874,564	37.0

In order to further prove the excellent effect of improved algorithm, we show the PR curve diagram of YOLOv3 and YOLO-MXANet, as well as the AP value of each category and the mAP value of all categories, which are shown in Figure 5. Compared with YOLOv3, the mAP value of YOLO-MXANet increases by 3.6 percentage points, and the AP value of each category of YOLO-MXANet has increased. Specifically, the AP value of category "Pedestrian" increases by 4.6 percentage points, the AP value of category "Car" increases by 1.1 percentage points, and the AP value of category "Cyclist" increases by 5 percentage points.

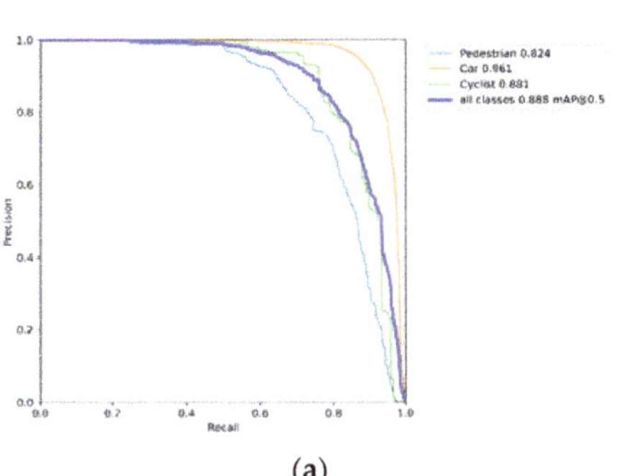

(a)

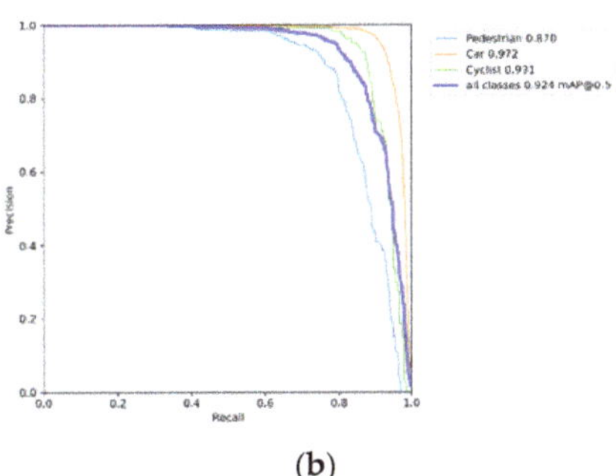

(b)

Figure 5. The PR curve diagram of YOLOv3 and YOLO-MXANet on the KITTI dataset. (a) YOLOv3. (b) YOLO-MXANet.

3.2. Comparison Experiments with Other Algorithms on the KITTI Dataset

In order to further verify the performance of improved algorithm, comparative experiments are conducted between YOLO-MXANet and other algorithms on the KITTI dataset, and the comparison results are shown in Table 2. The experimental results show that YOLO-MXANet has higher detection accuracy compared with YOLOv5s, and YOLO-MXANet has less parameters than YOLOv5m while keeping slightly better accuracy. Compared with YOLOv3 and YOLOv3-SPP, YOLO-MXANet has fewer parameters and more significant advantages in detection accuracy. Compared with the lightweight algorithm YOLOv3-tiny, although the number of parameters of YOLO-MXANet is a little more than that of YOLOv3-tiny, the mAP 0.5 value of YOLO-MXANet is 23.2 percentage points higher than that of YOLOv3-tiny, while the F1 value of YOLO-MXANet is 21.5 percentage points higher than that of YOLOv3-tiny. Compared with the latest lightweight algorithm YOLOv4-tiny, the mAP 0.5 value of YOLO-MXANet is 16.2 percentage points higher than that of YOLOv4-tiny, while the F1 value of YOLO-MXANet is 22.2 percentage points higher than that of YOLOv4-tiny.

Table 2. The comparison results of the algorithms on the KITTI dataset.

Algorithm \ Indicator	P	R	F1	mAP 0.5	Params (M)
YOLOv3	0.923	0.765	0.837	0.888	61.5
YOLOv3-SPP	0.923	0.783	0.847	0.894	62.6
YOLOv5s	0.922	0.781	0.846	0.889	7
YOLOv5m	0.899	0.862	0.880	0.923	21.1
YOLOv3-tiny	0.763	0.598	0.670	0.692	8.7
YOLOv4-tiny	0.589	0.761	0.663	0.762	5.9
YOLO-MXANet	0.930	0.844	0.885	0.924	13.8

The actual detection results of YOLO-MXANet and YOLOv3 (baseline) on the KITTI dataset are compared in Figure 6. As can be seen from the comparison figures, YOLO-MXANet can detect complex objects that YOLOv3 cannot detect, such as small objects and occluded objects. Specifically, it can be seen from the first group of images that both YOLOv3 and YOLO-MXANet can detect three objects, but the confidence of bounding box of YOLO-MXANet is higher. As can be seen from the second group of images, YOLOv3 detects three objects "Car", and YOLO-MXANet can detect four objects "Car", in other words, YOLO-MXANet detects one smaller object with low light than YOLOv3, and the other three object's bounding boxes detected by YOLO-MXANet have higher confidence. Similarly, it can be seen from the third group and the fourth group that YOLO-MXANet can also detect smaller objects and each bounding box detected by YOLO-MXANet has a higher confidence. This is because YOLO-MXANet can effectively enhance the characteristic information of object and suppress environmental interference.

3.3. Comparison Experiments with Other Algorithms on the CCTSDB Dataset

In order to further verify the performance of YOLO-MXANet, the comparative experiments with other advanced algorithms are conducted on the CCTSDB dataset. In the experiment, we select 3105 images from the CCTSDB dataset and use the dataset partition algorithm to randomly divide the CCTSDB dataset into train set and validation set and test set, and the 8:1:1 ratio is also adopted in this study. CCTSDB dataset is classified into three types of objects, namely, warning, prohibitory, and mandatory. The comparison results between YOLO-MXANet and the latest object detection algorithm on the CCTSDB dataset are shown in Table 3. Compared with YOLOv5m, YOLO-MXANet has fewer parameters and has more tremendous advantages in terms of detection accuracy. Compared with the lightweight algorithm YOLOv3-tiny, although the number of parameters of YOLO-MXANet is a little more than that of YOLOv3-tiny, the mAP 0.5 value of YOLO-MXANet is 6.8 percentage points higher than that of YOLOv3-tiny, and the F1 value of improved algorithm is 5.6 percentage points higher than that of YOLOv3-tiny. Compared with the lightweight algorithm YOLOv4-tiny, the mAP 0.5 value of YOLO-MXANet is 2.2 percentage points higher than that of YOLOv4-tiny, and the F1 value of improved algorithm is 7.7 percentage points higher than that of YOLOv4-tiny. Therefore, YOLO-MXANet is more suitable for object detection in traffic scenes.

(a)

(b)

Figure 6. The detection results of YOLOv3 and YOLO-MXANet on the KITTI dataset. (**a**) YOLOv3. (**b**) YOLO-MXANet.

Table 3. The comparison results of the algorithms on the CCTSDB dataset.

Indicator Algorithm	P	R	F1	mAP 0.5
YOLOv3	0.910	0.894	0.902	0.928
YOLOv3-SPP	0.929	0.877	0.902	0.937
YOLOv5m	0.968	0.939	0.953	0.966
YOLOv3-tiny	0.911	0.873	0.892	0.905
YOLOv4-tiny	0.795	0.964	0.871	0.951
YOLO-MXAet	0.930	0.967	0.948	0.973

The comparisons of actual detection results between YOLO-MXANet and YOLOv3 (baseline) on the CCTSDB dataset are shown in Figure 7. In the first set of images, YOLO-MXANet can detect small and dim objects. As can be seen from the second group, YOLO-MXANet can detect the "warning" objects that YOLOv3 cannot detect, and the bounding box detected by YOLO-MXANet has a higher confidence. As we can see from the third pictures, YOLOv3 misses two objects, while YOLO-MXANet detects all of them. As can be seen from the fourth group of pictures, although YOLOv3 can detect large objects with dim light, the confidence of bounding box detected by YOLOv3 is not as high as that detected by YOLO-MXANet, and the detection ability of YOLOv3 is not as good as that of YOLO-MXANet in terms of small objects.

(a)   (b)

Figure 7. The detection results of YOLOv3 and YOLO-MXANet on the CCTSDB dataset. (**a**) YOLOv3. (**b**) YOLO-MXANet.

4. Conclusions

Based on general one-stage object detection algorithms, we propose a small object detection algorithm in traffic scenes (named YOLO-MXANet), which not only solves the problem that original algorithm is not high in detecting small-scale objects but also reduces the number of parameters from 61.5 M to 13.8 M and improves the detection speed. Therefore, YOLO-MXANet balances the detection accuracy, inference speed, and model complexity. We utilize CIoU to improve the loss function of YOLOv3 and improve the positioning accuracy of small objects. A lightweight backbone network (named MobileNeXt) is used to reduce the number of parameters and amount of computation, which can improve the detection speed of the model. However, the light weight will reduce the accuracy of the model to a certain extent. To further enhance the feature extraction capability of MobileNeXt, we present SA-MobileNeXt based on the Shuffle Channel and Spatial Attention module as the backbone network. In order to make up for the loss of precision caused by light weight, we use Mosaic and Mixup to train the model, which can enhance the ability of small object detection and thus improve the robustness of the model. To further enhance the characteristics of the small object, we add a Down-top path and fuse the four-fold subsampled feature maps from the backbone network. At the same time, to reduce the number of parameters without weakening the feature extraction ability of the network, we utilize the idea of CSPNet to combine convolution blocks. We perform ablation experiments on the KITTI dataset to demonstrate the effectiveness of each improved method. In addition, we conduct comparative experiments with other advanced algorithms on the KITTI and CCTSDB datasets, and the experimental results show that our algorithm has certain advantages in terms of detection accuracy, detection speed, and model complexity. Although our algorithm has achieved some improvements in accuracy and model complexity, there is still a long way to go before it can be deployed on mobile devices. Therefore, the next step is to further balance the detection accuracy, speed, and model complexity to provide excellent theoretical basis and practical value for intelligent transportation and unmanned driving.

Author Contributions: Conceptualization, X.H. and R.C.; methodology, X.H. and R.C.; software, X.H. and R.C.; validation, X.H. and R.C.; formal analysis, X.H. and R.C.; investigation, R.C., X.H. and Z.W.; resources, X.H. and Z.Z.; data curation, R.C. and X.H.; writing—original draft preparation, X.H. and R.C.; writing—review and editing, X.H. and R.C.; visualization, X.H. and R.C.; supervision, X.H.; project administration, X.H. and Z.Z.; funding acquisition, X.H. and Z.Z. All authors have read and agreed to the published version of the manuscript.

Funding: This research was funded by the National Natural Science Foundation of China (NSFC): 61572023; Natural Science Foundation of Zhejiang Province: Z22F023843; National Natural Science Foundation of China (NSFC): 61672467.

Institutional Review Board Statement: Not applicable.

Informed Consent Statement: Not applicable.

Data Availability Statement: Some or all data, models, or code generated or used during the study are available from the corresponding author by request.

Conflicts of Interest: The authors declare no conflict of interest.

References

1. Girshick, R.; Donahue, J.; Darrelland, T.; Malik, J. Rich feature hierarchies for accurate object detection and semantic segmentation. In Proceedings of the IEEE Conference on Computer Vision and Pattern Recognition, Columbus, OH, USA, 24–27 June 2014.
2. Girshick, R. Fast R-CNN. In Proceedings of the IEEE Conference on Computer Vision and Pattern Recognition, Boston, MA, USA, 8–10 June 2015; pp. 1440–1448.
3. Ren, S.; He, K.; Girshick, R.; Sun, J. Faster R-CNN: Towards real-time object detection with region proposal networks. *IEEE Trans. Pattern Anal. Mach. Intell.* **2017**, *39*, 1137–1149. [CrossRef] [PubMed]
4. Liu, W.; Anguelov, D.; Erhan, D.; Szegedy, C.; Reed, S.; Fu, C.Y. SSD: Single shot multi box detector. In Proceedings of the Europe Conference on Computer Vision, Amsterdam, The Netherlands, 8–16 October 2016; pp. 21–37.
5. Fu, C.Y.; Liu, W.; Ranga, A.; Tyagi, A.C.; Berg, A. DSSD: Deconvolutional Single Shot Detector. *arXiv* **2017**, arXiv:1701.06659. Available online: https://arxiv.org/abs/1701.06659 (accessed on 23 January 2017).
6. Redmon, J.; Divvala, S.; Girshick, R.; Farhadi, A. You only look once: Unified, real-time object detection. In Proceedings of the IEEE Conference on Computer Vision and Pattern Recognition, Las Vegas, NV, USA, 27–30 June 2016; pp. 779–788.
7. Redmon, J.; Farhadi, A. YOLO9000: Better, Faster, Stronger. In Proceedings of the IEEE Conference on Computer Vision and Pattern Recognition, Honolulu, HI, USA, 22–25 July 2017; pp. 6517–6525.
8. Redmon, J.; Farhadi, A. YOLOv3: An Incremental Improvement. *arXiv* **2018**, arXiv:1804.02767v1. Available online: https://arxiv.org/abs/1804.02767 (accessed on 8 April 2018).
9. Zhou, X.; Zhuo, J.; Krahenbuhl, P. Bottom-up object detection by grouping extreme and center points. In Proceedings of the IEEE Conference on Computer Vision and Pattern Recognition, Long Beach, CA, USA, 16–20 June 2019.
10. Duan, K.; Bai, S.; Xie, L.; Qi, H.; Huang, Q.; Tian, Q. CenterNet: Keypoint triplets for object detection. In Proceedings of the IEEE/CVF International Conference on Computer Vision, Seoul, Korea, 27 October–2 November 2019.
11. Tian, Z.; Shen, C.; Chen, H.; He, T. FCOS: Fully convolutional one-stage object detection. In Proceedings of the 2019 IEEE/CVF International Conference on Computer Vision (ICCV), Seoul, Korea, 27 October–2 November 2019; pp. 9626–9635.
12. Zhu, C.; He, Y.H.; Savvides, M. Feature selective anchor-free module for single-shot object detection. In Proceedings of the IEEE Computer Vision and Pattern Recognition, Long Beach, CA, USA, 16–20 June 2019.
13. Que, L.; Zhang, T.; Guo, H.; Jia, C.; Gong, Y.; Chang, L.; Zhou, J. A Lightweight Pedestrian Detection Engine with Two-StageLow-Complexity Detection Network and Adaptive Region Focusing Technique. *Sensors* **2021**, *21*, 5851. [CrossRef] [PubMed]
14. Yang, X.; Liu, Q. Scale-sensitive feature reassembly network for pedestrian detection. *Sensors* **2021**, *21*, 4189. [CrossRef] [PubMed]
15. Ma, L.; Gong, X.; Ouyang, H. Improvement of Tiny YOLOv3 object detection. *Opt. Precis. Eng.* **2020**, *28*, 988–995.
16. Guo, F.; Zhang, Y.; Tang, J.; Li, W. YOLOv3-A: A traffic sign detection network based on attention mechanism. *J. Commun.* **2021**, *42*, 87–99.
17. Liu, C.; Wang, Q.; Bi, X. Research on Multi-target and Small-scale Vehicle Target Detection Method. *Control Decis.* **2021**, *36*, 2707–2712.
18. Zhu, Z.; Liang, D.; Zhang, S.; Huang, X.; Li, B.; Hu, S. Traffic-sign detection and classification in the wild. In Proceedings of the IEEE Conference on Computer Vision and Pattern Recognition, Las Vegas, NV, USA, 26 June–1 July 2016; pp. 2110–2118.
19. Wang, H.; Wang, J.; Bai, K.; Sun, Y. Centered Multi-Task Generative Adversarial Network for Small Object Detection. *Sensors* **2021**, *21*, 5194. [CrossRef]
20. Bosquet, B.; Mucientes, M.; Brea, V. STDnet-ST: Spatio-temporal ConvNet for small object detection. *Pattern Recognit.* **2021**, *116*, 107929. [CrossRef]
21. Lian, J.; Yin, Y.; Li, L.; Wang, Z.; Zhou, Y. Small Object Detection in Traffic Scenes Based on Attention Feature Fusion. *Sensors* **2021**, *21*, 3031. [CrossRef] [PubMed]

22. Zhang, C.; Zhang, X.; Tu, D.; Wang, Y. Small object detection using deep convolutional networks: Applied to garbage detection system. *J. Electron. Imaging* **2021**, *30*, 043013. [CrossRef]
23. Liu, C.; Li, S.; Chang, F.; Wang, Y. Machine Vision Based Traffic Sign Detection Methods: Review, Analyses and Perspectives. *IEEE Access* **2019**, *7*, 86578–86596. [CrossRef]
24. Chen, R.-C.; Saravanarajan, V.S.; Hung, H.-T. Monitoring the behaviours of pet cat based on YOLO model and raspberry Pi. *Int. J. Appl. Sci. Eng.* **2021**, *18*, 1–12. [CrossRef]
25. Zheng, Z.; Wang, P.; Ren, D.; Liu, W.; Ye, R.; Hu, Q.; Zuo, W. Enhancing Geometric Factors in Model Learning and Inference for Object Detection and Instance Segmentation. *arXiv* **2020**, arXiv:2005.03572.
26. Daquan, Z.; Hou, Q.; Chen, Y.; Feng, J.; Yan, S. Rethinking Bottleneck Structure for Efficient Mobile Network Design. *arXiv* **2020**, arXiv:2007.02269. Available online: https://arxiv.org/abs/2007.02269 (accessed on 27 November 2020).
27. Bochkovskiy, A.; Wang, C.Y.; Liao, H.Y.M. YOLOv4: Optimal Speed and Accuracy of Object detection. *arXiv* **2020**, arXiv:2004.10934. Available online: https://arxiv.org/abs/2004.10934 (accessed on 23 April 2020).
28. Zhang, H.; Gisse, M.; Dauphin, Y.N.; Lopez-Paz, D. mixup: Beyond Empirical Risk Minimization. *arXiv* **2017**, arXiv:1710.09412. Available online: https://arxiv.org/abs/1710.09412 (accessed on 27 April 2018).
29. Wang, C.Y.; Mark Liao, H.Y.; Wu, Y.H.; Chen, P.Y.; Hsieh, J.W.; Yeh, I.H. CSPNet: A new backbone that can enhance learning capability of cnn. In Proceedings of the IEEE/CVF Conference on Computer Vision and Pattern Recognition Workshops, Glasgow, UK, 23–28 August 2020; pp. 390–391.
30. He, K.; Zhang, X.; Ren, S.; Sun, J. Deep residual learning for image recognition. In Proceedings of the 2016 IEEE Conference on Computer Vision and Pattern Recognition (CVPR), Las Vegas, NV, USA, 27–30 June 2016; pp. 770–778.
31. Lin, T.Y.; Dollár, P.; Girshick, R.; He, K.; Hariharan, B.; Belongie, S. Feature Pyramid Networks for Object Detection. In Proceedings of the IEEE Conference on Computer Vision and Pattern Recognition, Honoluu, HI, USA, 22–25 July 2017; pp. 2117–2125.
32. Howard, A.; Zhu, M.; Chen, B.; Kalenichenko, D.; Wang, W.; Weyand, T.; Andreetto, M.; Adam, H. Mobilenets: Efficient convolutional neural networks for mobile vision applications. *arXiv* **2017**, arXiv:1704.04861. Available online: https://arxiv.org/abs/1704.04861 (accessed on 17 April 2017).
33. Sandler, M.; Howard, A.; Zhu, M.; Zhmoginov, A.; Chen, L. MobileNetV2: Inverted residuals and linear bottlenecks. In Proceedings of the 2018 IEEE/CVF Conference on Computer Vision and Pattern Recognition, Salt Lake City, UT, USA, 18–23 June 2018.
34. Howard, A.; Sandler, M.; Chen, B.; Wang, W.; Chen, L.-C.; Tan, M.; Chu, G.; Vasudevan, V.; Zhu, Y.; Pang, R.; et al. Searching for MobileNetv3. In Proceedings of the 2019 IEEE/CVF International Conference on Computer Vision (ICCV), Seoul, Korea, 27 October–2 November 2019; pp. 1314–1324.
35. Zhang, X.; Zhou, X.; Lin, M.; Sun, J. ShuffleNet: An extremely efficient convolutional neural network for mobile devices. In Proceedings of the 2018 IEEE/CVF Conference on Computer Vision and Pattern Recognition, Salt Lake City, UT, USA, 18–23 June 2018; pp. 6848–6856.
36. Ma, N.; Zhang, H.; Zheng, H.; Sun, J. ShuffleNetv2: Practical Guidelines for Efficient CNN Architecture Design. *arXiv* **2018**, arXiv:1807.11164. Available online: https://arxiv.org/abs/1807.11164 (accessed on 30 July 2018).
37. Landola, F.N.; Moskewicz, M.W.; Ashraf, K.; Han, S.; Dally, W.J.; Keutzer, K. SqueezeNet: AlexNet-level accuracy with 50x fewer parameters and <1 MB model size. *arXiv* **2016**, arXiv:1602.07360. Available online: https://arxiv.org/abs/1602.07360 (accessed on 4 November 2016).
38. Hu, J.; Shen, L.; Sun, G. Squeeze-and-excitation networks. In Proceedings of the 2018 IEEE Conference on Computer Vision and Pattern Recognition, Salt Lake City, UT, USA, 18–22 June 2018.
39. Hou, Q.; Zhou, D.; Feng, J. Coordinate Attention for Efficient Mobile Network Design. *arXiv* **2021**, arXiv:2103.02907. Available online: https://arxiv.org/abs/2103.02907 (accessed on 4 March 2021).
40. Liu, S.; Qi, L.; Qin, H.; Shi, J.; Jia, J. Path Aggregation Network for Instance Segmentation. In Proceedings of the IEEE Conference on Computer Vision and Pattern Recognition, Salt Lake City, UT, USA, 18–23 June 2018; pp. 8759–8768.
41. Elfwing, S.; Uchibe, E.; Doya, K. Sigmoid-weighted linear units for neural network function approximation in reinforcement learning. *Neural Netw.* **2018**, *107*, 3–11. [CrossRef] [PubMed]
42. Dewi, C.; Chen, R.-C.; Liu, Y.-T.; Jiang, X.; Hartomo, K.D. Yolo V4 for Advanced Traffic Sign Recognition With Synthetic Training Data Generated by Various GAN. *IEEE Access* **2019**, *7*, 97228–97242.
43. Cheng, R.; He, X.; Zheng, Z.; Wang, Z. Multi-Scale Safety Helmet Detection Based on SAS-YOLOv3-Tiny. *Appl. Sci.* **2021**, *11*, 3652. [CrossRef]
44. Rezatofighi, H.; Tsoi, N.; Gwak, J.; Sadeghian, A.; Reid, I.; Savarese, S. Generalized intersection over Union: A metric and a loss for bounding box regression. In Proceedings of the 32nd IEEE/CVF Conference on Computer Vision and Pattern Recognition (CVPR), Long Beach, CA, USA, 16–21 June 2019; pp. 658–666.
45. Zheng, Z.; Wang, P.; Liu, W.; Li, J.; Ye, R.; Ren, D. Distance-IoU Loss: Faster and Better Learning for Bounding Box Regression. *arXiv* **2019**, arXiv:1911.08287. Available online: https://arxiv.org/abs/1911.08287 (accessed on 9 March 2020). [CrossRef]

sensors

MDPI

Article

A Hybrid Deep Learning and Visualization Framework for Pushing Behavior Detection in Pedestrian Dynamics

Ahmed Alia [1,2,3], Mohammed Maree [4,*] and Mohcine Chraibi [1,*]

[1] Institute for Advanced Simulation, Forschungszentrum Jülich, 52425 Jülich, Germany; a.alia@fz-juelich.de
[2] Computer Simulation for Fire Protection and Pedestrian Traffic, Faculty of Architecture and Civil Engineering, University of Wuppertal, 42285 Wuppertal, Germany
[3] Department of Management Information Systems, Faculty of Engineering and Information Technology, An-Najah National University, Nablus, Palestine
[4] Department of Information Technology, Faculty of Engineering and Information Technology, Arab American University, Jenin, Palestine
* Correspondence: mohammed.maree@aaup.edu (M.M.); m.chraibi@fz-juelich.de (M.C.)

Abstract: Crowded event entrances could threaten the comfort and safety of pedestrians, especially when some pedestrians push others or use gaps in crowds to gain faster access to an event. Studying and understanding pushing dynamics leads to designing and building more comfortable and safe entrances. Researchers—to understand pushing dynamics—observe and analyze recorded videos to manually identify when and where pushing behavior occurs. Despite the accuracy of the manual method, it can still be time-consuming, tedious, and hard to identify pushing behavior in some scenarios. In this article, we propose a hybrid deep learning and visualization framework that aims to assist researchers in automatically identifying pushing behavior in videos. The proposed framework comprises two main components: (i) Deep optical flow and wheel visualization; to generate motion information maps. (ii) A combination of an EfficientNet-B0-based classifier and a false reduction algorithm for detecting pushing behavior at the video patch level. In addition to the framework, we present a new patch-based approach to enlarge the data and alleviate the class imbalance problem in small-scale pushing behavior datasets. Experimental results (using real-world ground truth of pushing behavior videos) demonstrate that the proposed framework achieves an 86% accuracy rate. Moreover, the EfficientNet-B0-based classifier outperforms baseline CNN-based classifiers in terms of accuracy.

Keywords: deep learning; convolutional neural network; EfficientNet-B0-based classifier; image classification; crowd behavior analysis; pushing behavior detection; motion information maps; deep optical flow

Citation: Alia, A.; Maree, M.; Chraibi, M. A Hybrid Deep Learning and Visualization Framework for Pushing Behavior Detection in Pedestrian Dynamics. *Sensors* **2022**, *22*, 4040. https://doi.org/10.3390/s22114040

Academic Editors: KWONG Tak Wu Sam, Yun Zhang, Xu Long and Tiesong Zhao

Received: 22 April 2022
Accepted: 23 May 2022
Published: 26 May 2022

Publisher's Note: MDPI stays neutral with regard to jurisdictional claims in published maps and institutional affiliations.

Copyright: © 2022 by the authors. Licensee MDPI, Basel, Switzerland. This article is an open access article distributed under the terms and conditions of the Creative Commons Attribution (CC BY) license (https://creativecommons.org/licenses/by/4.0/).

1. Introduction

In entrances of large-scale events, pedestrians either follow the social norm of queuing or force some pushing behavior to gain faster access to the events [1]. Pushing behavior in this context is an unfair strategy that some pedestrians use to move quickly and enter an event faster. This behavior involves pushing others and moving forward quickly by using one's arms, shoulders, elbows, or upper body, as well as using gaps among crowds to overtake and gain faster access [2,3]. Pushing behavior, as opposed to queuing behavior, can increase the density of crowds [4]. Consequently, such behavior may lead to threatening the comfort and safety of pedestrians, resulting in dangerous situations [5]. Thus, understanding pushing behavior, what causes it, and the consequences are crucial, especially when designing and constructing comfortable and safe entrances [1,6]. Conventionally, researchers have attempted to study pushing behavior manually by observing and identifying pushing cases among video recordings of crowded events. For instance, Lügering et al. [3] proposed a rating system on forward motions in crowds to understand when, where, and why pushing behavior appears. The system relies on two trained observers to classify the behaviors of pedestrians over time in a video (the behavior is classified into either pushing or non-pushing categories). In this context, each category includes two gradations: mild and strong for pushing, and falling behind and just walking for non-pushing. For more details on this system, we refer the reader to [3]. To carry out their tasks, the observers analyzed top-view video recordings using pedestrian trajectory data and PeTrack software [7]. However, this manual rating procedure is time-consuming, tedious, and requires a lot of effort by observers, making it hard to identify pushing behavior, specifically when the number of videos and pedestrians in each video increase [3]. Consequently, there is a pressing demand to develop an automatic and reliable framework to identify when and where pushing behavior appears in videos. This article's main motivation is to help social psychologists and event managers identify pushing behavior in videos. However, automatic pushing behavior detection is highly challenging due to several factors, including diversity in pushing behavior, the high similarity and overlap between pushing and non-pushing behaviors, and the high density of crowds at event entrances.

According to a computer vision perspective, automatic pushing behavior detection belongs to the video-based abnormal human behavior detection field [8]. Several human behaviors have been addressed, including walking in the wrong direction [9], running away [10], sudden people grouping or dispersing [11], human falls [12], suspicious behavior, violent acts [13], abnormal crowds [14], hitting, pushing, and kicking [15]. It is worth highlighting that pushing as defined in [15] is different from the "pushing behavior" term in this article. In [15], pushing is a strategy used for fighting, and the scene contains only up to four persons. To the best of our knowledge, no previous studies have automatically identified pushing behavior for faster access from videos.

With the rapid development in deep learning, CNN has achieved remarkable performance in animal [16,17] and human [13,18] behavior detection. The main advantage of CNN is that it directly learns the useful features and classification from data without any human effort [19]. However, CNN requires a large training dataset to build an accurate classifier [20,21]. Unfortunately, this requirement is unavailable in most human behaviors. To alleviate this limitation, several studies have used a combination of CNN and handcrafted feature descriptors [22,23]. The hybrid-based approaches use descriptors to extract valuable information. Then, CNN automatically models abnormal behavior from the extracted information [24,25]. Since labeled data for pushing behavior are scarce, the hybrid-based approaches could be more suitable for automatic pushing behavior detection. Unfortunately, the existing approaches are inefficient for pushing behavior detection [22]. Their main limitations are: (1) their descriptors do not work well to extract accurate information from dense crowds due to occlusions, or they cannot extract the needed information for pushing behavior representation [22,26]; (2) Some used CNN architectures are not efficient enough to deal with the high similarity between pushing and non-pushing behaviors (high inter-class similarity) and the increased diversity in pushing behavior (intra-class variance), leading to misclassification [25,26].

To address the above limitations, we propose a hybrid deep learning and visualization framework for automatically detecting pushing behavior at the patch level in videos. The proposed framework exploits video recordings of crowded entrances captured by a top-view static camera, and comprises two main components: (1) motion information extraction aims to generate motion information maps (MIMs) from the input video. A MIM is an image that contains useful information for pushing behavior representation. This component divides each MIM into several MIM patches, making it easier to see where pedestrians are pushing. For this purpose, recurrent all-pairs field transforms (RAFT) [27] (one of the newest and most promising deep optical flow methods) and the wheel visualization method [28,29] are combined; (2) The pushing patch annotation adapts the EfficientNet-B0-based CNN architecture (the EfficientNet-B0-based CNN [30] is an effective and simple architecture in the EfficientNet family proposed by Google in 2019, achieving the highest accuracy in the ImageNet dataset [31]) to build a robust classifier, which aims to select the relevant features from the MIM patches and label them into pushing and non-pushing categories. We utilized a false reduction algorithm to enhance the classifier's predictions. Finally, the component outputs pushing the annotated video showed when and where the pushing behaviors appeared.

We summarize the main contributions of this article as follows:

1. To the best of our knowledge, we proposed the first framework dedicated to automatically detecting when and where pushing occurs in videos.
2. An integrated EfficientNet-B0-based CNN, RAFT, and wheel visualization within a unique framework for pushing behavior detection.
3. A new patch-based approach to enlarge the data and alleviate the class imbalance problem in the used video recording datasets.
4. To the best of our knowledge, we created the first publicly available dataset to serve this field of research.
5. A false reduction algorithm to improve the accuracy of the proposed framework.

The rest of this paper is organized as follows: Section 2 reviews the related work of video-based abnormal human behavior detection. In Section 3, we introduce the proposed framework. A detailed description of dataset preparation is given in Section 4. Section 5 discusses experimental results and comparisons. Finally, the conclusion and future work are summarized in Section 6.

2. Related Works

Existing video-based abnormal human behavior detection methods can be generally classified into object-based and holistic-based approaches [25,26]. Object-based methods consider the crowd as an aggregation of several pedestrians and rely on detecting and tracking each pedestrian to define abnormal behavior [32]. Due to occlusions, these approaches face difficulties in dense crowds [33,34]. Alternatively, holistic-based approaches deal with crowds as single entities. Thus, they analyze the crowd itself to extract useful information and detect abnormal behaviors [24,25,34]. In this section, we briefly review some holistic-based approaches related to the context of this research. Specifically, the approaches are based on CNN or a hybrid of CNN and handcrafted feature descriptors.

Tay et al. [35] presented a CNN-based approach to detect abnormal actions from videos. The authors trained the CNN on normal and abnormal behaviors to learn the features and classification. As mentioned before, this type of approach requires a large dataset with normal and abnormal behaviors. To address the lack of large datasets with normal and abnormal behaviors, some researchers applied a one-class classifier using datasets of normal behaviors. Obtaining or preparing a dataset with only normal behaviors is easier than a dataset with normal and abnormal behaviors [34,36]. The main idea of the one-class classifier is to learn from the normal behaviors only; to define a class boundary between the normal and not defined (abnormal) classes. Sabokrou et al. [36] utilized a new pre-trained CNN to extract the motion and appearance information from crowded scenes. Then, they used a one-class Gaussian distribution to build the classifier from datasets with normal behaviors. In the same way, the authors of [34,37] used datasets of normal behaviors to develop their one-class classifiers. Xu et al. used a convolutional variational autoencoder to extract features in [34]. Then, multiple Gaussian models were employed to predict abnormal behavior. Ref. [37] adopted a pre-trained CNN model for feature extraction and a one-class support vector machines to predict abnormal behavior. In another work, Ilyas et al. [24] used pre-trained CNN along with a gradient sum of the frame difference to extract relevant features. Afterward, three support vector machines were trained on normal behavior to detect abnormal behavior. In general, the one-class classifier is popular when the abnormal behavior or target behavior class is rare or not well-defined [38]. In contrast, the pushing behavior is well-defined and not rare, especially in high-density and competitive scenarios. Moreover, this type of classifier considers the new normal behavior as abnormal.

In order to overcome the drawback of CNN-based approaches and one-class classifier approaches, several studies used a hybrid-based approach with a multi-class classifier. Duman et al. [22] employed the classical Farnebäck optical flow method [23] and CNN to identify abnormal behavior. The authors used Farnebäck and CNN to extract the direction and speed information. Then, they applied a convolutional long short-term memory network for building the classifier. In [39], the authors used a histogram of gradient and CNN to extract the relevant features, while a least-square support vector was employed for classification. In a similar line of the hybrid approaches, Direkoglu [25] combined the Lucas–Kanade optical flow method and CNN to extract the relevant features and detect "escape and panic behaviors". Almazroey et al. [26] employed mainly a Lucas–Kanade optical flow, pre-trained CNN, and feature selection (neighborhood component analysis) methods to select the relevant features. The authors then applied a support vector machine to generate a trained classifier. Zhou et al. [40] presented a CNN method for detecting and localizing anomalous activities. The study integrated optical flow with a CNN for feature extraction and it used a CNN for the classification task.

In summary, hybrid-based approaches have shown better accuracy than CNN-based approaches on small datasets [41]. Unfortunately, the reviewed hybrid-based approaches are inefficient for dense crowds and pushing behavior detection due to (1) their feature extraction parts being inefficient for dense crowds; (2) The reviewed approaches cannot extract all of the required information for pushing behavior representation; (3) Their classifiers are not efficient enough toward pushing behavior detection. Hence, the proposed framework combines the power of supervised EfficientNet-B0-based CNN, RAFT, and wheel visualization methods to solve the above limitations. The RAFT method works well for estimating optical flow vectors from dense crowds. Moreover, the integration of RAFT and wheel visualization helps to simultaneously extract the needed information for pushing behavior representation. Finally, the adapted EfficientNet-B0-based binary classifier detects distinct features from the extracted information and identifies pushing behavior at the patch level.

3. The Proposed Framework

This section describes the proposed framework for automatic pushing behavior detection at the video patch level. As shown in Figure 1, there are two main components: motion information extraction and pushing patches annotation. The first component extracts motion information from input video recordings, which is further exploited by the pushing patch annotation component to detect and localize pushing behavior, producing pushing annotated video. The following subsections discuss both components in more detail.

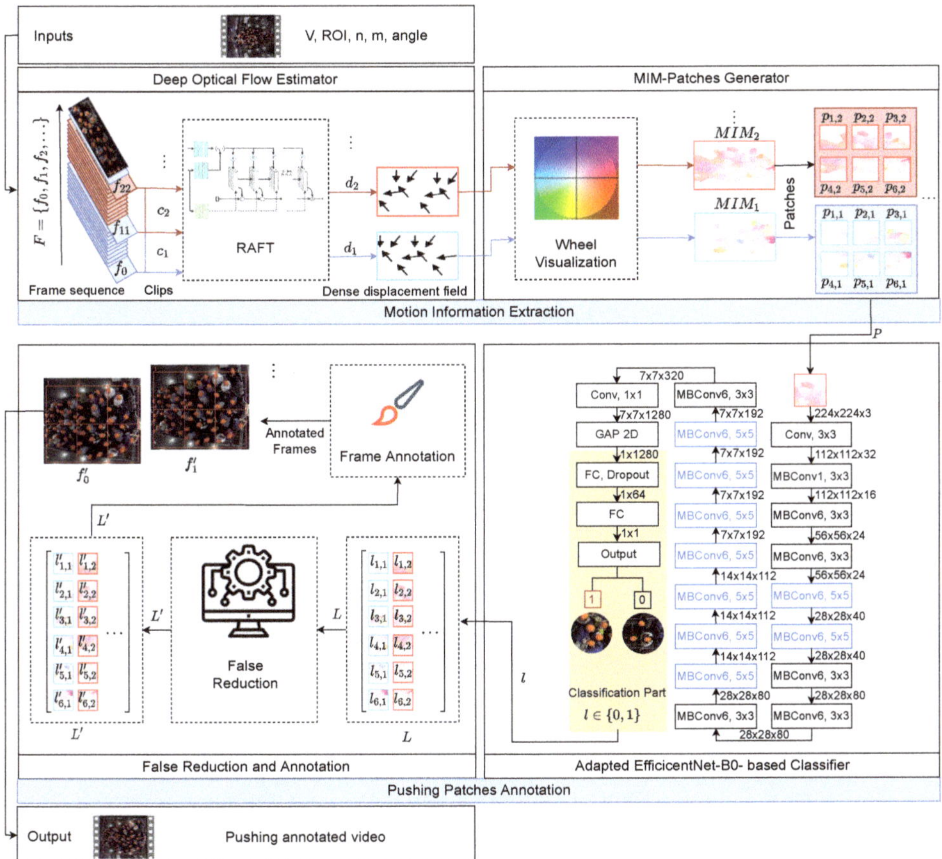

Figure 1. The architecture of the proposed automatic deep learning framework. n and m are two rows and three columns, respectively, for patching. Clip size s is 12 frames. *MIM*: motion information map. P: patch sequence. L: a matrix of all patches labels. L': an updated L by false reduction algorithm. V: the input video. *ROI*: region of interest (entrance area). *angle*: the rotation angle of the input video.

3.1. Motion Information Extraction

This component employs RAFT and wheel visualization to estimate and visualize the crowd motion from the input video at the patch level. The component has two modules, a deep optical flow estimator and a MIM patch generator.

The deep optical flow estimator relies on RAFT to calculate the optical flow vectors for all pixels between two frames. RAFT was introduced in 2020; it is a promising approach for dense crowds because it reduces the effect of occlusions on optical flow estimation [27]. RAFT is based on a composition of CNN and recurrent neural network architectures. Moreover, RAFT has strong cross-dataset generalization and its pre-trained weights are publicly available. For additional information about RAFT, we refer the reader to [27]. This module is based on the RAFT architecture with its pre-trained weights along with three inputs, which are a video of crowded event entrances, the rotation angle of the input video, and the region of interest (ROI) coordinates. To apply RAFT, firstly, we determine the bounding box of the entrance area (ROI) in the input video V. This process is based on user-defined left–top and bottom–right coordinates of the ROI in the pixel unit. Then, we extract the frame sequence $F = \{f_t \mid t = 1, 2, 3, \ldots, T\}$ with ROI only from V, where $f_t \in \mathbb{R}^{w \times h \times 3}$, w and h are the f_t width and height, respectively, 3 is the number of channels,

t is the order of the frame f in V, and T is the total number of frames in V. After that, we rotate the frames (based on the user-defined *angle*) in F to meet the baseline direction of the crowd flow that is used in the classifier, which is from left to right. The rotation process is essential to improve the classifier accuracy because the classifier will be built by training the adapted EfficientNet-B0 on the crowd flow from left to right. Next, we construct from F the sequence of clips $C = \{c_i \mid i = 1, 2, 3, \dots\}$ and c_i is defined as

$$c_i = \{f_{(i-1)\times(s-1)+1}, f_{(i-1)\times(s-1)+2}, \cdots, f_{(i-1)\times(s-1)+s}\},$$ (1)

where s is the clip size. Finally, RAFT is applied on c_i, to calculate the dense displacement field d_i between $f_{(i-1)\times(s-1)+1}$ and $f_{(i-1)\times(s-1)+s}$. The output of RAFT of each pixel location $\langle x, y \rangle$ in c_i is a vector, as shown in.

$$\langle u_{\langle x,y \rangle}, v_{\langle x,y \rangle} \rangle_{c_i} = RAFT(\langle x, y \rangle_{c_i}),$$ (2)

where u and v are horizontal and vertical displacements of a pixel at the $\langle x, y \rangle$ location in c_i, respectively. This means d_i is a matrix of the vector values for the entire c_i, as described in

$$d_i = \left\{ \langle u_{\langle x,y \rangle}, v_{\langle x,y \rangle} \rangle_{c_i} \right\}_{(x,y)=(1,1)}^{(w,h)}$$ (3)

In summary, d_i is the output of this module and will act as the input of the MIM patch generator module.

The second module, the MIM patch generator, employs the wheel visualization to infer the motion information from each d_i. Firstly, the wheel visualization calculates the magnitude and the direction of each motion vector at each pixel $\langle x, y \rangle$ in d_i. Equations (4) and (5) are used to calculate the motion direction and magnitude, respectively. Then, from the calculated information, the wheel visualization generates MIM_i, where $MIM_i \in \mathbb{R}^{w \times h \times 3}$. In MIM, the color refers to the motion direction and the intensity of the color represents the motion magnitude or speed. Figure 2 shows the color wheel scheme (b) and an example of MIM (MIM_{37}) (c) that is generated from c_{37}, whose first and last frames are f_{397} and f_{408}, respectively (a). c_{37} is taken from the experiment 270 [42].

$$\theta(\langle x, y \rangle)_{c_i} = \pi^{-1} \arctan\left(\frac{v_{\langle x,y \rangle}}{u_{\langle x,y \rangle}}\right)$$ (4)

$$mag(\langle x, y \rangle)_{c_i} = \sqrt{u_{\langle x,y \rangle}^2 + v_{\langle x,y \rangle}^2}$$ (5)

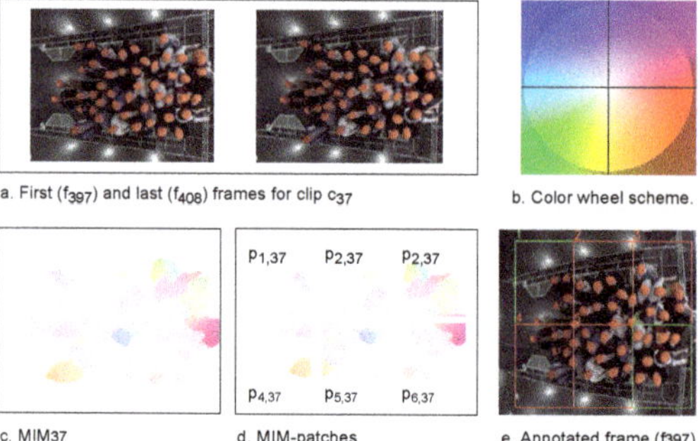

a. First (f_{397}) and last (f_{408}) frames for clip c_{37}

b. Color wheel scheme.

c. MIM37

d. MIM-patches

e. Annotated frame ($f397$)

Figure 2. An illustration of two frames (experiment 270 [42]), color wheel scheme [29], MIM, MIM patches, and annotated frame. In sub-figure (**e**), red boxes refer to pushing patches, while green boxes represent non-pushing patches.

To detect pushing behavior at the patch level, the MIM patch generator divides each MIM_i into several patches. A user-defined row (n) and column (m) are used to split MIM_i into patches $\{p_{k,i} \in \mathbb{R}^{(w/m) \times (h/n) \times 3} \mid k = 1, 2, \ldots, n \times m\}$, where k is the order of the patch in MIM_i. Afterward, each $p_{k,i}$ is resized to a dimension of $224 \times 224 \times 3$, which is the input size of the second component of the framework. For example, MIM_{37} in Figure 2c represents an entrance with dimensions 5×3.4 m on the ground, and it is divided into 2×3 patches $\{p_{k,37} \mid k \leq 6\}$ as given in Figure 2d. These patches are equal in pixels, whereas the area that is covered by them is not necessarily equal. The far patches from the camera cover a larger viewing area compared to close patches; because the far-away object has fewer pixels per m than a close object [43]. In Figure 2d, the average width and height of the $p_{k,37}$ are approximately 1.67×1.7 m.

In summary, the output of the motion information extraction component can be described as $P = \{p_{k,i} \in \mathbb{R}^{224 \times 224 \times 3} \mid k \leq n \times m \ \& \ i \leq |C|\}$, and will serve as input for the second component of the framework.

3.2. Pushing Patches Annotation

This component localizes the pushing patches in $c_i \in C$, annotates the patches in the first frame ($f_{(i-1) \times (s-1)+1}$) of each c_i, and stacks the annotated frame sequence $F' = \{f'_i \mid i = 1, 2, \ldots, |C|\}$ as a video. The Adapted EfficientNet-B0-based classifier and false reduction algorithm are the main modules of this component. In the following, we provide a detailed description.

The main purpose of the first module is to classify each $p_{k,i} \in P$ as pushing or non-pushing. The module is based on EfficientNet-B0 and real-world ground truth of pushing behavior videos. Unfortunately, the existing effective and simple EfficientNet-B0 is unsuitable for detecting pushing behavior because its classification is not binary. However, binary classification is required in our scenario. Therefore, we modify the classification part in EfficientNet-B0 to support a binary classification. The module in Figure 1 shows the architecture of the adapted EfficientNet-B0. Firstly, it executes a 3×3 convolution operation on the input image with dimensions of $224 \times 224 \times 3$. Afterwards, the next 16 mobile inverted bottleneck convolutions are used to extract the feature maps. The final stacked feature maps $\in \mathbb{R}^{7 \times 7 \times 1280}$, where 7 and 7 are the dimensions of each feature map, and 1280 is the number of feature maps. The following global average pooling2D (GAP) layer reduces the dimensions of the stacked feature maps into $1 \times 1 \times 1280$. For the binary classification, we employed a fully connected (FC) layer with a ReLU activation function and a dropout rate of 0.5 [44] before the final FC. The final layer operates as output with a sigmoid activation function to find the probability δ of the class of each $p_{k,i} \in P$.

In order to generate the trained classifier, we trained the adapted EfficientNet-B0 with pushing and non-pushing MIM patches. The labeled MIM patches were extracted from a real-world ground truth of pushing behavior videos, where the ground truth was manually created. In Sections 4 and 5.1, we show how to prepare the labeled MIM patches and train the classifier, respectively. Overall, after several empirical experiments (Section 5.2), the trained classifier on MIM patches of 12 frames produces the best accuracy results. Therefore, our framework uses 12 frames for the clip size (s). Moreover, the classifier uses the threshold for determining the label $l_{k,i}$ of the input $p_{k,i}$ as:

$$l_{k,i} = \begin{cases} 1 \ (\text{pushing class}) & \text{if } \delta \geq 0.5 \\ 0 \ (\text{non-pushing class}) & \text{if } \delta < 0.5 \end{cases} \tag{6}$$

Finally, the output of this module can be described as $L = \{l_{k,i} \in 0, 1 \mid k \leq n \times m \ \& \ i \leq |C|\}$ and will perform as the input of the next module.

In the second module, the false reduction algorithm aims to reduce the number of false predictions in L, which improves the overall accuracy of the proposed framework. Comparing the predictions (L) with the ground truth pushing, we notice that the time interval of the same behavior of each patch region could help improve the accuracy of the framework. We assume a threshold value of $\frac{34}{25}$ second. This value is based on visual inspection.

The example in Figure 3 visualizes the $\{l_{k,i} \mid k \leq 3 \ \& \ i \leq 4\}$ on the first frame of c_1, c_2, c_3, and c_4 in the video. Each c_i represents $\frac{12}{25}$ second. c_1 (Figure 3a) contains one false non-pushing, $p_{2,1}$, while the same region of the patch in $\{c_2, c_3, c_4\}$ is true pushing (Figure 3b–d). This means, we have two time intervals for $\{p_{2,i} \mid i \leq 4\}$. The first has one clip (c_1) (Figure 3a) with a duration of $\frac{12}{25}$ second, which is lesser than the defined threshold. The second time interval contains three clips ($\{c_2, c_3, c_4\}$), with durations equal to the threshold. Then the algorithm changes the prediction of $p_{2,1}$ to "pushing", while it confirms the predictions of $p_{2,2}$, $p_{2,3}$, and $p_{2,4}$. Algorithm 1 presents the pseudocode of the false reduction algorithm. Lines 2–8 show how to reduce the false predictions of the patches in $\{c_i \mid i \leq |c| - 2\}$ Then, lines 9–16 recheck the first two clips (c_1, c_2) to discover the false predictions that are not discovered by lines 2–8. After that, lines 17–32 focus on the last two clips $\{c_{|C|-1}, c_{|C|}\}$. Finally, the updated L is stored in L', which can be described as $L' = \{l'_{k,i} \in 0, 1 \mid k \leq n \times m \ \& \ i \leq |C|\}$.

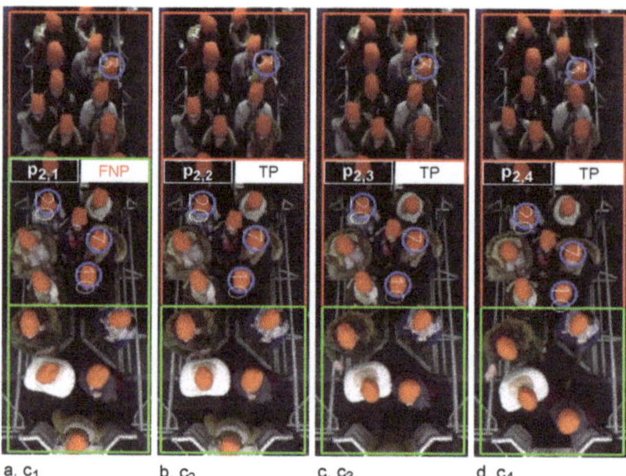

Figure 3. Examples of the visualized classifier predictions with ground truth pushing. The images represent the first frames $\{f_1, f_{12}, f_{23}, f_{34}\}$ of $\{c_1, c_2, c_3, c_4\}$ in a video, respectively; the video is for experiment 110 [42]. Red boxes: pushing patches. Green boxes: non-pushing patches. Blue circles: ground truth pushing. FNP: false non-pushing. TP: true pushing.

After applying the false reduction algorithm, the pushing patch annotation component based on L' identifies the regions of pushing patches on the first frame for each c_i to generate the annotated frame sequence F'. Finally, all annotated frames are stacked as a video, which is the final output of the proposed framework.

Algorithm 1 False Reduction.

Input:
    $matrix[N, M] \leftarrow L$
Output:
    L'

1: **for** $i \leftarrow 0, 1, \ldots, N$ **do**
2:    **for** $j \leftarrow 0, 1, \ldots, M - 2$ **do**
                                          ▷ Excepting the last two clips
3:        **if** $\mathrm{matrix}[i, j] \neq \mathrm{matrix}[i, j + 1]$ **then**
4:          **if** $\mathrm{count}(\mathrm{matrix}[i, j]$ in $\mathrm{matrix}[i, j + 2 \text{ to } j + 4]) > 1$ **then**
5:            $\mathrm{matrix}[i, j + 1] \leftarrow \text{not } \mathrm{matrix}[i, j + 1]$
6:          **end if**
7:        **end if**
8:    **end for**
                                          ▷ Recheck the first two clips
9:    **if** $\mathrm{matrix}[i, 0 \text{ to } 2]$ is not identical **then**
10:      **if** $\mathrm{matrix}[i, 1]$ is not in $\mathrm{matrix}[i, 2 \text{ to } 4]$ **then**
11:        $\mathrm{matrix}[i, 1] \leftarrow \text{not } \mathrm{matrix}[i, 1]$
12:      **end if**
13:      **if** $\mathrm{matrix}[i, 0]$ not in $\mathrm{matrix}[i, 1 \text{ to } 3]$ **then**
14:        $\mathrm{matrix}[i, 0] \leftarrow \text{not } \mathrm{matrix}[i, 0]$
15:      **end if**
16:    **end if**
                                          ▷ For the last two clips
17:    **if** $\mathrm{matrix}[i, M - 1] \neq \mathrm{matrix}[i, M - 2]$ **then**
18:      **if** $\mathrm{matrix}[i, M - 1] \neq \mathrm{matrix}[i, M - 3]$ **then**
19:        $\mathrm{matrix}[i, M - 1] \leftarrow \text{not } \mathrm{matrix}[i, M - 1]$
20:      **end if**
21:    **end if**
22:    **if** $\mathrm{matrix}[i, M - 1] \neq \mathrm{matrix}[i, M - 2]$ **then**
23:      **if** $\mathrm{matrix}[i, M - 1] = \mathrm{matrix}[i, M - 3]$ **then**
24:        $\mathrm{matrix}[i, M - 2] \leftarrow \text{not } \mathrm{matrix}[i, M - 2]$
25:      **end if**
26:    **end if**
27:    **if** $\mathrm{matrix}[i, M - 1] = \mathrm{matrix}[i, M - 2]$ **then**
28:      **if** $\mathrm{matrix}[i, M - 1]$ not in $\mathrm{matrix}[i, M - 5 \text{ to } M - 3]$ **then**
29:        $\mathrm{matrix}[i, M - 1] \leftarrow \text{not } \mathrm{matrix}[i, M - 1]$
30:        $\mathrm{matrix}[i, M - 2] \leftarrow \text{not } \mathrm{matrix}[i, M - 2]$
31:      **end if**
32:    **end if**
33: **end for**
34: $L' \leftarrow matrix$

---

4. Datasets Preparation

This section prepares the required datasets for training and evaluating our classifier. In the following, firstly, four MIM-based datasets are prepared. Then, we present a new patch-based approach for enlarging the data and alleviating the class imbalance problem in the MIM-based datasets. Finally, the patch-based approach is applied to the datasets.

4.1. MIM-Based Datasets Preparation

In this section, we prepare four MIM-based datasets using two clip sizes, Farnebäck and RAFT optical flow methods. Two clip sizes (12 and 25 frames) are used to study the impact of the period of motion on the classifier accuracy. Selecting a small clip size (s) for the MIM sequence (MIM^{Q_s}) leads to redundant and irrelevant information, while a large size leads to a few samples. Consequently, we chose 12 and 25 frames as the two clip sizes. The four datasets can be described as RAFT-$\mathrm{MIM}^{Q_{12}}$, RAFT-$\mathrm{MIM}^{Q_{25}}$, Farnebäck-$\mathrm{MIM}^{Q_{12}}$, and Farnebäck-$\mathrm{MIM}^{Q_{25}}$. For more clarity, the "RAFT-$\mathrm{MIM}^{Q_{12}}$" term means that a combination of RAFT and wheel visualization is used to generate the $\mathrm{MIM}^{Q_{12}}$. As mentioned before, the EfficientNet-B0 learns from MIM sequences generated based on RAFT. Therefore, RAFT-$\mathrm{MIM}^{Q_{12}}$-based and RAFT-$\mathrm{MIM}^{Q_{25}}$-based datasets play the primary role in training and evaluating the proposed classifier. Moreover, we create Farnebäck-$\mathrm{MIM}^{Q_{12}}$-based and Farnebäck-$\mathrm{MIM}^{Q_{25}}$-based datasets to evaluate the impact of RAFT on the classifier accuracy. The pipeline for preparing the datasets (Figure 4) is illustrated below.

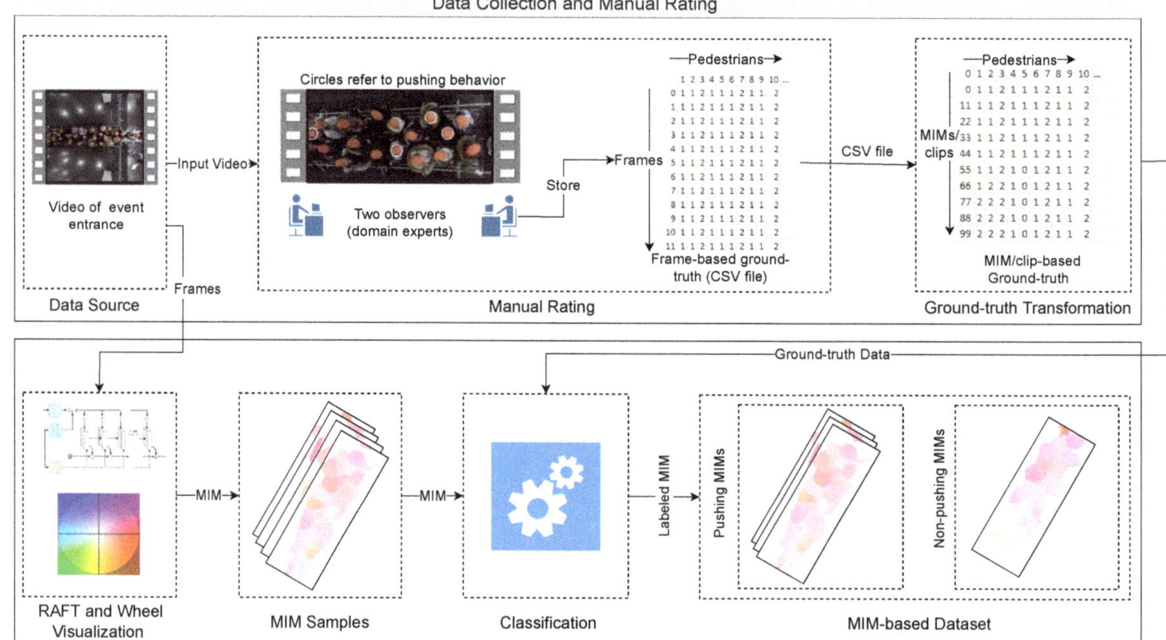

Figure 4. The pipeline of MIM-based dataset preparation.

4.1.1. Data Collection and Manual Rating

In this section, we discuss the data source and the manual rating methodology for the datasets. Five experiments were selected from the data archive hosted by the Forschungszentrum Jülich under CC Attribution 4.0 International license [42]. The experiments mimicked the crowded event entrances. The videos were recorded by a top-view static camera with a frame rate of 25 frames per second and 1920×1440 pixels resolution. In addition to the videos, parameters for video undistortion and trajectory data are available. In Figure 5, the left part sketches the experimental setup and Table 1 shows the different characteristics of the selected experiments.

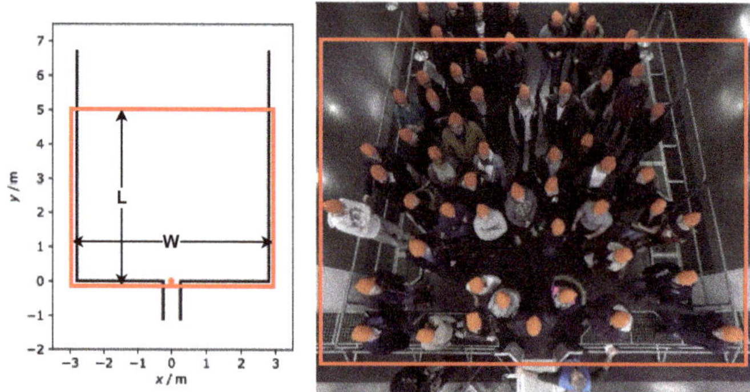

Figure 5. ROI in the entrance. (**Left**) experimental setup with the red dot indicating the coordinate origin [42], (**right**) overhead view of an exemplary experiment. The original frame in the right image is from [42]. The entrance gate width is 0.5 m. The rectangle indicates the entrance area (ROI). L: length of ROI in m. According to the experiment, the width of the ROI (w) varies from 1.2 to 5.6 m.

Table 1. Characteristics of the selected experiments.

Experiment *	Width (m)	Pedestrians	Direction	Frames **
110	1.2	63	Left to right	1285
150	5.6	57	Left to right	1408
170	1.2	25	Left to right	552
270	3.4	67	Right to left	1430
280	3.4	67	Right to left	1640

* The same names as reported in [42]; ** The number of frames that contain pedestrians in the ROI.

Experts performing the manual rating are social psychologists who developed the corresponding rating system [3]. PeTrack [7] was used to track each pedestrian one-by-one, over every frame in the video experiments. Pedestrian ratings are annotated for the first frame when the respective participant becomes visible in the video. The first rating can be extended to the whole video and every frame if that pedestrian does not change his/her behavior. If there is a behavioral change during the experiment, then the rating is also changed. Likewise, it can be extended to the rest of the frames if there is no additional change in the behavior. The rating process is finished after every frame is filled with ratings for every pedestrian. The behaviors of pedestrians are labeled with numbers $\in \{0, 1, 2\}$; 0 indicates that a corresponding pedestrian does not appear in the clip, while 1 and 2 represent non-pushing and pushing behaviors, respectively. Two ground truth files ($\text{MIM}^{Q_{12}}$ and $\text{MIM}^{Q_{25}}$) for each experiment were produced for this paper. Further information about the manual rating can be found in [3].

4.1.2. MIM Labeling and Dataset Creation

Three steps are required to create the labeled MIM-based datasets. In the first step, we generated the samples from the videos; the samples were: RAFT-$\text{MIM}^{Q_{12}}$, RAFT-$\text{MIM}^{Q_{25}}$, Farnebäck-$\text{MIM}^{Q_{12}}$, and Farnebäck-$\text{MIM}^{Q_{25}}$ sequences. The MIM represents the crowd motion in the ROI, which is presented by the rectangle in Figure 5. It is worth mentioning that the directions of the crowd flows in the videos are not similar. This difference could influence building an efficient classifier because changing the direction is one candidate feature for pushing behavior representation. To address this problem, we unified the direction in all videos from left to right before extracting the samples. Additionally, to improve the efficiency of the datasets, we discarded roughly the first seconds from each video to guarantee that all pedestrians started to move forward.

Based on the ground truth files, the second step labels MIMs in the four MIM sequences into pushing and non-pushing. Each MIM that contains at least one pushing pedestrian is classified as pushing; otherwise, it is labeled as non-pushing.

Finally, we randomly split each dataset into three distinct sets: 70% for training, 15% for validation, and 15% for testing. The 70%-15%-15% split ratio is one of the most common ratios in the deep learning field [45]. The information about the number of pushing and non-pushing samples in the training, validation and test sets for the four MIM-based datasets is given in Table 2. As can be seen from Table 2, our MIM-based datasets suffer from two main limitations: lack of data and a class imbalance problem, since less than 20% of samples are non-pushing.

Table 2. Number of labeled samples in training, validation, and test sets for each MIM-based dataset.

Dataset		Experiment										All		
		110		150		170		270		280				
		P	NP	P	NP	P	NP	P	NP	P	NP	P	NP	Total
RAFT-MIM$^{Q_{12}}$	Training	66	16	76	14	28	5	61	29	86	11	317	75	392
	Validation	13	3	15	3	5	1	13	6	18	2	64	15	79
	Test	13	3	15	3	5	1	13	6	18	2	64	15	79
	Total	92	22	106	20	38	7	87	41	122	15	445	105	550
RAFT-MIM$^{Q_{25}}$	Training	30	6	35	6	13	1	29	13	40	4	147	30	177
	Validation	6	2	7	1	3	1	6	2	8	1	30	7	37
	Test	6	2	7	1	3	1	6	2	8	1	30	7	37
	Total	42	10	49	8	19	3	41	17	56	6	207	44	251
Farnebäck-MIM$^{Q_{12}}$	It has the same samples as the RAFT$^{Q_{12}}$ sets while they are generated using Farnebäck.													
Farnebäck-MIM$^{Q_{25}}$	It has the same samples as the RAFT$^{Q_{25}}$ sets while they are generated using Farnebäck.													

P: pushing samples. NP: non-pushing samples. All: all experiments. 110, 150, 170, 270, and 280: names of the video experiments.

4.2. The Proposed Patch-Based Approach

In this section, we propose a new patch-based approach to alleviate the limitations of the MIM-based datasets. The general idea behind our approach is to enlarge the small pushing behavior dataset by dividing each MIM into several patches. After that, we label each patch into "pushing" or "non-pushing" to create a patch-based MIM dataset. The patch should cover a region that can contain a group of pedestrians, where the motion information of the group is essential for pushing behavior representation. Section 5.2 investigates the impact of the patch area on the classifier accuracy. To further clarify the idea of the proposed approach, we take an example of a dataset with one pushing MIM and one non-pushing MIM, as depicted in Figure 6. After applying our idea with 2×3 patches on the dataset, we obtain a patch-based MIM dataset with four pushing, six non-pushing, and two empty MIM patches. The empty patches are discarded. In conclusion, the dataset is enlarged from two images into ten images. The methodology of our approach, as shown in Figure 7 and Algorithm 2, consists of four main phases: automatic patches labeling, visualization, manual revision, and patch-based MIM dataset creation. The following paragraphs discuss the inputs and the workflow of the approach.

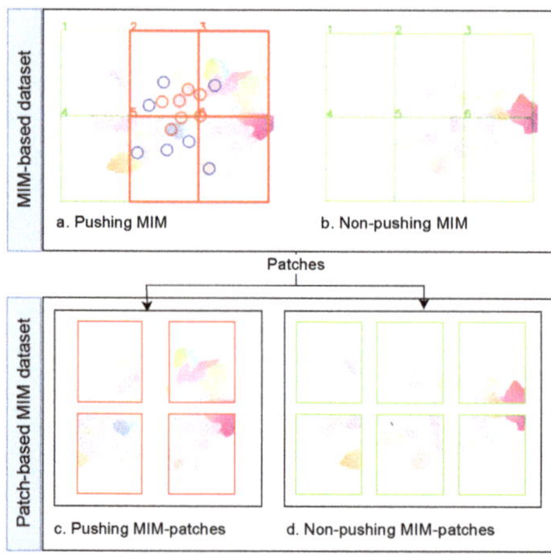

Figure 6. A simple example of the patch-based approach idea. Circles: ground truth pushing. Red boxes: pushing patches. Green boxes: non-pushing patches.

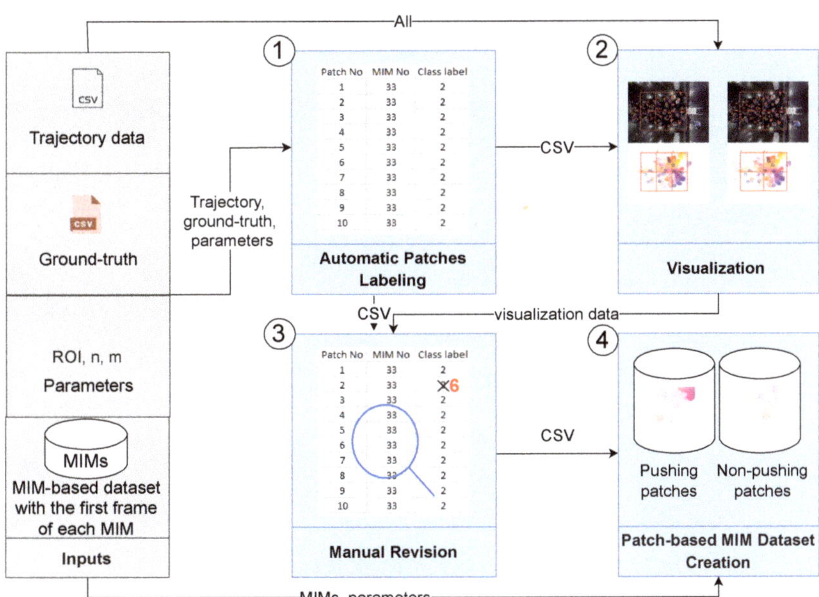

Figure 7. The flow diagram of the proposed patch-based approach. n and m: the numbers of rows and columns, respectively, that are used to divide ROI into n × m regions.

Our approach relies on four inputs (Algorithm 2 and Figure 7, inputs part): (1) MIM-based dataset, which contains a collection of MIMs with the first frame of each MIM; the frames are used in the visualization phase; (2) ROI, n and m, parameters that aim to identify the regions for patches; (3) Pedestrian trajectory data to find the pedestrians in each patch; (4) Manual rating information (ground truth file) helps to label the patches.

The first phase, automatic patch labeling, identifies and labels the patches in each MIM (Algorithm 2, lines 1–33 and Figure 7, first phase). The phase contains two steps: (1) Finding the regions of the patches. For this purpose, we find the coordinates of the regions that are generated from dividing the ROI area into $n \times m$ parts. The extracted regions can be described as $\{a_k | k = 1, 2, \ldots, n \times m\}$, where a_k represents a patch sequence $\{p_{k,i} \in \mathbb{R}^{(w/m) \times (h/n) \times 3} | i = 1, 2, \ldots, |MIM^Q|\}$, w and h are the ROI width and height, respectively, see Algorithm 2, lines 1–15. We should point out that identifying the regions is performed on at least two levels; to avoid losing any useful information. For example, in Figure 8, we first split ROI by 3 × 3 regions (Algorithm 2, lines 2–8), while in the second level, we reduce the number of regions (2 × 2) to obtain larger patches (Algorithm 2, lines 9–15) containing the missing pushing behaviors (pushing behaviors are divided between the patches) in the first level; (2) Labeling the patches is executed according to the pedestrians' behavior in each patch $p_{k,i}$. Firstly, we find all pedestrians who appear in MIM_i (Algorithm 2, lines 18 and 19). Then, we label each $p_{k,i}$ as pushing if it contains at least one pushing behavior; otherwise, it is labeled as non-pushing (Algorithm 2, lines 20–28). Finally, we store k, i, and the label of $p_{k,i}$ in a CSV-file (Algorithm 2, lines 29 and 30).

Algorithm 2 Patch-Based Approach.

Inputs:
dataset ← collection of MIMs with the first frame of each MIM
ROI ← $matrix[left\_top : [x\_coordinate, y\_coordinate], right\_bottom : [x\_coordinate, y\_coordinate]]$
n, m ← the numbers of rows and columns that are used to divide ROI into $n \times m$ regions.
trajectory ← CSV file, each row represents $\langle order\,of\ frame(f_t), pedestrian\,no., pixel\,x - coordinate, pixel\,y - coordinate \rangle$
ground_truth ← CSV file, each row represents $\langle order\,of\ c_i\ or\ MIM, behavior\,of\,pedestrian\,1, behavior\,of\,pedestrian\,2, \ldots, behavior\ of\ last\ pedestrian \rangle$

Outputs:
pushing_folder, non-pushing_folder
1: $region \leftarrow matrix[[]]$ ▷ Automatic patches labeling
2: $patch\_width \leftarrow (ROI[1,0] - ROI[0,0])/m$
3: $patch\_height \leftarrow (ROI[1,1] - ROI[0,1])/n$
4: **for** $i \leftarrow 0, 1, \ldots, n-1$ **do**
5:     **for** $j \leftarrow 0, 1, \ldots, m-1$ **do**
6:         $region.append([ROI[0,0] + j \times patch\_width, ROI[0,1] + i \times patch\_height, ROI[0,0] + (j+1) \times patch\_width, ROI[0,1] + (i+1) \times patch\_height])$
7:     **end for**
8: **end for**
9: $patch\_width \leftarrow (ROI[1,0] - ROI[0,0])/(m-1)$
10: $patch\_height \leftarrow (ROI[1,1] - ROI[0,1])/(n-1)$
11: **for** $i \leftarrow 0, 1, \ldots, n-2$ **do**
12:     **for** $j \leftarrow 0, 1, \ldots, m-2$ **do**
13:         $region.append([ROI[0,0] + j \times patch\_width, ROI[0,1] + i \times patch\_height, ROI[0,0] + (j+1) \times patch\_width, ROI[0,1] + (i+1) \times patch\_height])$
14:     **end for**
15: **end for**
16: $file \leftarrow CSV\,file$
17: **for each** $MIM \in dataset$ **do**
18:     $frame\_order \leftarrow MIM\,name$
19:     $ped \leftarrow Filter(trajectory.frame\_order)[1]$
20:     $patch\_no \leftarrow 1$
21:     **for each** $patch\_region \in region$ **do**
22:         $behavior \leftarrow 1$ //non-pushing
23:         **for each** $ped \in patch\_region$ **do**
24:             **if** $Filter(ground\_truth.frame\_order\,\&\,ped) == 2$ **then**
25:                 $behavior \leftarrow 2$ //pushing
26:                 break
27:             **end if**
28:         **end for**
29:         $record \leftarrow [patch\_no, frame\_order, behavior]$
30:         $file.write(record)$
31:         $patch\_no \leftarrow patch\_no + 1$
32:     **end for**
33: **end for**

                                     ▷ Visualization
34: **for each** $frame \in dataset$ **do**
35:     $frame\_order \leftarrow frame\,name$
36:     $ped \leftarrow Filter(trajectory.frame\_order)[1]$
37:     **for each** $person \in ped$ **do**
38:         $behavior \leftarrow Filter(ground\_truth.frame\_order\,\&\,person)$
39:         **if** behavior $==2$ **then**
40:             draw a circle around the position $\langle person[2], person[3] \rangle$ of pedestrian $person[1]$ over $frame$
41:         **end if**
42:     **end for**
43:     **for** $patch\_no \leftarrow 1, 2, \ldots, len(region)$ **do**
44:         **if** $Filter(file.frame\_order\,\&\,patch\_no)[2] == 2$ **then**
45:             draw a red rectangle around $region[patch\_no - 1]$ over $frame$
46:         **else**
47:             draw a green rectangle around $region[patch\_no - 1]$ over $frame$
48:         **end if**
49:     **end for**
50: **end for**

                                     ▷ Manual revision
51: **for each** $frame \in dataset$ **do**
52:     **for each** $patch\_region \in region$ **do**
53:         manual revision of $patch\_region$ in $frame$
54:         **if** $patch\_region$ contains only a part of one pushing behavior and its label is 2 **then**
55:             manually updating the label of the patch_region in $file$ to 6, where 6 means unknown patch
56:         **end if**
57:     **end for**
58: **end for**

                              ▷ Patch-based MIM dataset creation
59: **for each** $MIM \in dataset$ **do**
60:     $MIM\_order \leftarrow MIM\,name$
61:     **for** $patch\_no \leftarrow 1, 2, \ldots, len(region)$ **do**
62:         $patch \leftarrow MIM[region[patch\_no - 1, 1] : region[patch\_no - 1, 3], [region[patch\_no - 1, 0] : region[patch\_no - 1, 2]]$
63:         **if** $Filter(file.MIM\_order\,\&\,patch\_no)[2] == 2$ **then**
64:             save $patch$ to pushing_folder under name "$MIM\_order - patch\_no$"
65:         **else if** $Filter(file.MIM\_order\,\&\,patch\_no)[2] == 1$ **then**
66:             save $patch$ to non-pushing_folder under name "$MIM\_order - patch\_no$"
67:         **end if**
68:     **end for**
69: **end for**

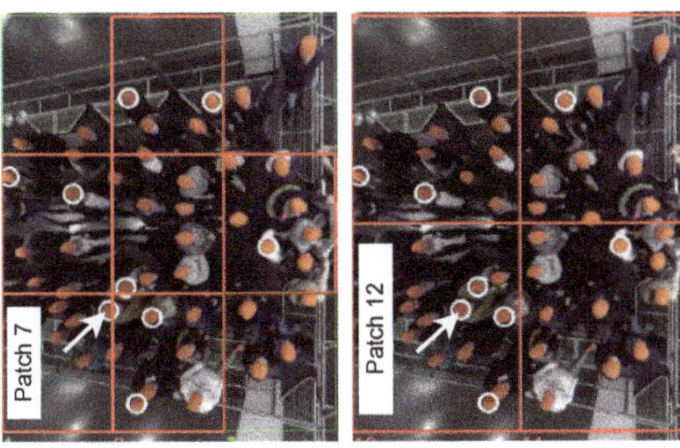

Figure 8. An example of identifying patches and the visualization process. The original frames are from [42]. Red boxes: pushing patches. Green boxes: non-pushing patches. White circles: ground truth pushing.

Despite the availability of the pedestrian trajectories, the automatic patch labeling phase is not 100% accurate, affecting the quality of the dataset. The automatic way fails to label some of the patches that only contain a part of one pushing behavior. Therefore, manual revision is required to improve the dataset quality. To ease this process and make it more accurate, the visualization phase (Algorithm 2, lines 34–50 and Figure 7, second phase) visualizes the ground truth pushing (Algorithm 2, lines 36–42), and the label of each $p_{k,i}$ (Algorithm 2, lines 43–49) on the first frame of MIM_i. Figure 8 is an example of the visualization process.

The manual revision phase ensures that each $p_{k,i}$ takes the correct label by manually revising the visualization data (Algorithm 2, lines 51–58 and Figure 7, third phase). The criteria used in the revision are as follows: if $p_{k,i}$ only has a part of one pushing behavior, we change the labels to unknown labels in the CSV-file generated by the first phase; otherwise, the label of $p_{k,i}$ is not changed. The unknown patches do not offer complete information about pushing behavior or non-pushing behavior. Therefore, the final phase in our approach will discard them. A good example of an unknown patch is patch 7, Figure 8a. This patch contains a part of one pushing behavior, as highlighted by the arrow. On the other hand, patch 12 in the aforementioned example (b) contains the whole pushing behavior that we lose in discarding patch 7.

In the final phase (Algorithm 2, lines 59–69 and Figure 7, fourth phase), the patch-based MIM dataset creation is responsible for creating the labeled patch-based MIM dataset, containing two groups of MIM patches, pushing and non-pushing. Firstly, we crop $p_{k,i}$ from MIM_i (Algorithm 2, line 62). Next, and according to the labels of the patches, the pushing patches are stored in the first group (Algorithm 2, lines 63 and 64), while the second group archives the non-pushing patches (Algorithm 2, lines 65 and 66).

4.3. Patch-Based MIM Dataset Creation

In this section, we aimed to create several patch-based MIM datasets using the proposed patch-based approach and the MIM-based datasets. The main purposes of the created datasets are: (1) to build and evaluate our classifier; (2) examine the influence of the patch area and clip size on classifier accuracy.

In order to study the impact of the patch area on classifier accuracy, we used two different areas. As we mentioned before, the regions covered by the patches should be enough to house a group of pedestrians. Therefore, according to the ROIs of the experiments, we selected the two patch areas as follows: 1 m × (1 to 1.2) m and 1.67 m × (1.2 to 1.86) m. The dimensions of each area refer to the length x width of patches. Due to the width difference between the experiment setups, there is a variation in the width between the experiments. Table 1 shows the width of each experiment's setup, while the length of the

ROI area in all experiment setups was 5 m (Figure 5, left part). For the sake of discussion, we name the 1 m × (1 to 1.2) m patch area as the small patch, and 1.67 m × (1.2 to 1.86) m as the medium patch. Moreover, the small and medium patching with the used levels are illustrated in Figure 9.

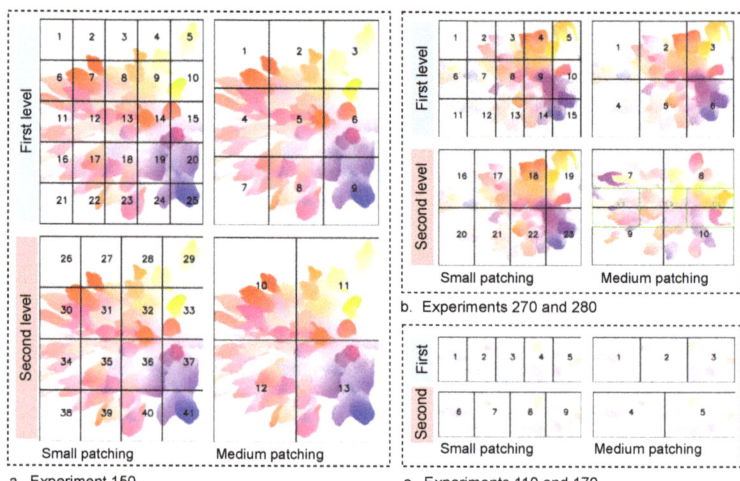

Figure 9. The visualization of patching for the experiments. Numbers represent the patch order in each experiment and level.

The patch-based approach is performed on the RAFT-MIM-based training sets to generate patch-based RAFT-MIM training sets, while it creates patch-based RAFT-MIM validation sets from the RAFT-MIM-based validation sets. The created patch-based RAFT-MIM datasets with their numbers of labeled samples are presented in Table 3. The table and Figure 10 demonstrate that the proposed approach enlarges the RAFT-MIM-based training and validation sets in both small and medium patching. The approach roughly duplicates the MIM-based training and validation sets 13 times in small patching. While in medium patching, each MIM-based training and validation set is duplicated 8 times. Moreover, our approach decreases the class imbalance issue significantly.

Table 3. Number of labeled MIM patches in training and validation sets for each patch-based MIM dataset.

Dataset		Experiment												
		110		150		170		270		280		All		
		P	NP	P	NP	P	NP	P	NP	P	NP	P	NP	Total
Patch-based small RAFT-MIM$^{Q}_{12}$	Training	350	279	523	932	121	97	528	784	634	806	2156	2898	5054
	Validation	67	53	89	161	20	21	91	169	108	162	375	566	941
	Total	417	332	612	1093	141	118	619	953	742	968	2531	3464	5995
Patch-based small RAFT-MIM$^{Q}_{25}$	Training	156	124	249	419	53	42	236	379	324	354	1018	1318	2336
	Validation	33	26	35	82	9	12	56	53	67	89	200	262	462
	Total	189	150	284	501	62	54	292	432	391	443	1218	1580	2798
Patch-based medium RAFT-MIM$^{Q}_{12}$	Training	237	131	298	354	95	38	540	439	698	326	1868	1288	3156
	Validation	45	26	55	64	16	8	98	105	126	81	340	284	624
	Total	282	157	353	418	111	46	638	544	824	407	2208	1572	3780
Patch-based medium RAFT-MIM$^{Q}_{25}$	Training	107	58	142	151	42	14	242	219	338	146	871	585	1459
	Validation	22	14	20	37	8	6	56	27	68	32	174	116	290
	Total	129	72	162	188	50	20	298	246	406	178	1045	704	1749

P: pushing samples. NP: non-pushing samples. All: all experiments. 110, 150, 170, 270, and 280: names of the video experiments.

The approach reduces the difference percentage between the pushing and non-pushing classes in the patch-based MIM training and validation sets as follows: patch-based small

RAFT-MIM$^{Q_{12}}$, from 62% to 16%. Patch-based medium RAFT-MIM$^{Q_{12}}$, from 62% to 17%. Patch-based small RAFT-MIM$^{Q_{25}}$, from 65% to 13%. Patch-based medium RAFT-MIM$^{Q_{25}}$, from 65% to 20%.

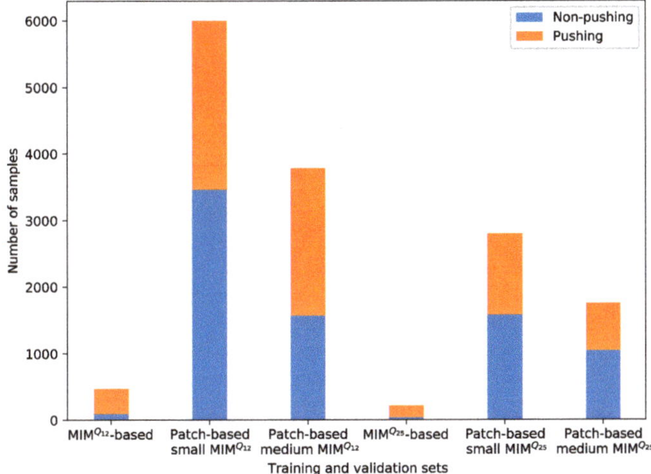

Figure 10. The visualization of the number of pushing and non-pushing samples for the training and validation sets.

Despite these promising results, we can only assess the efficiency of our approach when the CNN-based classifier is trained and tested on our patch-based RAFT-MIM datasets. For this important process, we generate four patch-based RAFT-MIM test sets. The patch-based approach applies the first level of patching on RAFT-MIM-based test sets (Table 2) to generate the patch-based RAFT-MIM test sets. We apply the first level in the small and medium patching (because we need to evaluate our classifier for detecting pushing behavior at the small and medium patches). Table 4 shows the number of labeled MIM patches in the patch-based RAFT-MIM test sets and their experiments. In Section 5.3, we discuss the impact of the patch-based approach on the accuracy of CNN-based classifiers.

Table 4. Number of labeled MIM patches in patch-based test sets.

Test Set	Experiment												
	110		150		170		270		280		All		
	P	NP	P	NP	P	NP	P	NP	P	NP	P	NP	Total
Patch-based small RAFT-MIM$^{Q_{12}}$ test	40	28	47	99	9	13	59	112	61	108	216	360	576
Patch-based small RAFT-MIM$^{Q_{25}}$ test	18	15	19	44	7	8	28	54	25	36	97	157	254
Patch-based medium RAFT-MIM$^{Q_{12}}$ test	26	16	25	47	8	6	47	41	50	40	156	150	306
Patch-based medium RAFT-MIM$^{Q_{25}}$ test	13	8	8	26	5	5	22	19	20	18	68	76	144

P: pushing samples. NP: non-pushing samples. All: all experiments. 110, 150, 170, 270, and 280: names of the video experiments.

5. Experimental Results

This section presents the parameter setup and performance metrics used in the evaluation. Then, it trains and evaluates our classifier and studies the impact of the patch area and clip size on the classifier performance. After that, we investigate the influence of the patch-based approach on the classifier performance. Next, the effect of RAFT on the classifier is discussed. Finally, we evaluate the performance of the proposed framework on the distorted videos.

5.1. Parameter Setup and Performance Metrics

For the training process, the RMSProp optimizer with a binary cross-entropy loss function was used. The batch size and epochs were set to 128 and 100, respectively. Moreover, when the validation accuracy did not increase for 20 epochs, the training process was automatically terminated. In the RAFT and Farnebäck methods, we used the default parameters.

The implementations in this paper were performed on a personal computer running the Ubuntu operating system with an Intel(R) Core(TM) i7-10510U CPU @ 1.80 GHz (8 CPUs) 2.3 GHz and 32 GB RAM. The implementation was written in Python using PyTorch, Keras, TensorFlow, and OpenCV libraries.

In order to evaluate the performance of the proposed framework and our classifier, we used accuracy and F1 score metrics. This combination was necessary since we had imbalanced datasets. Further information on the evaluation metrics can be found in [46].

5.2. Our Classifier Training and Evaluation, the Impact of Patch Area and Clip Size

In this section, we have two objectives: (1) training and evaluating the adapted EfficientNet-B0-based classifier. (2) Investigating the impact of the clip size and patch area on the performance of the classifier.

We compare the adapted EfficientNet-B0-based classifier with three well-known CNN-based classifiers (MobileNet [47], InceptionV3 [48], and ResNet50 [49]) to achieve the above objectives. The classification part in the well-known CNN architectures is modified to be binary. The four classifiers train from scratch on the patch-based RAFT-MIM training and validation sets. Then we evaluate the trained classifiers on patch-based RAFT-MIM test sets to explore their performance.

From the results in Table 5 and Figure 11, it is seen that our trained classifier on the patch-based medium RAFT-MIM$^{Q_{12}}$ dataset achieves better accuracy and F1 scores than other classifiers. More specifically, the EfficientNet-B0-based classifier has 88% accuracy and F1 scores. Furthermore, the medium patches help all classifiers to obtain better performances than small patches. At the same time, MIM$^{Q_{12}}$ is better than MIM$^{Q_{25}}$ for training the four classifiers in terms of accuracy and F1 score.

Table 5. Comparison with well-known CNN-based classifiers on patch-based MIM datasets.

CNN-Based Classifier	Patch-Based MIM Dataset							
	Medium RAFT-MIM$^{Q_{12}}$		Small RAFT-MIM$^{Q_{12}}$		Medium RAFT-MIM$^{Q_{25}}$		Small RAFT-MIM$^{Q_{25}}$	
	Accuracy%	F1 Score%	Accuracy%	F1 Score%	Accuracy%	F1 Score%	Accuracy%	F1 Score%
MobileNet	87	87	79	78	85	85	77	74
EfficientNet-B0	**88**	**88**	**81**	**80**	**87**	**87**	**78**	**78**
InceptionV3	85	85	76	75	80	80	76	74
ResNet50	80	80	70	70	74	73	71	69

Bold: best results in each dataset. Gray highlight: Best results among all datasets.

The patch area influences the classifier performance significantly. For example, medium patches improve the EfficientNet-B0-based classifier accuracy and F1 scores by 7% and 8%, respectively, compared to the small patches. On the other hand, the effect of the MIM sequence (clip size) on the classifier performance is lesser than the influence of the patch area. Compared to medium MIM$^{Q_{25}}$, medium MIM$^{Q_{12}}$ enhances the accuracy and F1 score by 1% in the EfficientNet-B0-based classifier.

In summary, the trained adapted EfficientNet-B0-based classifier on the patch-based medium RAFT-MIM$^{Q_{12}}$ dataset achieves the best performance.

5.3. The Impact of the Patch-Based Approach

We evaluated the impact of the proposed patch-based approach on the performance of the trained classifiers on patch-based medium RAFT-MIM$^{Q_{12}}$ training and validation sets. To achieve that, we trained the four classifiers on RAFT-MIM$^{Q_{12}}$-based training and validation sets (Table 2). Then the trained classifiers were evaluated on patch-based medium RAFT-MIM$^{Q_{12}}$ test sets (Table 4).

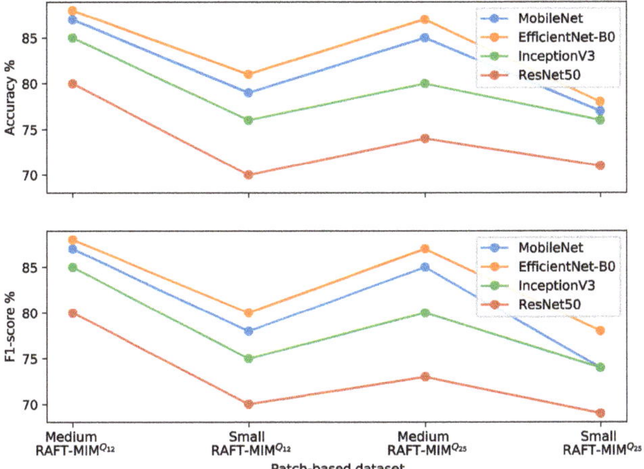

Figure 11. Comparisons of four classifiers over all patch-based RAFT-MIM sets.

Table 6 represents the performance of MIM-based classifiers. The comparison between patch-based classifiers and MIM-based classifiers is visualized in Figure 12. We can see that the EfficientNet-B0-based classifier (MIM-based classifier) achieves the best performance, which is a 78% accuracy and F1 score. In comparison, the corresponding patch-based classifier achieves an 88% accuracy and F1 score. This means that the patch-based approach improves the accuracy and F1 score of the EfficientNet-B0-based classifier by 10%. Similarly, in other classifiers, the patch-based approach increases the accuracy and F1 score by at least 15% for each.

Table 6. MIM -based classifier evaluation.

CNN-Based Classifier	Patch-Based Classifier		MIM-Based Classifier	
	Accuracy%	F1 Score%	Accuracy%	F1 Score%
MobileNet	87	87	71	69
EfficientNet-B0	88	88	78	78
InceptionV3	85	85	51	34
ResNet50	80	80	51	34

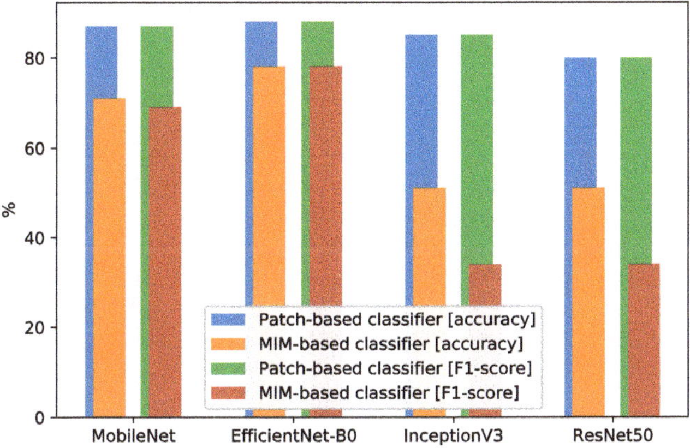

Figure 12. Comparison between MIM-based classifiers and patch-based classifiers.

5.4. The Impact of RAFT

In order to study the impact of RAFT on our classifier, we trained it using the patch-based medium Farnebäck-MIM$^{Q_{12}}$ dataset. Farnebäck is one of the most popular optical flow methods used in human action detection. Firstly, we created patch-based medium training and validation and test sets from the Farnebäck-MIM$^{Q_{12}}$-based dataset (Table 2). The training and validation sets were used to train the EfficientNet-B0-based classifier (Farnebäck-based classifier), while the test set was used to evaluate the classifier. Finally, we compared the performance of the classifier based on RAFT with the classifier based on Farnebäck. As shown in Table 7 and Figure 13, we find that RAFT improves the classifier performance in all classifiers compared to Farnebäck. In particular, RAFT enhances the EfficientNet-B0-based classifier performance by 8%.

Table 7. Comparison between RAFT-based classifiers and Farnebäck-based classifiers.

Classifier	RAFT-Based Classifier		Farnebäck-Based Classifier	
	Accuracy%	F1 Score%	Accuracy%	F1 Score%
MobileNet	87	87	81	81
EfficientNet-B0	88	88	80	80
InceptionV3	85	85	79	79
ResNet50	80	80	74	73

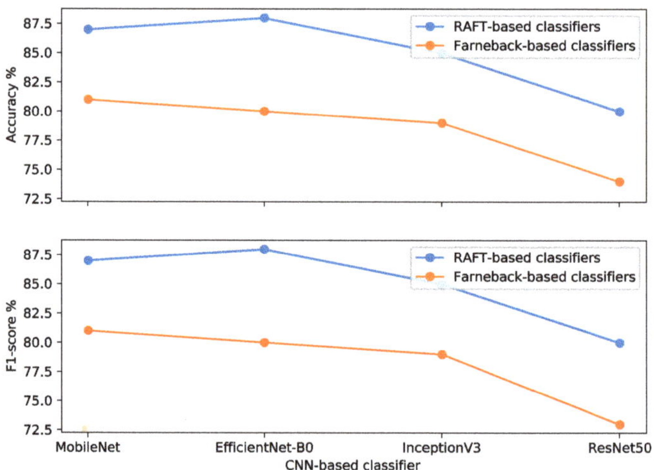

Figure 13. Comparison between the RAFT-based classifier and the Farnebäck-based classifier.

5.5. Comparison between the Proposed Classifier and the Customized CNN-Based Classifiers in Related Works

In this section, we evaluate our classifier by comparing it with two of the most recent customized CNN architectures (CNN-1 [25] and CNN-2 [35]) in the video-based abnormal human behavior detection field. Customized CNNs have simple architectures; CNN-1 used 75 × 75 pixels as an input image, three convolutional layers followed by batch normalization and max pooling operations. Finally, a fully connected layer with a softmax activation function was employed for classification. On the other hand, CNN-2 resized the input images into 28 × 28 pixels, then employed three convolutional layers with three max pooling layers (each max pooling layer with strides of 2 pixels). Moreover, it used two fully connected layers for predictions; the first layer was based on a ReLU activation function, while the second layer used a softmax activation function. For more details on CNN-1 and CNN-2, we refer the reader to [25,35], respectively.

The three classifiers were trained and evaluated based on the patch-based medium RAFT-MIM$^{Q_{12}}$ dataset. As shown in Table 8 and Figure 14, CNN-1 and CNN-2 obtained low accuracy and F1 scores (less than 61%), while our classifier achieved an 88% accuracy and F1 score.

Table 8. Comparisons to the customized CNN-based classifiers in the related works.

Classifier	Accuracy%	F1 Score%
EfficientNet-B0 (our classifier)	88	88
CNN-1 [25]	60	54
CNN-2 [35]	54	35

In summary, and according to Figure 15, the reviewed customized CNN architectures are simple and not enough to detect pushing behaviors because the differences between pushing and non-pushing behaviors are not clear in many cases. To address this challenge, we need an efficient classifier (such as the proposed classifier).

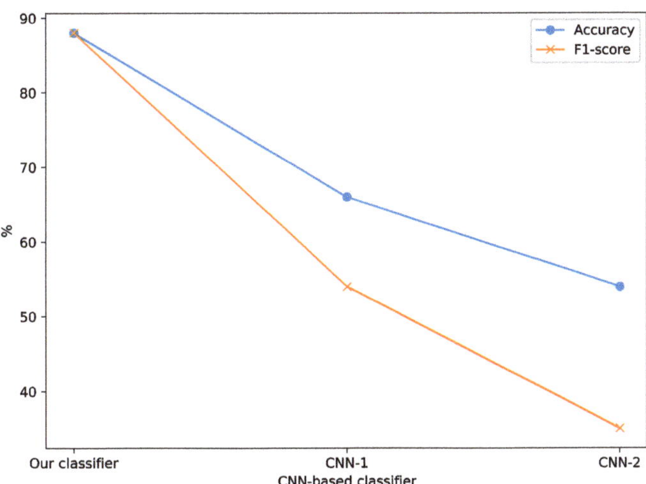

Figure 14. Comparison between our classifier, CNN-1 [25] and CNN-2 [35] based on the patch-based medium RAFT-MIM$^{Q_{12}}$ dataset.

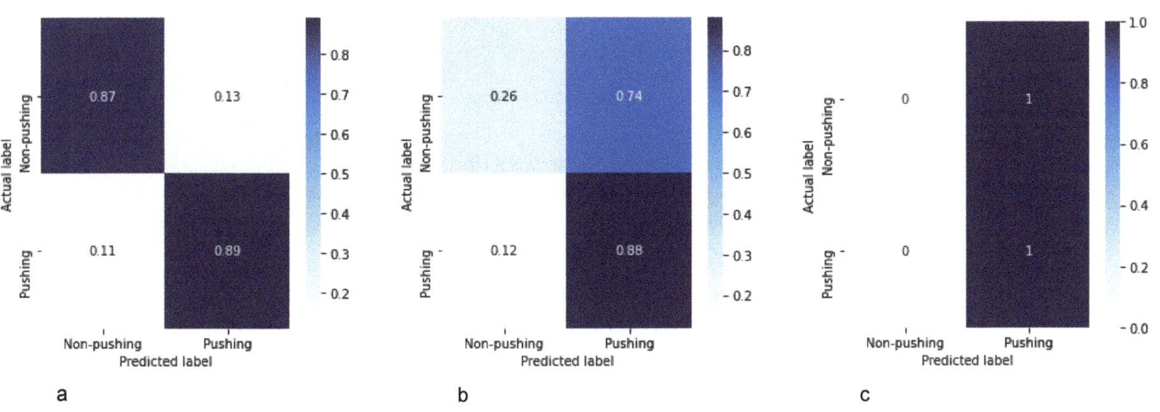

Figure 15. Confusion matrices for our classifier (**a**), CNN-1 [25] (**b**) and CNN-2 [35] (**c**) based on the patch-based medium RAFT-MIM$^{Q_{12}}$ dataset.

5.6. Framework Performance Evaluation

Optical imaging systems often suffer from distortion artifacts [50]. According to [51], distortion is "a deviation from the ideal projection considered in a pinhole camera model, it is a form of optical aberration in which straight lines in the scene do not remain straight in an image". The distortion leads to inaccurate trajectory data [52]. Therefore, PeTrack corrects the distorted videos before extracting the accurate trajectory data, whereas the required information for the correction is not often available. Unfortunately, training our classifier on undistorted videos could decrease the framework performance on distorted videos. Therefore, in this section, we evaluated the proposed framework performance on the distorted videos and studied the impact of the false reduction algorithm on the framework performance. To achieve both goals, firstly, we evaluated the framework's performance without the algorithm on the distorted videos. Then, the framework with the algorithm was evaluated. Finally, we compared both performances.

A qualitative methodology was used in both evaluations; the methodology consisted of four steps: (1) we applied the framework to annotate distorted clips corresponding to MIMs in the RAFT-MIM$^{Q_{12}}$-based test set (Figure 16); the bottom image is an example of an annotated distorted clip; (2) Unfortunately, we could not visualize the ground truth pushing on the distorted frames because the trajectory data were inaccurate. Therefore, we visualized ground truth pushing on the first frame of the corresponding undistorted clips to the distorted clips, Figure 16, top image. Then, we manually identified pushing behaviors on the distorted clips based on the corresponding annotated undistorted clips; This process is highlighted by arrows in Figure 16. (3) We manually calculated the number of true pushing, false pushing, true non-pushing, and false non-pushing. Note that the empty patches were discarded. Non-empty patches containing more than half of the pushing behaviors are labeled as pushing; otherwise, they are labeled as non-pushing. Half of the pushing behavior means that more than half of the visible pedestrian body contributes to pushing; (4) Finally, we measured the accuracy and F1 score metrics.

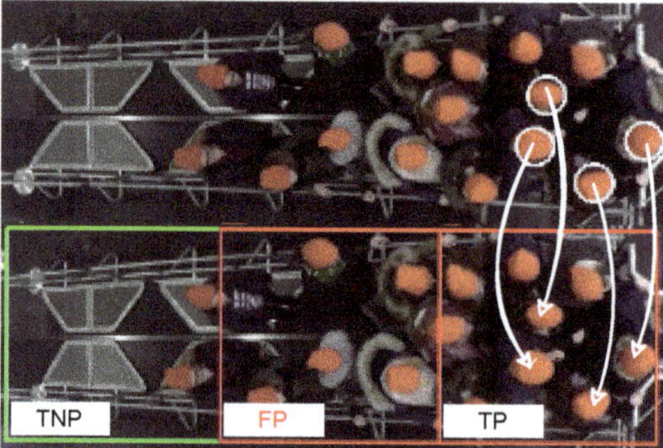

Figure 16. An example of the used qualitative methodology. (**Top**) the first frame of an undistorted clip; (**Bottom**) the first frame of a distorted clip. White arrows: connecting the pushing locations in both undistorted and distorted clips. TP: true pushing. FP: false pushing. TNP: true non-pushing. White circles: ground truth pushing. Red boxes: predicted pushing patches. Green boxes: predicted non-pushing patches.

From Table 9, we can see that our framework with the false reduction algorithm can achieve an 86% accuracy and F1 score on the distorted videos. Moreover, the false reduction improves the performance by 2%.

Table 9. The performance of the framework with and without false reduction on distorted videos.

Framework	Accuracy%	F1 Score%
Without false reduction	84	84
With false reduction	86	86

6. Conclusions, Limitations, and Future Work

This paper proposed a hybrid deep learning and visualization framework for automatic pushing behavior detection at the patch level, particularly from top-view video recordings of crowded event entrances. The framework mainly relied on the power of EfficientNet-B0-based CNN, RAFT, and wheel visualization methods to overcome the high complexity of pushing behavior detection. RAFT and wheel visualization are combined to extract crowd motion information and generate MIM patches. After that, the combination of the EfficientNet-B0-based classifier and false reduction algorithm detects the pushing MIM patches and produces the pushing annotated video. In addition to the proposed framework, we introduced an efficient patch-based approach to increase the number of samples and alleviate the class imbalance issue in pushing datasets. The approach aims to improve the accuracy of the classifier and the proposed framework. Furthermore, we created new datasets using a real-world ground truth of pushing behavior videos and the proposed patch-based approach for evaluation. The experimental results show that: (1) the patch-based medium RAFT-MIM$^{Q_{12}}$ dataset is the best compared to the other generated datasets for training the CNN-based classifiers; (2) Our classifier outperformed the baseline well-known CNN architectures in image classification as well as customized CNN architectures in the related works; (3) Compared to Farnebäck, RAFT improved the accuracy of the proposed classifier by 8%; (4) The proposed patch-based approach helped to enhance our classifier accuracy from 78% to 88%; (5) Overall, the proposed adapted EfficientNet-B0-based classifier obtained 88% accuracy on the patch-based medium RAFT-MIM$^{Q_{12}}$ dataset; (6) The above results were based on undistorted videos, while the proposed framework obtained 86% accuracy on the distorted videos; (7) The developed false reduction algorithm improved the framework accuracy on distorted videos from 84% to 86%. The main reason behind decreasing the framework accuracy on distorted videos was training the classifier based on undistorted videos.

The main limitations of the proposed framework cannot be applied in real time. Additionally, it does not work well with recorded videos from a moving camera. Moreover, the framework was evaluated only on specific scenarios of crowded event entrances.

In future work, we plan to evaluate our framework in more scenarios of crowded event entrances. Additionally, we plan to optimize the proposed framework to allow real-time detection.

Author Contributions: Conceptualization, A.A., M.M. and M.C.; methodology, A.A., M.M. and M.C.; software, A.A.; validation, A.A.; formal analysis, A.A., M.M. and M.C.; investigation, A.A., M.M. and M.C.; data curation, A.A.; writing—original draft preparation, A.A.; writing—review and editing, A.A., M.M. and M.C.; supervision, M.M. and M.C. All authors have read and agreed to the published version of the manuscript.

Funding: This work was funded by the German Federal Ministry of Education and Research (BMBF: funding number 01DH16027) within the Palestinian-German Science Bridge project framework, and partially by the Deutsche Forschungsgemeinschaft (DFG, German Research Foundation)—491111487.

Institutional Review Board Statement: The experiments were conducted according to the guidelines of the Declaration of Helsinki, and approved by the ethics board at the University of Wuppertal, Germany.

Informed Consent Statement: Informed consent was obtained from all subjects involved in the experiments.

Data Availability Statement: All videos and trajectory data used in generating the datasets were obtained from the data archive hosted by the Forschungszentrum Jülich under CC Attribution 4.0 International license [42]. The undistorted videos, trained CNN-based classifiers, test sets, results, codes (framework; building, training and evaluating the classifiers) generated or used in this paper are publicly available at: https://github.com/PedestrianDynamics/DL4PuDe (accessed on 10 April 2022). The training and validation sets are available from the author upon request.

Acknowledgments: The authors are thankful to Armin Seyfried for the many helpful and constructive discussions. They would also like to thank Anna Sieben, Helena Lügering, and Ezel Üsten for developing the rating system and annotating the pushing behavior in the video experiments. Additionally, the authors would like to thank Maik Boltes and Tobias Schrödter for valuable technical discussions.

Conflicts of Interest: The authors declare no conflict of interest.

References

1. Adrian, J.; Boltes, M.; Sieben, A.; Seyfried, A. Influence of Corridor Width and Motivation on Pedestrians in Front of Bottlenecks. In *Traffic and Granular Flow 2019*; Springer: Berlin, Germany, 2020; pp. 3–9.
2. Adrian, J.; Seyfried, A.; Sieben, A. Crowds in front of bottlenecks at entrances from the perspective of physics and social psychology. *J. R. Soc. Interface* **2020**, *17*, 20190871. [CrossRef] [PubMed]
3. Lügering, H.; Üsten, E.; Sieben, A. Pushing and Non-Pushing Forward Motion in Crowds: A Systematic Psychological Method for Rating Individual Behavior in Pedestrian Dynamics. 2022, *Manuscript submitted for publication*.
4. Haghani, M.; Sarvi, M.; Shahhoseini, Z. When 'push' does not come to 'shove': Revisiting 'faster is slower' in collective egress of human crowds. *Transp. Res. Part A Policy Pract.* **2019**, *122*, 51–69. [CrossRef]
5. Sieben, A.; Schumann, J.; Seyfried, A. Collective phenomena in crowds—Where pedestrian dynamics need social psychology. *PLoS ONE* **2017**, *12*, e0177328. [CrossRef] [PubMed]
6. Adrian, J.; Boltes, M.; Holl, S.; Sieben, A.; Seyfried, A. Crowding and queuing in entrance scenarios: Influence of corridor width in front of bottlenecks. *arXiv* **2018**, arXiv:1810.07424.
7. Boltes, M.; Seyfried, A.; Steffen, B.; Schadschneider, A. Automatic extraction of pedestrian trajectories from video recordings. In *Pedestrian and Evacuation Dynamics 2008*; Springer: Berlin, Germany, 2010; pp. 43–54.
8. Nayak, R.; Pati, U.C.; Das, S.K. A comprehensive review on deep learning-based methods for video anomaly detection. *Image Vis. Comput.* **2021**, *106*, 104078. [CrossRef]
9. Roshtkhari, M.J.; Levine, M.D. An on-line, real-time learning method for detecting anomalies in videos using spatio-temporal compositions. *Comput. Vis. Image Underst.* **2013**, *117*, 1436–1452. [CrossRef]
10. Singh, G.; Khosla, A.; Kapoor, R. Crowd escape event detection via pooling features of optical flow for intelligent video surveillance systems. *Int. J. Image Graph. Signal Process.* **2019**, *10*, 40. [CrossRef]
11. George, M.; Bijitha, C.; Jose, B.R. Crowd panic detection using autoencoder with non-uniform feature extraction. In Proceedings of the 8th International Symposium on Embedded Computing and System Design (ISED), Cochin, India, 13–15 December 2018; pp. 11–15.
12. Santos, G.L.; Endo, P.T.; Monteiro, K.H.D.C.; Rocha, E.D.S.; Silva, I.; Lynn, T. Accelerometer-based human fall detection using convolutional neural networks. *Sensors* **2019**, *19*, 1644. [CrossRef]
13. Mehmood, A. LightAnomalyNet: A Lightweight Framework for Efficient Abnormal Behavior Detection. *Sensors* **2021**, *21*, 8501. [CrossRef]
14. Zhang, X.; Zhang, Q.; Hu, S.; Guo, C.; Yu, H. Energy level-based abnormal crowd behavior detection. *Sensors* **2018**, *18*, 423. [CrossRef] [PubMed]
15. Kooij, J.F.; Liem, M.C.; Krijnders, J.D.; Andringa, T.C.; Gavrila, D.M. Multi-modal human aggression detection. *Comput. Vis. Image Underst.* **2016**, *144*, 106–120. [CrossRef]
16. Gan, H.; Xu, C.; Hou, W.; Guo, J.; Liu, K.; Xue, Y. Spatiotemporal graph convolutional network for automated detection and analysis of social behaviours among pre-weaning piglets. *Biosyst. Eng.* **2022**, *217*, 102–114. [CrossRef]
17. Gan, H.; Ou, M.; Huang, E.; Xu, C.; Li, S.; Li, J.; Liu, K.; Xue, Y. Automated detection and analysis of social behaviors among preweaning piglets using key point-based spatial and temporal features. *Comput. Electron. Agric.* **2021**, *188*, 106357. [CrossRef]
18. Vu, T.H.; Boonaert, J.; Ambellouis, S.; Taleb-Ahmed, A. Multi-Channel Generative Framework and Supervised Learning for Anomaly Detection in Surveillance Videos. *Sensors* **2021**, *21*, 3179. [CrossRef]
19. Yamashita, R.; Nishio, M.; Do, R.K.G.; Togashi, K. Convolutional neural networks: An overview and application in radiology. *Insights Imaging* **2018**, *9*, 611–629. [CrossRef]
20. Li, L.; Zhang, S.; Wang, B. Apple leaf disease identification with a small and imbalanced dataset based on lightweight convolutional networks. *Sensors* **2021**, *22*, 173. [CrossRef]
21. Wang, P.; Fan, E.; Wang, P. Comparative analysis of image classification algorithms based on traditional machine learning and deep learning. *Pattern Recognit. Lett.* **2021**, *141*, 61–67. [CrossRef]
22. Duman, E.; Erdem, O.A. Anomaly detection in videos using optical flow and convolutional autoencoder. *IEEE Access* **2019**, *7*, 183914–183923. [CrossRef]
23. Farnebäck, G. Two-frame motion estimation based on polynomial expansion. In Proceedings of the 13th Scandinavian Conference on Image Analysis, Gothenburg, Sweden, 29 June–2 July 2003; pp. 363–370.
24. Ilyas, Z.; Aziz, Z.; Qasim, T.; Bhatti, N.; Hayat, M.F. A hybrid deep network based approach for crowd anomaly detection. *Multimed. Tools Appl.* **2021**, *80*, 1–15. [CrossRef]

25. Direkoglu, C. Abnormal crowd behavior detection using motion information images and convolutional neural networks. *IEEE Access* **2020**, *8*, 80408–80416. [CrossRef]
26. Almazroey, A.A.; Jarraya, S.K. Abnormal Events and Behavior Detection in Crowd Scenes Based on Deep Learning and Neighborhood Component Analysis Feature Selection. In Proceedings of the International Conference on Artificial Intelligence and Computer Vision (AICV2020), Cairo, Egypt, 8–10 April 2020; pp. 258–267.
27. Teed, Z.; Deng, J. Raft: Recurrent all-pairs field transforms for optical flow. In Proceedings of the European Conference on Computer Vision, Glasgow, UK, 23–28 August 2020; pp. 402–419.
28. Tom Runia, D.F. Optical Flow Visualization. Available online: https://github.com/tomrunia/OpticalFlow_Visualization (accessed on 2 April 2020).
29. Baker, S.; Scharstein, D.; Lewis, J.; Roth, S.; Black, M.J.; Szeliski, R. A database and evaluation methodology for optical flow. *Int. J. Comput. Vis.* **2011**, *92*, 1–31. [CrossRef]
30. Tan, M.; Le, Q. Efficientnet: Rethinking model scaling for convolutional neural networks. In Proceedings of the International Conference on Machine Learning, Long Beach, CA, USA, 9–15 June 2019; pp. 6105–6114.
31. Deng, J.; Dong, W.; Socher, R.; Li, L.J.; Li, K.; Fei-Fei, L. Imagenet: A large-scale hierarchical image database. In Proceedings of the 2009 IEEE Conference on Computer Vision and Pattern Recognition, Miami, FL, USA, 20–25 June 2009; pp. 248–255.
32. Coşar, S.; Donatiello, G.; Bogorny, V.; Garate, C.; Alvares, L.O.; Brémond, F. Toward abnormal trajectory and event detection in video surveillance. *IEEE Trans. Circuits Syst. Video Technol.* **2016**, *27*, 683–695. [CrossRef]
33. Jiang, J.; Wang, X.; Gao, M.; Pan, J.; Zhao, C.; Wang, J. Abnormal behavior detection using streak flow acceleration. *Appl. Intell.* **2022**, 1–18. [CrossRef]
34. Xu, M.; Yu, X.; Chen, D.; Wu, C.; Jiang, Y. An efficient anomaly detection system for crowded scenes using variational autoencoders. *Appl. Sci.* **2019**, *9*, 3337. [CrossRef]
35. Tay, N.C.; Connie, T.; Ong, T.S.; Goh, K.O.M.; Teh, P.S. A robust abnormal behavior detection method using convolutional neural network. In *Computational Science and Technology*; Springer: Berlin, Germany, 2019; pp. 37–47.
36. Sabokrou, M.; Fayyaz, M.; Fathy, M.; Moayed, Z.; Klette, R. Deep-anomaly: Fully convolutional neural network for fast anomaly detection in crowded scenes. *Comput. Vis. Image Underst.* **2018**, *172*, 88–97. [CrossRef]
37. Smeureanu, S.; Ionescu, R.T.; Popescu, M.; Alexe, B. Deep appearance features for abnormal behavior detection in video. In Proceedings of the International Conference on Image Analysis and Processing, Catania, Italy, 11–15 September 2017; pp. 779–789.
38. Khan, S.S.; Madden, M.G. One-class classification: Taxonomy of study and review of techniques. *Knowl. Eng. Rev.* **2014**, *29*, 345–374. [CrossRef]
39. Hu, Y. Design and implementation of abnormal behavior detection based on deep intelligent analysis algorithms in massive video surveillance. *J. Grid Comput.* **2020**, *18*, 227–237. [CrossRef]
40. Zhou, S.; Shen, W.; Zeng, D.; Fang, M.; Wei, Y.; Zhang, Z. Spatial–temporal convolutional neural networks for anomaly detection and localization in crowded scenes. *Signal Process. Image Commun.* **2016**, *47*, 358–368. [CrossRef]
41. Zhang, C.; Xu, Y.; Xu, Z.; Huang, J.; Lu, J. Hybrid handcrafted and learned feature framework for human action recognition. *Appl. Intell.* **2022**, 1–17. [CrossRef]
42. Adrian, J.; Seyfried, A.; Sieben, A. Crowds in Front of Bottlenecks from the Perspective of Physics and Social Psychology. Available online: http://ped.fz-juelich.de/da/2018crowdqueue (accessed on 2 April 2020).
43. Hollows, G.; James, N. Understanding Focal Length and Field of View. *Retrieved Oct.* **2016**, *11*, 2018.
44. Srivastava, N.; Hinton, G.; Krizhevsky, A.; Sutskever, I.; Salakhutdinov, R. Dropout: A simple way to prevent neural networks from overfitting. *J. Mach. Learn. Res.* **2014**, *15*, 1929–1958.
45. Genc, B.; Tunc, H. Optimal training and test sets design for machine learning. *Turk. J. Electr. Eng. Comput. Sci.* **2019**, *27*, 1534–1545. [CrossRef]
46. Ismael, S.A.A.; Mohammed, A.; Hefny, H. An enhanced deep learning approach for brain cancer MRI images classification using residual networks. *Artif. Intell. Med.* **2020**, *102*, 101779. [CrossRef]
47. Howard, A.G.; Zhu, M.; Chen, B.; Kalenichenko, D.; Wang, W.; Weyand, T.; Andreetto, M.; Adam, H. Mobilenets: Efficient convolutional neural networks for mobile vision applications. *arXiv* **2017**, arXiv:1704.04861.
48. Szegedy, C.; Vanhoucke, V.; Ioffe, S.; Shlens, J.; Wojna, Z. Rethinking the inception architecture for computer vision. In Proceedings of the IEEE Conference on Computer Vision and Pattern Recognition, Las Vegas, NV, USA, 27–30 June 2016; pp. 2818–2826.
49. He, K.; Zhang, X.; Ren, S.; Sun, J. Deep residual learning for image recognition. In Proceedings of the IEEE Conference on Computer Vision and Pattern Recognition, Las Vegas, NV, USA, 27–30 June 2016; pp. 770–778.
50. Van der Jeught, S.; Buytaert, J.A.; Dirckx, J.J. Real-time geometric lens distortion correction using a graphics processing unit. *Opt. Eng.* **2012**, *51*, 027002. [CrossRef]
51. Stankiewicz, O.; Lafruit, G.; Domański, M. Multiview video: Acquisition, processing, compression, and virtual view rendering. In *Academic Press Library in Signal Processing*; Elsevier: Amsterdam, The Netherlands, 2018; Volume 6, pp. 3–74.
52. Vieira, L.H.; Pagnoca, E.A.; Milioni, F.; Barbieri, R.A.; Menezes, R.P.; Alvarez, L.; Déniz, L.G.; Santana-Cedrés, D.; Santiago, P.R. Tracking futsal players with a wide-angle lens camera: Accuracy analysis of the radial distortion correction based on an improved Hough transform algorithm. *Comput. Methods Biomech. Biomed. Eng. Imaging Vis.* **2017**, *5*, 221–231. [CrossRef]

Article

Deepfake Detection Using the Rate of Change between Frames Based on Computer Vision

Gihun Lee [†] **and Mihui Kim** *[,†]

Department of Computer Science & Engineering, Computer System Institute, Hankyong National University, Jungang-ro, Anseong-si 17579, Gyeonggi-do, Korea; comb1001@hknu.ac.kr
* Correspondence: mhkim@hknu.ac.kr; Tel.: +82-31-670-5167
† Current address: School of Computer Engineering & Applied Mathematics, Computer System Institute, Hankyong National University, Jungang-ro, Anseong-si 17579, Gyeonggi-do, Korea.

Abstract: Recently, artificial intelligence has been successfully used in fields, such as computer vision, voice, and big data analysis. However, various problems, such as security, privacy, and ethics, also occur owing to the development of artificial intelligence. One such problem are deepfakes. Deepfake is a compound word for deep learning and fake. It refers to a fake video created using artificial intelligence technology or the production process itself. Deepfakes can be exploited for political abuse, pornography, and fake information. This paper proposes a method to determine integrity by analyzing the computer vision features of digital content. The proposed method extracts the rate of change in the computer vision features of adjacent frames and then checks whether the video is manipulated. The test demonstrated the highest detection rate of 97% compared to the existing method or machine learning method. It also maintained the highest detection rate of 96%, even for the test that manipulates the matrix of the image to avoid the convolutional neural network detection method.

Keywords: deepfake; computer vision; the rate of change

Citation: Lee, G.; Kim, M. Deepfake Detection Using the Rate of Change between Frames Based on Computer Vision. *Sensors* **2021**, *21*, 7367. https://doi.org/10.3390/s21217367

Academic Editors: Yun Zhang, KWONG Tak Wu Sam, Xu Long and Tiesong Zhao

Received: 21 September 2021
Accepted: 1 November 2021
Published: 5 November 2021

Publisher's Note: MDPI stays neutral with regard to jurisdictional claims in published maps and institutional affiliations.

Copyright: © 2021 by the authors. Licensee MDPI, Basel, Switzerland. This article is an open access article distributed under the terms and conditions of the Creative Commons Attribution (CC BY) license (https://creativecommons.org/licenses/by/4.0/).

1. Introduction

Deepfake is a technology that uses artificial intelligence to synthesize another person's face with the face of a person appearing in a video and manipulate the target person's doing or saying things [1]. Deepfake technology has gradually developed and created videos that human eyes cannot distinguish (see Figure 1).

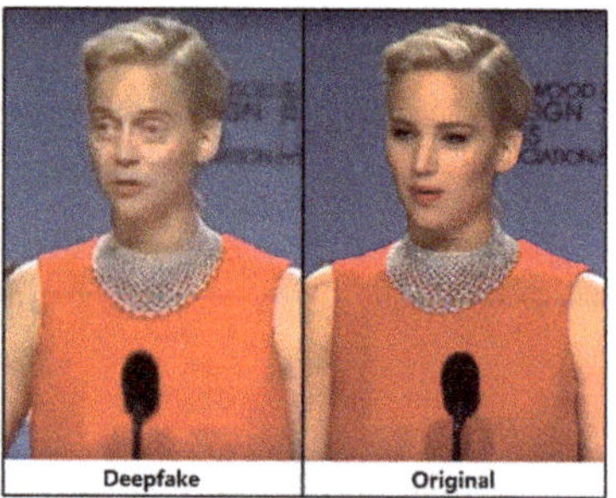

Figure 1. Deepfake image and original image [2].

The development of deepfake technology poses a significant threat to digital content explosion owing to the development of smartphones and social networks. Particularly, problems include creating confusion in the stock market owing to false news, producing malicious effects on election campaigns, and generating regional political tensions between countries. Facial manipulation has been developed from modifying the lip motion of a person to synthesizing non-existent faces or manipulating the real face of one person [1]. Nowadays, Autoencoder and generative adversarial network (GAN) artificial intelligence have appeared. As a result, deepfake videos can be made easily for identity swapping.

Accordingly, various methods for detecting deepfakes have been proposed. Afchar et al. [3] proposed detection with a deep neural network using tiny noises in an image using convolutional neural network (CNN). Güera et al. [4] proposed detection using long short term memory (LSTM) by extracting features of the frame image of a video using a CNN. Li et al. [5] proposed extracting eye blinks using CNNs and detecting them using LSTM. Li et al. [6] proposed detection using the disparity of a distorted face using ResNet50 and the VGG16 model based on CNN. Yang et al. [7] proposed a method for extracting 68 landmarks from face images and detecting them using SVMs. Agarwal et al. [8] proposed detection using the dynamics of the mouth shape using a CNN. In most proposed methods, deepfakes are detected by extracting features from video frames using a CNN. However, CNNs are vulnerable to changes in metrics, such as blur, brightness, contrast, noise, and angle. Because a CNN has a convolutional filter of a specific size to extract features while moving around the image, if factors, such as blur, brightness, contrast, noise, and angle, change, differences from previously learned features occur. Test data with these changed factors have a lower detection rate in the learned CNN [9]. Therefore, in this study, computer vision features were extracted from the frames of videos without using a CNN, and then the rate of change of features between frames was calculated. We propose a method to detect deepfakes using the distribution of the data. The proposed method can detect manipulated digital content irrespective of changes in factors, such as blur, brightness, contrast, noise, and angle. In addition, a CNN must learn additional learning data by creating images with changed angles or contrasts to increase the detection rate. However, the proposed method can minimize these costs. Conversely, a CNN can detect manipulated digital content by extracting features from a single image, but the proposed method requires more than a certain number of frames to determine.

The contribution of this work is summarized as follows: First, we propose a method detecting deepfake video without a convolutional neural network. Usually, CNN learns a representation by embedding a vector in a hypersphere from an image. Then, it is used as the classifier's input. In contrast, we extracted computer vision features first and used just a fully connected layer for classification. Second, we focus on detecting deepfake videos. Autoencoder and GAN make deepfake videos by manipulating frame by frame. We used unnatural differences between frames that can be made during manipulating. Thus, we calculated the rate of change between frames and used this for detecting deepfake videos. Third, we have many benefits because we do not use CNN. We can have comparable performance without data augmentation. Moreover, training time is saved because of the smaller parameter of the network and smaller datasets. Most importantly, our method is robust in regards to adversarial attacks or CNN's weakness.

The remainder of this paper is organized as follows. Section 2 introduces the deepfake technology and existing deepfake detection methods. Section 3 describes the proposed deepfake detection method. Section 4 shows the feasibility of the proposed method by evaluating its performance and comparing it with other mechanisms.

2. Related Works

2.1. Deepfake Creation

Deepfake is a technology that synthesizes the face of a character in a video into the face of a specific target using artificial intelligence technology. The artificial intelligence technologies used are primarily autoencoders [10] and the generative adversarial network [11]. Figure 2 illustrates the deepfake creation process using an autoencoder. An autoencoder comprises an encoder and a decoder. The goal of the encoder is to extract features from the image through dimensional reduction, and the goal of the decoder is to restore the original image as much as possible using the extracted features. Two autoencoders are used for learning to create a deepfake. The encoders, shown in Figure 2a,b, are trained using the same encoder. Therefore, the encoder learns common features that appear in face A (Figure 2a) and face B (Figure 2b). Examples of features include the position of the eyes, nose, and mouth. The decoders, depicted in Figure 2a,b, are trained separately. Figure 2c illustrates the deepfake creation process. After extracting the features of face A using an encoder, an image is generated using what the decoder learned, as shown in Figure 2b. FaceApp [2] is an example of deepfake production using an autoencoder.

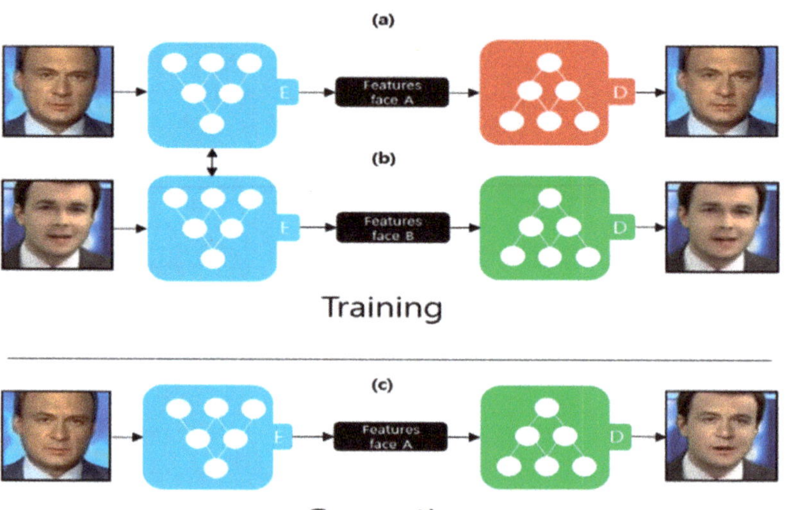

Figure 2. Deepfake creation process using an autoencoder [2]. (**a**) Autoencoder trained by face A; (**b**) Autoencoder trained by face B; (**c**) Deepfake creation process.

Figure 3 illustrates the deepfake creation process using a GAN. A GAN comprises a discriminator and a generator. The generator, as depicted in Figure 3a, receives the source and target images to be synthesized as the input data. The generator creates a new image using the input data. The discriminator, as shown in Figure 3b, learns to distinguish between the real and generated fake images. As depicted in Figure 3c, this process repeats until the discriminator cannot distinguish between the generated fake image and the original image. StarGAN is an example of creating a deepfake using a GAN [12].

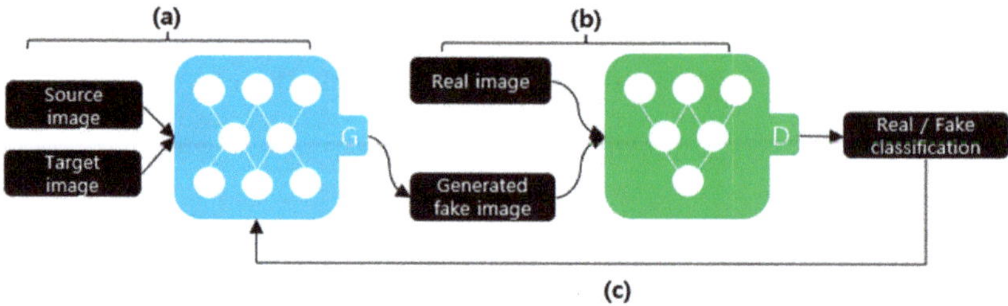

Figure 3. Deepfake creation process using a GAN. (**a**) Training the generator; (**b**) Training the discriminator; (**c**) Repeat.

2.2. Deepfake Detections

Table 1 summarizes the methods proposed for deepfake detection in the past three years. Each proposed method can be classified as a key feature and architecture.

Table 1. Deepfake detection method proposed in the past three years.

Methods	Key Features	Architecture	Published
Microscopic analyses [3]	Mesoscopic properties of images	MesoNet (based on CNN)	2018
Temporal inconsistencies [4]	Frame level temporal features	CNN + LSTM	2018
Eye blinking [5]	Temporal patterns of eye blinking	CNN + LSTM	2018
Face warping [6]	Inconsistencies in warped face and surrounding area	VGG16, ResNet50 (based on CNN)	2019
Discrepancy [7]	Temporal discrepancies across frames	CNN + RNN	2019
Spoken phoneme mismatches [8]	Mismatches between the dynamics of the mouth shape	CNN	2020

Afchar et al. [3] extracted features by analyzing mesoscopic noise from a single image using a CNN and then detected deepfakes using this feature. Microscopic analyses based on image noise cannot be applied in a compressed video context in which the image noise is strongly degraded.

Güera et al. [4] used a CNN and LSTM. The CNN extracts a feature vector of 2048 dimensions in units of frames. The LSTM receives the feature vector and detects the deepfake by searching for features with temporal significance between multiple frames.

Li et al. [5] used a CNN and LSTM. The CNN extracts the blinking patterns of the eyes. Using these extracted features, LSTM detects deepfakes by determining features with temporal significance between frames. The synthesized fake videos did not efficiently exhibit a physiological signal.

Li et al. [6] used VGG16 and ResNet50 models. These two neural networks are CNN-based neural networks. The CNN extracts the landmarks of the face to compute the transform matrices to align the faces to a standard configuration. The deepfake is detected by comparing the inconsistencies in the generated face areas and their surrounding regions. When creating a deepfake, matrix transformation occurs because limited images are used.

Yang et al. [7] used a CNN and an RNN. The CNN extracts features from each frame. The RNN detects the inconsistencies between frames from the extracted features. When creating a deepfake, inconsistencies may occur between frames because images are synthesized in units of frames.

Agarwal et al. [8] used a CNN. The CNN focuses on the visemes associated with words having the sound M, B, and P, in which the mouth must completely close to pronounce these phonemes. Deepfakes are detected using the inconsistencies between what is actually said and the shape of the mouth. Manipulated videos are occasionally inconsistent with spoken phonemes.

The deepfake detection methods proposed for the past three years detect deep fakes using a CNN. A CNN is a model that exhibits high performance, particularly related to image recognition, among artificial intelligence technologies [9]. Figure 4 illustrates a convolutional filter process that moves around the image by one space to create a feature map of the image. The convolutional filter is the core of CNN. This process results in the locality of pixel dependencies. It efficiently determines the small features of the image [13].

However, the performance is highly dependent on several factors in the image. When metrics, such as blur, brightness, contrast, noise, and angle, change, the detection rate of CNN drops significantly [9]. Malicious users can use this problem. Usually, an artificial network is trained by a dataset that has general representations. Malicious users could put just one filter to control with uncommon conditions in video. The eye can not feel the difference in people, but the pretrained network model cannot work properly in this image. In contrast, our method extracts computer vision features. Extracted features will change obviously. Nevertheless, our method focuses on the rate of change between frames. Each frame has the same condition change. Therefore, it is not critical for our method. These benefits make our method more robust in regards to CNN problems and adversarial attacks. Moreover, our model can be trained faster in a DFDC dataset that considers different acquisition scenarios, light conditions, distance from the camera, and pose variation.

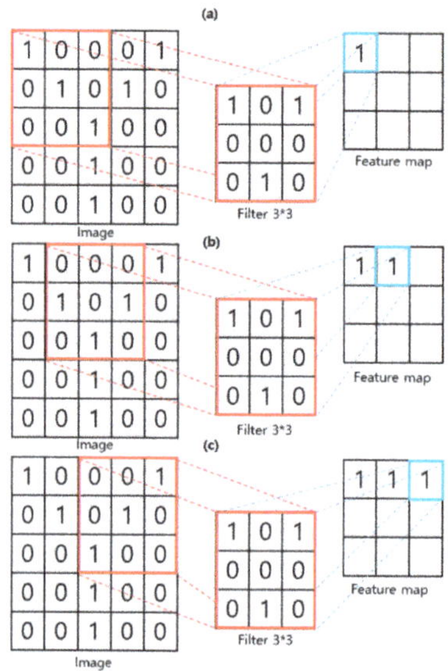

Figure 4. Feature map extraction process using a convolutional filter. (**a**) First step for feature extraction; (**b**) Next step (stride 1); (**c**) Next step (stride 1).

Figure 5 demonstrates an example in which the deepfake detection model using a CNN cannot detect. Figure 5a shows an image that can be detected as the frame of a general manipulated video. However, the remaining samples were not detected. Figure 5b shows the application of Gaussian noise in the manipulated frame. Figure 5c depicts changes in the brightness in the manipulated frame. Figure 5d shows the application of salt and pepper noise in the manipulated frame. Figure 5e depicts changes in the angle in the manipulated frame. The disadvantage of being undetectable owing to such a change in metrics can be used to avoid the CNN-based detection method [9,14].

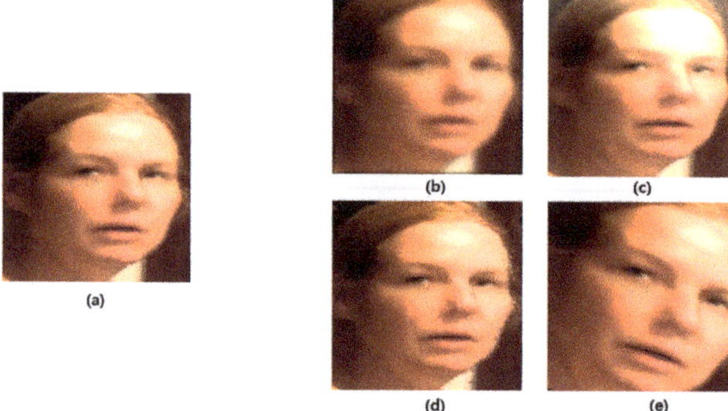

Figure 5. Example of undetectable image [15]. (**a**) Detectable deepfake image; (**b**) Undetectable deepfake image owing to Gaussian noise; (**c**) Undetectable deepfake image owing to brightness change; (**d**) Undetectable deepfake image owing to salt and pepper noise; (**e**) Undetectable deepfake image owing to angle change.

3. Proposed System

Figure 6 demonstrates the proposed system structure. The method is divided into preprocessing and classification processes. The preprocessing process extracts a face image from a frame image, extracts computer vision features, and then extracts the difference between the frames. The classification process detects a deepfake using a DNN by obtaining the variance of a certain number of frames from the preprocessed data.

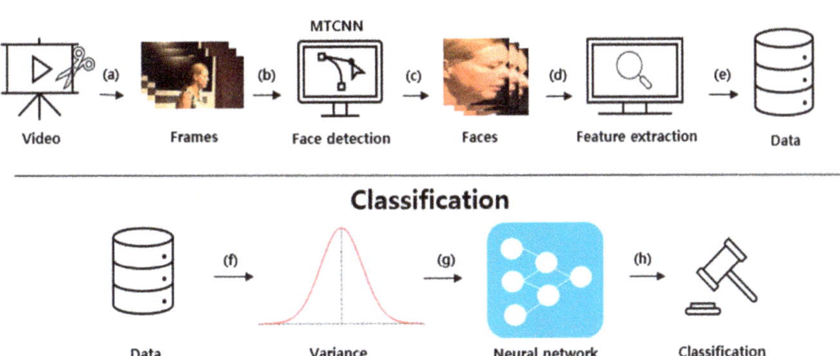

Figure 6. Proposed system structure [15]. (**a**) Extracting frames from video; (**b**) Face detection using MTCNN from each frames; (**c**) Crop detected faces; (**d**) Feature extraction from cropped faces; (**e**) Collecting extracted features; (**f**) Calculate variance from data; (**g**) Using neural network with data; (**h**) Classification from neural network.

3.1. Preprocessing

First, the video was divided into frames, as shown in Figure 6a. Then, the face part was detected and cut using MTCNN [16] in each frame, as depicted in Figure 6b. MTCNN is a Python module that improves the accuracy of face detection by 95% accuracy compared to a CNN. By only extracting the face and measuring the amount of change, it can focus more on the transformation of the face in computer vision. The extracted face image frames were arranged, as demonstrated in Figure 6c. Subsequently, various computer vision features were extracted from the face image, as illustrated in Figure 6d. A feature vector was generated by extracting computer vision features from the aligned face images using computation, clustering, and filtering.

The extracted features are presented in Table 2. The mean squared error (mse) measures the similarity of an image using the difference in the intensity of pixels between two images. The peak signal-to-noise ratio (psnr) evaluates the loss information for the image quality. psnr focuses on numerical differences rather than human visual differences. Because psnr is calculated using mse, when mse is 0, psnr is also set to 0. The structural similarity index measure (ssim) evaluates the temporal difference felt by humans in terms of luminance, contrast, and structural aspects. Red, green, blue (rgb), and the hue, saturation, and value (hsv) represent the color space of an image. The histogram represents the distribution of hues in the images. The luminance represents the average total brightness of the image. The variance represents the variance of the image brightness values. edge_density is the ratio of the edge components of all the pixels. The discrete cosine transform (dct) refers to the sharpness of an image. Because the deepfake production method synthesizes the target image for each frame, it may cause unnatural changes to various computer vision features. In addition, when creating a deepfake, the target image is obtained with limited resolution, and the size is changed as transformation matrices to fit the source image. Therefore, the sharpness is often inferior. In addition, distortion and blurring occur. The selected features greatly influence the deepfake creation process.

Figure 7 demonstrates frames with a significant change rate value for each computer vision feature among data obtained by preprocessing from a single deepfake video. Figure 6e takes the absolute value after calculating the difference between the extracted computer vision features of the i-th frame from the $i + 1$-th frame. The degree of change in the computer vision features was different for each video. Therefore, the rate of change was calculated by dividing the change by the average value of the change between all video frames.

Table 2. Extracted computer vision features.

Attribute	Explanation
mse	The average squared difference between the estimated values and the actual value
psnr	The ratio between the maximum possible power of a signal and the power of corrupting noise
ssim	The perceived quality of digital television and cinematic pictures
rgb	The percentage of each red, green, and blue color of the image
hsv	The percentage of each hue, saturation, and value of the image
histogram	The histogram plots the number of pixels in the image with a particular brightness or tonal value
luminance	The mean of the total brightness of the image
variance	Image variance of the image
edge_density	The ratio of edge pixels to the total pixels of in the image
dct	DCT bias of the image

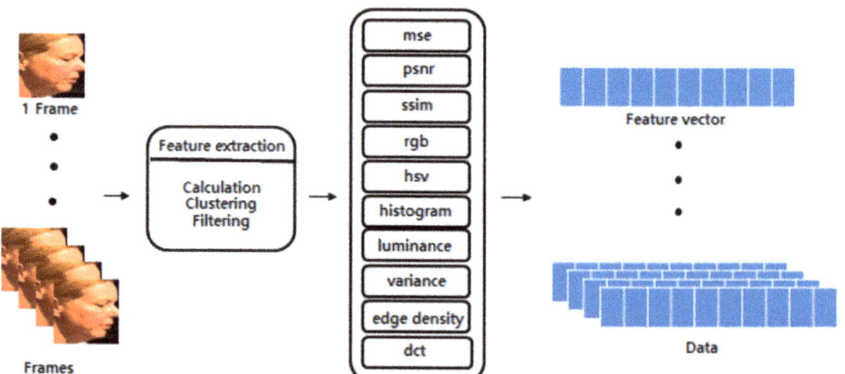

Figure 7. Computer vision feature extraction process [15].

Each feature is calculated using Equation (1). f_i denotes the feature value of the i-th frame. f_{i+1} denotes the feature value of the $i + 1$-th frame. mean(f) denotes the average of the feature values of all frames obtained from one video.

$$f_i = \frac{\mathrm{abs}(f_{i+1} - f_i)}{mean(f)},$$

(1)

Figure 8 shows the extraction of the frame with the most significant change in each feature from one deepfake video.

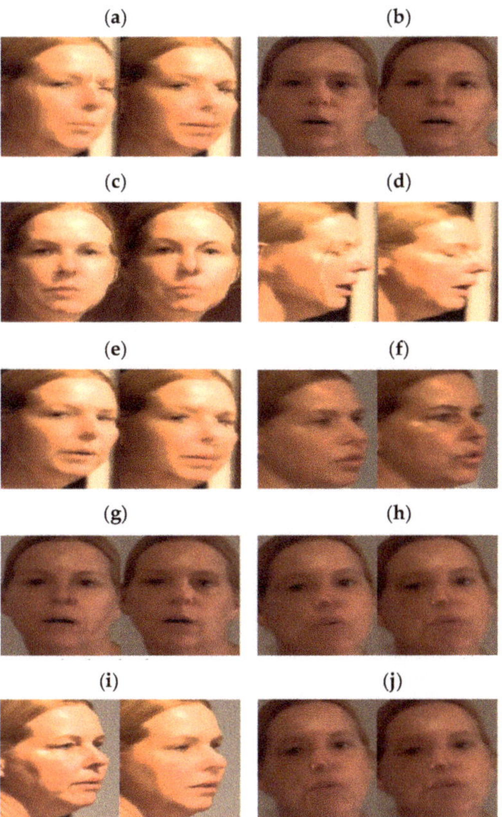

Figure 8. Frames showing a significant rate of change [15]. (**a**) mse; (**b**) psnr; (**c**) ssim; (**d**) rgb; (**e**) hsv; (**f**) histogram; (**g**) luminance; (**h**) variance; (**i**) edge density; (**j**) dct.

3.2. Classification

The variance for each feature was calculated by grouping the rate of change between certain extracted frame numbers in Figure 6f. The calculated variance of each feature was used as the data for DNN learning. A dependent variable indicating whether the data is a deepfake video was attached. Finally, these data were learned by the DNN and used to detect deepfakes. The final data were calculated using Equation (2). $Data_i$ denotes one feature value of the i-th a data sample used for DNN learning. d_i denotes the *i*-th data obtained by preprocessing. $\bar{d}$ denotes the average value of n data obtained by preprocessing.

$$Data_i = \frac{1}{n} \sum_{i*n}^{n} \left(d_i - \bar{d} \right)^2 \tag{2}$$

3.3. Modeling

Table 3 presents the accuracy by calculating the variance of the extracted adjacent frame change rate by a certain number. The highest accuracy of 95.22% was obtained when the DNN was trained by calculating the variance with 20 pieces of data.

Table 4 presents the accuracy by changing the optimizer function and the number of hidden layers to determine the appropriate hyperparameter. The Keras module was used for the learning. Image feature extraction was performed using OpenCV [17]. Binary cross-entropy was used as the loss function of the DNN. The highest accuracy of 97.39% was obtained when the DNN used the Adam optimizer function and five hidden layers.

Table 3. Accuracy by the number of distributed data.

Count	Accuracy
5	90.78%
10	92.33%
20	95.22%
30	86.67%
50	76.67%

Table 4. Hyperparameters—model performance.

Optimizer	# Hidden Layers	Loss	Accuracy
SGD	3	0.5560	67.83
	5	0.4146	78.26
	8	0.3439	81.74%
AdaGrad	3	0.6577	60.43%
	5	0.6672	55.22%
	8	0.6494	62.83
Adam	3	0.1608	94.35%
	5	0.0722	97.39%
	8	0.1120	94.78%

When comparing our method and MesoNet using CNN, our model has 3–8 layers and has about 15,202 total parameters. On the other hand, MesoNet has 6–18 layers and has about 27,977 total parameters. Thus, our model has almost 50% fewer hyperparameters. Moreover, the training time is faster than Mesonet, by more than 30%, because it skips the data augmentation process.

4. Performance Evaluation

4.1. Dataset

A total of three datasets were used. The Face2Face and FaceSwap datasets are provided by FaceForensics++ [18]. This dataset contains more than 1000 videos. Kaggle provides the Deepfake Detection Challenge (DFDC) dataset [15]. This dataset is over 470 GB. The characters appearing in all datasets are composed of various races, genders, and various shooting environments. This study used 206 videos of Face2Face, 210 videos of FaceSwap, and 176 videos of DFDC for the experiment. Three hundred frames were extracted from one video, and the face size extracted using MTCNN was set to 160 × 160 pixels. Python 3 and the image processing library OpenCV were used to extract the computer vision features from each frame. To confirm the result was owing to the change in the metric in the frame, 15% of the frame images in the DFDC test dataset indicated a 10% metric change.

4.2. Evaluation

Each model was implemented in Python 3, and Keras was used for the machine learning model training. Table 5 lists the system specifications for the experiments. According to the dataset, the proposed methods, the Mesonet method using CNN, and the SVM method, were compared. Table 6 presents a comparison of the detection accuracy.

The Mesonet method using the Face2Face and FaceSwap datasets exhibited a higher than 90% detection accuracy. However, an experiment using the DFDC dataset with a changed metric showed a 77.71% detection accuracy. It could be inferred that the metric of the frame image was changed in the test data of the DFDC dataset, and the detection accuracy of the CNN was degraded.

Table 5. System specification for the experiment.

CPU	AMD Ryzen 7 3800X 8-Core Processor
RAM	32 GB DDR4
GPU	Nvidia GeForce GTX 1660 Ti
VRAM	6 GB GDDR6

Table 6. Deepfake detection performance comparison.

	Face2face	FaceSwap	DFDC
Proposed model	97.39%	95.65%	96.55%
Mesonet	93.21%	95.32%	77.71%
SVM	54.24%	53.46%	52.91%

The SVM [19] method for all datasets exhibited a detection accuracy of less than 60%. It could be inferred that detecting a deepfake video using only the rate of change between frames is difficult unless a major defect occurs when manipulating the image.

The proposed method using the Face2Face and FaceSwap datasets exhibited a detection accuracy of more than 95%. In addition, an experiment using the DFDC dataset with a changed metric exhibited 96.55% detection accuracy. Mesonet learned by creating a new image by changing metrics, such as the angle and contrast, of the training data. However, the proposed method exhibited a similar detection accuracy without additional learning. We used a similar amount of the dataset to other deepfake papers. However, the quality of the academic dataset is poor and not diverse. Therefore, if we use this method in a really good quality dataset, it will not be effective. Future studies will address these issues.

5. Conclusions

In this paper, to detect deepfake videos, we propose a method of extracting the rate of change of computer vision features between frames and using a DNN based on the variance of a certain number of frames. Unlike existing deepfake detection methods, the problem of avoiding detection methods owing to changes in various metrics was solved because a CNN was not used. In addition, the amount of training data was less than that of the existing CNN. The proposed method exhibited detection accuracies of 97.39% and 95.65% for the Face2face and FaceSwap datasets, respectively, and 96.55% for the DFDC dataset with the metric changed dataset.

Author Contributions: Conceptualization, G.L. and M.K.; methodology, G.L. and M.K.; software, G.L.; validation, G.L. and M.K.; formal analysis, G.L. and M.K.; investigation, G.L.; resources, M.K.; data curation, G.L.; writing—original draft preparation, G.L.; writing—review and editing, M.K.; visualization, G.L. and M.K.; supervision, M.K.; project administration, M.K.; funding acquisition, M.K. All authors have read and agreed to the published version of the manuscript

Funding: This research was supported by the National Research Foundation of Korea (NRF) grant funded by the Korea government (MSIT) [No.2018R1A2B6009620].

Institutional Review Board Statement: Not applicable.

Informed Consent Statement: Not applicable.

Conflicts of Interest: The authors declare no conflict of interest.

References

1. Ruben, T.; Ruben, V.R.; Julian, F.; Aythami, M.; Javier, O.G. DeepFakes and Beyond: A Survey of Face Manipulation and Fake Detection. *arXiv* **2020**, arXiv:2001.00179.
2. Faceswap. Available online: https://faceswap.dev (accessed on 3 May 2021).

3.   Afchar, D.; Nozick, V.; Yamagishi, J.; Echizen, I. MesoNet: A Compact Facial Video Forgery Detection Network. In Proceedings of the 2018 IEEE International Workshop on Information Forensics and Security (WIFS), Hong Kong, China, 11–13 December 2018; pp. 1–7.

4.   Güera, D.; Delp, E.J. Deepfake Video Detection Using Recurrent Neural Networks. In Proceedings of the 2018 15th IEEE International Conference on Advanced Video and Signal Based Surveillance (AVSS), Auckland, New Zealand, 27–30 November 2018; pp. 1–6.

5.   Li, Y.; Chang, M.-C.; Lyu, S. In Ictu Oculi: Exposing AI Created Fake Videos by Detecting Eye Blinking. In Proceedings of the 2018 IEEE International Workshop on Information Forensics and Security (WIFS), Hong Kong, China, 11–13 December 2018; pp. 1–7.

6.   Li, Y.; Lyu, S. Exposing DeepFake Videos by Detecting Face Warping Artifacts. *arXiv* **2019**, arXiv:1811.00656.

7.   Yang, X.; Li, Y.; Lyu, S. Exposing Deep Fakes Using Inconsistent Head Poses. *arXiv* **2018**, arXiv:1811.00661.

8.   Agarwal, S.; Farid, H.; Fried, O.; Agrawala, M. Detecting Deep-Fake Videos from Phoneme-Viseme Mismatches. In Proceedings of the 2020 IEEE/CVF Conference on Computer Vision and Pattern Recognition Workshops (CVPRW), Seattle, WA, USA, 14–19 June 2020; pp. 2814–2822.

9.   Grm, K.; Štruc, V.; Artiges, A.; Caron, M.; Ekenel, H.K. Strengths and weaknesses of deep learning models for face recognition against image degradations. *IET Biom.* **2018**, *7*, 81–89. [CrossRef]

10.   Hou, X.; Shen, L.; Sun, K.; Qiu, G. Deep Feature Consistent Variational Autoencoder. In Proceedings of the 2017 IEEE Winter Conference on Applications of Computer Vision (WACV), Santa Rosa, CA, USA, 24–31 March 2017; pp. 1133–1141.

11.   Goodfellow, I.J.; Pouget-Abadie, J.; Mirza, M.; Xu, B.; Warde-Farley, D.; Ozair, S.; Courville, A.; Bengio, Y. Generative Adversarial Networks. *arXiv* **2014**, arXiv:1406.2661. [CrossRef]

12.   Choi, Y.; Choi, M.; Kim, M.; Ha, J.-W.; Kim, S.; Choo, J. StarGAN: Unified Generative Adversarial Networks for Multi-Domain Image-to-Image Translation. In Proceedings of the 2018 IEEE/CVF Conference on Computer Vision and Pattern Recognition, Salt Lake City, UT, USA, 18–23 June 2018; pp. 8789–8797.

13.   Krizhevsky, A.; Sutskever, I.; Hinton, G.E. ImageNet classification with deep convolutional neural networks. *Commun. ACM* **2017**, *60*, 84–90. [CrossRef]

14.   Roy, P.; Ghosh, S.; Bhattacharya, S.; Pal, U. Effects of Degradations on Deep Neural Network Architectures. *arXiv* **2019**, arXiv:1807.10108.

15.   Deepfake Detection Challenge | Kaggle. Available online: https://www.kaggle.com/c/deepfake-detection-challenge (accessed on 3 May 2021).

16.   Zhang, K.; Zhang, Z.; Li, Z.; Qiao, Y. Joint face detection and alignment using multi-task cascaded convolutional networks. *IEEE Signal Process. Lett.* **2016**, *23*, 1499–1503. [CrossRef]

17.   OpenCV. Available online: https://opencv.org/ (accessed on 3 May 2021).

18.   Rössler, A.; Cozzolino, D.; Verdoliva, L.; Riess, C.; Thies, J.; Nießner, M. FaceForensics++: Learning to Detect Manipulated Facial Images. *arXiv* **2019**, arXiv:1901.08971.

19.   Schuldt, C.; Laptev, I.; Caputo, B. Recognizing Human Actions: A Local SVM Approach. In Proceedings of the 17th International Conference on Pattern Recognition, Cambridge, UK, 26–26 August 2004; pp. 32–36.

Article

A Timestamp–Independent Haptic–Visual Synchronization Method for Haptic-Based Interaction System

Yiwen Xu [1,2], Liangtao Huang [1], Tiesong Zhao [1], Ying Fang [1] and Liqun Lin [1,*]

[1] Fujian Key Lab for Intelligent Processing and Wireless Transmission of Media Information, College of Physics and Information Engineering, Fuzhou University, Fuzhou 350108, China; xu_yiwen@fzu.edu.cn (Y.X.); 211120101@fzu.edu.cn (L.H.); t.zhao@fzu.edu.cn (T.Z.); fangying@fzu.edu.cn (Y.F.)
[2] College of Zhicheng, Fuzhou University, Fuzhou 350108, China
* Correspondence: lin_liqun@fzu.edu.cn

Abstract: The booming haptic data significantly improve the users' immersion during multimedia interaction. As a result, the study of a Haptic-based Interaction System has attracted the attention of the multimedia community. To construct such a system, a challenging task is the synchronization of multiple sensorial signals that is critical to the user experience. Despite audio-visual synchronization efforts, there is still a lack of a haptic-aware multimedia synchronization model. In this work, we propose a timestamp-independent synchronization for haptic–visual signal transmission. First, we exploit the sequential correlations during delivery and playback of a haptic–visual communication system. Second, we develop a key sample extraction of haptic signals based on the force feedback characteristics and a key frame extraction of visual signals based on deep-object detection. Third, we combine the key samples and frames to synchronize the corresponding haptic–visual signals. Without timestamps in the signal flow, the proposed method is still effective and more robust in complicated network conditions. Subjective evaluation also shows a significant improvement of user experience with the proposed method.

Keywords: haptic-based interaction system; multimedia environment; human-centric multimedia; haptic–visual synchronization

Citation: Xu , Y.; Huang, L.; Zhao, T.; Fang, Y.; Lin, L. A Timestamp-Independent Haptic–Visual Synchronization Method for Haptic-Based Interaction System. *Sensors* **2022**, *22*, 5502. https://doi.org/10.3390/s22155502

Academic Editor: Stefano Berretti

Received: 17 June 2022
Accepted: 21 July 2022
Published: 23 July 2022

Publisher's Note: MDPI stays neutral with regard to jurisdictional claims in published maps and institutional affiliations.

Copyright: © 2022 by the authors. Licensee MDPI, Basel, Switzerland. This article is an open access article distributed under the terms and conditions of the Creative Commons Attribution (CC BY) license (https://creativecommons.org/licenses/by/4.0/).

1. Introduction

Recent developments in multimedia technology also require multimedia content that is more immersive. As an emerging multimedia signal, haptics provide newfangled and authentic user experiences beyond current audio-visual signals. Thus, a Haptic-based Interaction System (HIS) has garnered the attention of researchers [1–5].

An HIS has been used in a variety of applications. For example, Ilaria et al. [6] designed an immersive haptic VR system for rehabilitation training of children with motor neurological disorders which significantly improved the effect of rehabilitation training. Zhou et al. [7] proposed an approach with visual and haptic signals which helps physicians perform surgeries accurately and effectively and furthermore reduces their physical and cognitive burden during surgery. Chen et al. [8] designed a remote training system with force feedback for power grid operation training. It avoided the collision between the manipulator and steel bars, which helps guide operators reduce operational errors and complete tasks efficiently. Varun et al. [9] also introduced haptics into a VR-based training system to enhance training immersion, effectiveness and efficiency. The use of an HI for online shopping [10,11] can improve the realism of the shopping experience and help visually impaired patients enjoy the convenience of online shopping. An HIS can also be used in outdoor search and rescue scenarios to avoid collisions by providing tactile guidance [12]. In industry, an HI is usually used to enhance the operational ability of robots. For example, the work in [13] equipped a robot with bionic haptic manipulators to help it have more stable grasping ability in tele-operation tasks. In [14], the operator controls the robot to perform the tele-operation in real time by means of a pneumatic haptic feedback glove. Apparently, the HIS is widely used and worthy of further investigation.

In an HIS, similar to conventional audio-visual signals, the haptic signal can also be affected during network fluctuations or congestion. In a multimedia case, the haptic signal may lose synchronization with other signals, e.g., images and videos. Compared to video, audio or image, the transmission of haptics is more tolerant of data loss and bandwidth but has higher requirements for the latency between signals. To ensure more natural interactive operations, haptic-based multimedia signal transmission requires better inter-signal synchronization. As reported, the haptic–visual asynchronization greatly influences the user experience. Qi et al. [15] implemented several experiments to explore the impact of the delay between video and haptic signals on the quality of users' experience. The results showed that all the Mean Opinion Score (MOS) values decreased with the inter-flow synchronization error. The works from Aung et al. [16] also confirmed the above conclusion.

To address this issue, haptic–visual synchronization is needed. The system examines the synchronization status of signals in real time and adjusts the corresponding signals immediately when an asynchronization is found. However, to the best of our knowledge, current research on the synchronization of visual–haptic signals is mainly focused on studying the impact of visual–haptic asynchronization on user experience, while little research has been conducted on synchronization detection and adjustment of visual–haptic signals, and there is still room for improvement in this area.

The research on synchronization algorithms for audio-visual signals can be used as good references for the research on visual–haptic signals. In the state-of-the-art HIS systems, synchronization is achieved by the timestamp method [17,18] that was designed for generic signals. The timestamp-dependent method embeds the timestamps in the signal stream to avoid synchronization drift. The receiving-end detects the signal synchronization status based on the timestamps and the system clock. However, the timestamp-dependent method has its drawbacks. First, in the sending end, the timestamps are usually added after frame synchronization, format conversion or pre-processing, where the delay derived from these operations are not compensated [18]. Thus, this signal asynchronization in the sending end will take to and always exist in the receiving end. Second, as the sending and receiving ends have different system clocks (the same frequency), the initial delay and frequency offset caused by dynamic environments also lead to signal asynchronization. To solve these shortcomings, researchers have proposed some improvement algorithms. For example, the works in [19,20] utilized the correlation between audio-visual signals for synchronization detection. They extract lip pictures in video frames and then compare them with the features of an audio signal through a deep-learning-based model to determine the synchronization status of audio-visual signals. The limitation of this method is that the video frame must contain the lip region. Yang et al. [21] proposed a watermark-based method to keep the synchronization of the audio-visual signal. However, this method has a disadvantage in that the "watermark" is not well adapted to the video or audio signal when applying conversion, aspect ratio conversion or audio downmixing [18].

From the above analysis, we can make conclusions that:

i. Haptic–visual synchronization plays an important role in HIS. It is worthy of further investigation.
ii. The traditional timestamp-dependent method used in an HIS has some shortcomings. As a result, there is still room for research on the haptic–visual synchronization method.

Thus, In this paper, we propose a first-of-its-kind timestamp-independent synchronization method for haptic–visual signals. Our contributions are summarized as follows.

The sequential correlation between haptic–visual signals. We build a multimedia communication platform with both haptic and visual signals. Based on this platform, we observe a strong correlation between the two signals during haptic-aware interaction. This intrinsic correlation is further utilized to design our synchronization model.

The key sample/frame extraction during haptic–visual interaction. We exploit the statistical features of haptic–visual signals and then develop learning-based methods to extract key samples and key frames in haptic and visual signals, respectively.

The asynchronization detection and removal strategy. Combining the correlation and key samples/frames, we are able to detect and eliminate asynchronization when the registration delay is larger than a threshold. Experimental results with subjective evaluations validate the effectiveness of our method.

2. Motivation

In our opinion, there exists a strong sequential correlation between haptic–visual signals, which can help the judgment of the signal synchronization state without crystal oscillators or timestamps. Inspired by this, we propose a timestamp-independent haptic–visual synchronization model to detect and eliminate asynchronization phenomena in an HIS. In this section, we establish a haptic–visual simulation platform and subsequently confirm the correlation between haptic–visual signals via the platform.

As shown in Figure 1, we use a virtual interaction module to design a haptic–visual interaction scenario where a human user manipulates a virtual ball to push a virtual box. A Geomagic Touch is deployed to connect the real and virtual world: on one hand, it sends the human instructions to the virtual ball; on the other hand, it collects the force feedback of the virtual ball and sends the corresponding signals back to the human user. This haptic interaction is achieved with the kinesthetic signal, which is a major component of haptic information.

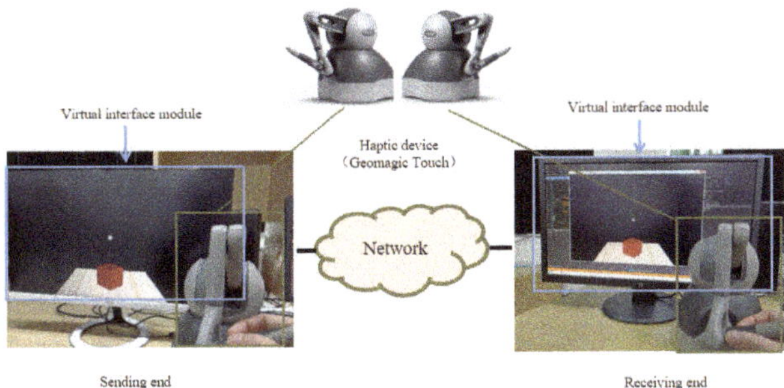

Figure 1. Our simulation platform for haptic–visual signal delivery.

In addition to the haptic signals captured by Geomagic Touch, the sending-end also records the visual contents of the virtual space, resulting in a high-definition video at a resolution of 1920×1080. Then the video is compressed by High Efficiency Video Coding (HEVC) and subsequently delivered with haptic signal by the network via User Datagram Protocol (UDP). Finally, the receiving-end combines both haptic and visual signals for a more immersive tele-presence, where another user can watch the scene in real time and also feel the haptic sensing via a haptic device.

The haptic and visual signals should be fully synchronized under normal conditions. Based on this simulation platform, we can observe the sequential correlation between haptic and visual signals. As shown in Figure 2, strong haptic signal fluctuations exist when the virtual hand (i.e., the ball) is on a collision course with another object. When the virtual hand visually touches the box, the force amplitude of the haptic changes simultaneously. As the two objects move closer, the force amplitude is also higher and vice versa. The force amplitude recovers to a constant when all objects are detached. These changes are also intuitive to the human users when operating a haptic-aware handle.

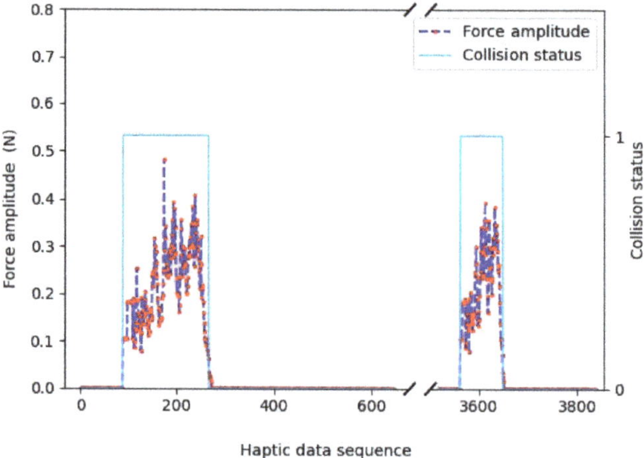

Figure 2. An example of haptic–visual correlations.

This intrinsic correlation inspires us to design a synchronization strategy. A sharp increase of force amplitude indicates a collision between the virtual hand and another object, while a sharp decrease implies a detachment between objects. If these deductions are inconsistent with the machine vision, we can conclude that there exists an asynchronization between haptic and visual signals and thus change the signal flows.

3. Proposed Method

Based on the above analysis, we propose the timestamp-independent synchronization method as shown in Figure 3. First, we extract the key samples in the haptic signal where the amplitude is intensively increased from near zero. Second, we extract the key frames in the visual signal where the visual collision happens. Third, we compare the time intervals of these key samples/frames to detect asynchronization phenomena. If a pair of time intervals (namely T_h and T_v) have a large difference, the haptic–visual asynchronization is found and further fixed. Note that here the object collision frequencies are low in the real world; therefore, we can easily identify different pairs of time intervals. In the following subsections, the key sample detection, key frame detection, threshold selection, asynchronization removal and the overall method are elaborated, respectively.

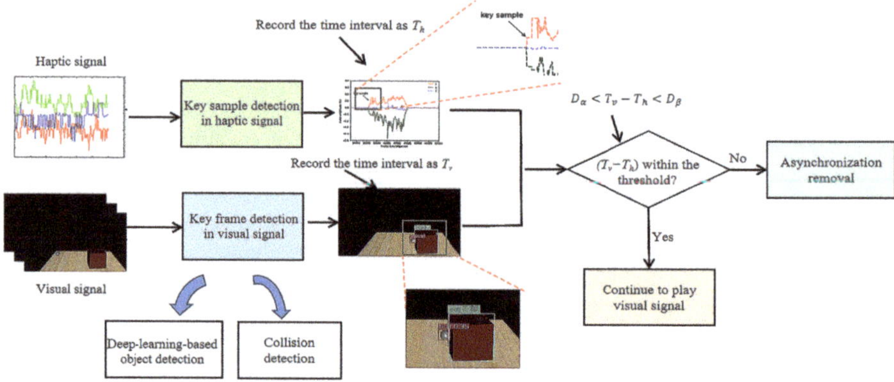

Figure 3. The flowchart of our proposed method.

3.1. Key Sample Detection in the Haptic Signal

For the haptic signal, the key samples are easily obtained for it consists of three one-dimensional signals (in x-axis, y-axis and z-axis). A sharp increase of force amplitude is found when its difference in any dimension is larger than a threshold (namely F_{th}). Through observations on a large number of samples, we found that the fluctuations of force amplitudes during non-collision are always below 0.01, and the force amplitudes of key samples are always above 0.07. Therefore, the F_{th} is empirically set as 0.05 in our work.

An example of this step is shown in Figure 4. An operation with force signals in three dimensions is presented, where all sharp increases are successfully detected and labeled as key samples. Correspondingly, their time intervals (i.e., T_h) are recorded for further comparison.

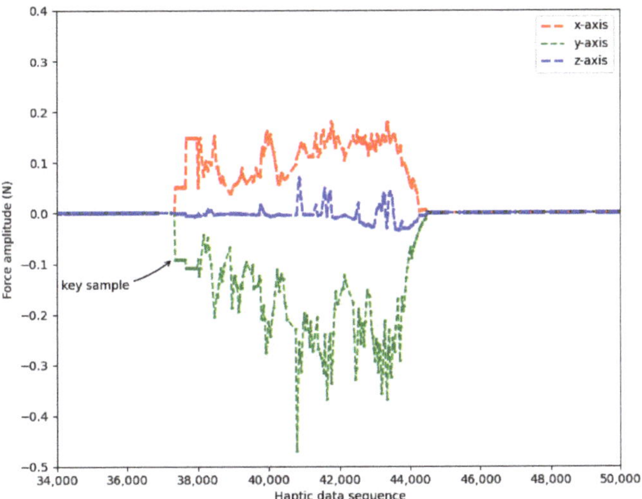

Figure 4. An example of key sample detection.

3.2. Key Frame Detection in the Visual Signal

The objective of key frame detection is to find the time intervals when the virtual hand touches the box. Essentially, it consists of two modules: object detection and collision detection. The first module identifies all objects, while the second module determines whether object collision occurs. Both modules are achieved by computer vision methods.

3.2.1. Object Detection

The commonly-used object detection algorithms are R-CNN [22], SPPNet [23], Fast R-CNN [24], Faster R-CNN [25], SSD [26] and YOLO [27,28]. Considering the efficiency, R-CNN, SPPNet and Fast R-CNN are not suitable for our scenario. Moreover, in our work, small object recognition, in which the performances of the Faster RCNN and SSD are not good enough, is needed. With a deep network, the YOLO network extracts the deep features of different objects and scenarios, thereby achieving object recognition with high accuracy. Consequently, we employ the V3 of YOLO network in our method [28].

We established our image database for training the YOLO V3 network. We acquired 1000 images from visual signals with an image size of 1600 × 900 pixels. Then the images were labeled via a label-making tool (the application software of labelImg). We used a rectangle to bound the balls in the images and labeled them as "ball" and accordingly, bound the boxes and labeled them as "box". All the labels were saved with xml files for using during training. The 800 images in this database are employed as the training set and the other 200 images are the test set.

The loss function plays an important role in the YOLO network. In this work, the position information of the ball and box is the target of the network. Therefore, the target's error of center coordinate in the form of squared difference is first taken into account in the loss function; then, to obtain the accurate bounding rectangle, the wide and high coordinate error in the form of cross-entropy is utilized; finally, as the detection of multiple categories of targets (ball and box) are involved, the category error in the form of cross-entropy must be considered. Hence, the loss function used in this work is:

$$
\begin{aligned}
Loss = {}& \lambda_{coord} \sum_{i=0}^{S^2} \sum_{j=0}^{B} I_{ij}^{obj}[(x_i - \hat{x}_i)^2 + (y_i - \hat{y}_i)^2] + \\
& \lambda_{coord} \sum_{i=0}^{S^2} \sum_{j=0}^{B} I_{ij}^{obj}[(\sqrt{w_i} - \sqrt{\hat{w}_i})^2 + (\sqrt{h_i} - \sqrt{\hat{h}_i})^2] - \\
& \sum_{i=0}^{S^2} \sum_{j=0}^{B} I_{ij}^{obj}[\hat{C}_i \log(C_i) + (1 - \hat{C}_i)\log(1 - C_i)] - \\
& \lambda_{noobj} \sum_{i=0}^{S^2} \sum_{j=0}^{B} I_{ij}^{noobj}[\hat{C}_i \log(C_i) + (1 - \hat{C}_i)\log(1 - C_i)] - \\
& \sum_{i=0}^{S^2} I_{ij}^{obj} \sum_{c \in classes} [\hat{P}_i \log(P_i) + (1 - \hat{P}_i)\log(1 - P_i)]
\end{aligned}
\tag{1}
$$

where the first row indicates the error of the center coordinates, S represents the grid size, B represents the bounding rectangle. I_{ij}^{obj} denotes whether targets are in the rectangle, and its value is one if there is a target in the bounding rectangle at grid (i, j), and zero vice versa. Here, x_i and y_i represent the true center coordinates; $\hat{x}_i$ and $\hat{y}_i$ represent the predicted center coordinates.

The second row represents the error of the width and height of the predicted rectangle in which w_i and h_i represent the true width and height and $\hat{w}_i$ and $\hat{h}_i$ represent the predicted width and height. The third and fourth rows indicate the error of the confidence level, where C_i denotes the true confidence level, and $\hat{C}_i$ denotes the predicted confidence level.

The fifth row denotes the error of classification, where P_i and $\hat{P}_i$ denote the true and the predicted categories, respectively; λ_{coord} and λ_{noobj} are the weights which will be trained as hyperparameters of the network.

The main hyperparameters used in training are set as shown in Table 1. Among them, the learning rate is set as cosine decay as follows:

$$
lr = \left\{ \frac{1}{2} \times [1 + cos(N_{trained} \times \frac{\pi}{N_{epoch}})] \times 0.95 + 0.05 \right\} \times 10^{-2},
\tag{2}
$$

where $N_{trained}$ denotes the number of epochs already trained, and N_{epoch} denotes the total number of training epochs.

Table 1. The hyperparameter settings in model training.

Epoch	Batchsize	λ_{coord}	λ_{noobj}	Learning Rate
300	16	0.5	0.5	cosine decay

With this method, the training module has a larger learning rate at the beginning to accelerate the training speed, and then the learning rate decreases with the increasing number of training epochs to more easily find the optimal solution.

After training, an example of a recognition result is shown in Figure 5 in which the virtual hand (i.e., the ball) and the box are detected, with their borders labeled by rectangular frames.

Figure 5. An example of object detection.

3.2.2. Collision Detection

We determine whether a collision happens based on the aforementioned rectangular frames. Let (X_1, Y_1) and (X_2, Y_2) denote the top-left locations of the virtual hand (i.e., the ball) and any object as the target in the 2D space, and (H_1, W_1) and (H_2, W_2) denote the sizes of the corresponding rectangular frames, the condition of no collision is:

$$(Y_1 + H_1 > Y_2)||(X_1 + W_1 < X_2)||(Y_1 < Y_2 + H_2)||(X_1 > X_2 + W_2). \tag{3}$$

Otherwise, the collision of objects is found. At the time of collision found, we extract the corresponding video frame as the key frame of the visual signal and record the time interval as T_v, which is further utilized for asynchronization detection.

3.3. The Synchronization Threshold

During haptic–visual delivery and playback, we can easily identify each key sample/frame pair considering the corresponding time intervals are usually very close to each other. For a pair of time intervals T_h and T_v, their difference is set as a criterion of haptic–visual asynchronization. A synchronization of signals is guaranteed if:

$$D_\alpha < (T_v - T_h) < D_\beta, \tag{4}$$

where D_α and D_β refer to the lower and upper bound of the perception threshold.

As results from a subjective test can be more consistent with users' perception experience, we designed a subjective test to determine D_α and D_β. Our test strictly follows the subjective test manual ITU-R BT.500 [29] with the following steps. First, we recruited 21 subjects without prior knowledge of haptic coding or delivery. Then, we used the two-alternative force choice method to perform the test. Each session of the test consisted of two randomly presented haptic–visual segments: with and without delay. The delay can be negative or positive with a range from -100 ms to 100 ms with an interval of 20ms. Each subject was asked to choose one segment where he/she could not feel delay between the two. Finally, for each session, the probability of correct choices, which is obtained by Equation (5), is recorded.

$$p_i = \frac{n_i}{N}, \tag{5}$$

where n_i denotes the number of subjects who have made a correct choice in the i-th delay, and N denotes the total number of subjects.

As shown in Figure 6, the probability of correct choices is around 0.5 when the delay of visual signals ranges from -60 ms to 80 ms. In other words, the human users cannot perceive the difference between delayed and non-delayed signals in this range. Therefore, we set the threshold of synchronization as $D_\alpha = -60$ ms, $D_\beta = 80$ ms.

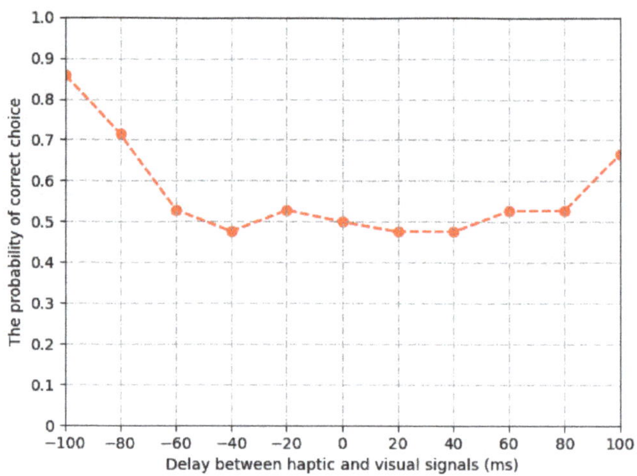

Figure 6. Subjective result of synchronization threshold.

3.4. Asynchronization Removal

To adjust the signal stream and remove asynchronization phenomena, a general method is to select a main stream and set the remaining as auxiliary streams. When asynchronization occurs, all auxiliary streams are adjusted to be synchronized with the main stream. As reported in [30], the human perception of haptic signals is very sensitive in that only haptic signals above 1 kHz provide smooth experience to users. This frequency is significantly higher than visual signals. Based on this fact, we utilize the haptic signal and the visual signal as the main stream and the auxiliary stream, respectively. For synchronization, the visual signal is moved to be consistent with the haptic signal.

In a multimedia communication system, the receiving-end usually sets a buffer zone to cache all multimedia data for a smooth display of them. Therefore, if the visual signal is delayed more than D_α, we will retrieve the correct video frame from the buffer zone. otherwise, if the visual signal is ahead by D_β, we will repeat the current frame until haptic–visual synchronization. Through this method, we are able to remove all asynchronization phenomena during haptic–visual delivery and playback.

3.5. The Overall Method

By summarizing Sections 3.1–3.4, the detailed steps of our method are presented as follows.

Step 1. Initialization. Set a buffer zone at the receiving end to cache haptic–visual data. Start the haptic–visual data delivery and playback. Go to Step 2.

Step 2. Key sample detection. Keep to detect the key samples of the haptic signals with the method in Section 3.1. If a key sample is found, set the time interval as T_h and go to Step 3.

Step 3. Key frame detection. Use the method in Section 3.2 to detect the corresponding key frames in the buffer and subsequent video of 1 s. If a key frame is found, set the time interval as T_v and go to Step 4; otherwise, the synchronization detection fails, go to Step 2.

Step 4. Asynchronization examination. If Equation (4) of Section 3.3 is true, go to Step 2 to check the following signals; otherwise go to Step 5.

Step 5. Asynchronization removal. Adjust the haptic–visual streams with the method shown in Section 3.4. Go to Step 2 to check the following signals.

4. Experimental Results

To examine the effectiveness of the proposed method, we implement it on the simulation platform shown in Section 2 and conduct both objective and subjective experiments. The frequencies of haptic and visual signals are set as 1000 Hz and 30 Hz, respectively. Due to the lack of a haptic–visual synchronization method, we compare our model with the original case only.

4.1. Estimation Accuracy of Synchronization Delay

The proposed method utilizes the synchronization delay $T_v - T_h$ to determine whether asynchronization happens. Therefore, the estimation accuracy of synchronization delay is critical in our method. We design the following experiment to examine the accuracy.

Based on the simulation platform, we randomly captured 100 haptic–visual clips, with the length of each clip as 30 s. In other words, there exist 30,000 haptic samples and 900 video frames in each clip; in total, 3 million haptic samples and 90,000 video frames exist). For each haptic–visual clip, we add a random delay on the visual signals. The delay is in the range of $(-330$ ms, 330 ms$)$ where the positive/negative values indicate the visual signal is ahead/behind the haptic signal. At the receiving-end, we employ our model to calculate the synchronization delay (namely $\hat{d}$) and compare it with the "actual" delay (namely d).

The Mean Absolute Error (MAE) and Maximum Absolute Error (MaxAE) are utilized to be assessment metrics. They are calculated by:

$$MAE = \frac{1}{M} \sum_{i=1}^{M} \left| \hat{d}_i - d_i \right|, \tag{6}$$

$$MaxAE = \max_{i \in \{1,2,3,...M\}} \left| \hat{d}_i - d_i \right|, \tag{7}$$

where M is the total number of samples.

The results are shown in Table 2. From the table, the MAE and MaxAE values are 7.3 ms and 15 ms, respectively. It is noted that the haptic–visual synchronization is unperceivable in $(-60$ ms, 80 ms$)$, where the ratio of MAE and MaxAE are only 5.2% and 10.7%, respectively. On the other hand, the frame length of each video frame is $\frac{1}{30}$ Hz = 33.3 ms, which is also significantly larger than the MAE/MaxAE values. Therefore, the estimation accuracy could fulfill the requirement in the practical applications of the haptic–visual system.

Table 2. The estimation accuracy of $T_v - T_h$.

Metrics	MAE (ms)	MaxAE (ms)
Results	7.3	15

4.2. Effectiveness of the Haptic–Visual Synchronization

To evaluate the effectiveness of our synchronization detection and removal method, we examine it on the same dataset presented in Section 4.

At the sending end, after sending random video frames (in the range of $(0, 100)$), we add a random delay (in the range of $(-330$ ms, 330 ms$)$ and denoted as t_n) on it. We repeat the above process until all the frames in each clip (totally 100 clips) are sent. Considering that the proposed asynchronization removal method adjusts the visual signal frame-by-frame, the interval of the above random delay is set the same as the frame interval of the visual signal (i.e., 33 ms). Therefore, the delay range of $(-330$ ms, 330 ms$)$ is equivalent to a delay random number (denoted as d_n) of video frames in the range of $(-10, 10)$. Taking a clip (900 frames) as example, the random numbers generated in the experiment are shown in Table 3. In the table, the values in the first column indicate that the visual signal is ahead of the haptic signal $7 \times 33 = 231$ ms, and the delay status lasts for $19 \times 33 = 627$ ms. The above random delay in the experiment is also intuitively shown in Figure 7 in which the vertical axis indicates the delay between visual and haptic signals and the horizontal axis indicates the order of the visual signal. From the figure, the delays are random and representative to evaluate our method.

Table 3. An example of random delay in the experiment.

d_n	7	−7	8	−8	8	9	−1	0	5	−8	−8	2	0	1	−1	−7
t_n	19	18	95	17	56	65	82	46	69	96	47	86	36	99	14	55

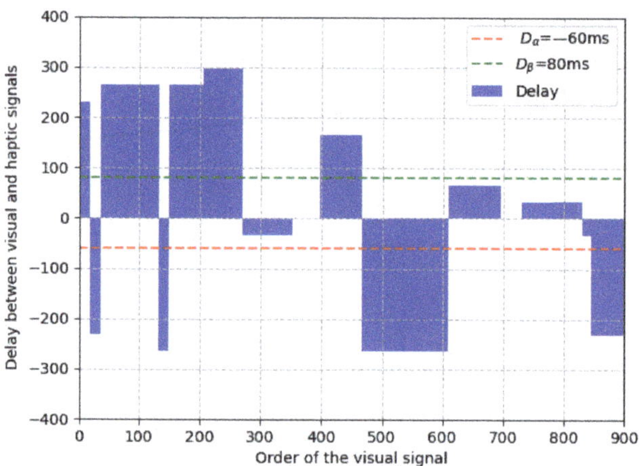

Figure 7. An example of random delay in the experiment.

At the receiving end, we compare the probabilities of successful synchronization with and without our method. The results are presented in Table 4. By using our model, the average probability of synchronization increases from 25.3% to 89.2%. It should be pointed out that our synchronization method is executed frame-by-frame. If the haptic–visual delay is larger than one frame, the signal is kept asynchronized during the synchronization process. That is the reason why there are still 10.8% signals asynchronized in Table 4. Even at this scenario with severe fluctuations, our method still achieves a high probability of 89.2%, which reveals the effectiveness and robustness of our method in haptic–visual synchronization. The utilization of our model guarantees the signal synchronization in most cases, thereby greatly improving the system performance of haptic–visual interaction.

Table 4. Probabilities of synchronization with and without our method.

	Without Our Method	**With Our Method**
Probabilities	25.3%	89.2%

4.3. Subjective Test on User Experience

In addition to objective evaluation, we also conducted a subjective test to evaluate the improvement of the user experience with our model. As mentioned in Section 1, the signal asynchronization is a critical factor to influence the user experience in haptic–visual interaction. Therefore, the improvement of user experience can be taken as circumstantial evidence of the effectiveness of our model.

We recruited 23 subjects to participate in this test, where all haptic–visual sequences are also the same to those in Section 4.1. The subjects' ages ranged from 17 to 26, and they have no exposure to the haptic-based system. To calculate the correlations, we introduce the delays that are evenly distributed from −10 to 10 frames (that is, ranged from −333 ms to

333 ms with the interval of 33.3 ms) and occasionally utilize the proposed synchronization method at the receiving end. However, whether or not we are using the synchronization is unknown for all subjects. As a result, a subject scores his/her experience based on real feelings and experiences. All scores are between 0 and 10 and their averaged value, the Mean Opinion Score (MOS), represents the average perceptions of human users.

The collected subjective test results were pre-processed to remove outliers based on the ITU subjective test regulations. We calculated the correlations, including the Pearson Linear Correlation Coefficient (PLCC) and the Spearman Rank-Order Correlation Coefficient (SROCC) [31], between each subject's score and the MOS. The results are shown in Figure 8. According to ITU-R BT.500 [29], a subject's score is considered as an outlier if the correlation between his/her score and the MOS is less than 0.7. Therefore, from Figure 8, the 12th and 18th subjects are considered as outliers and subsequently excluded in the final results.

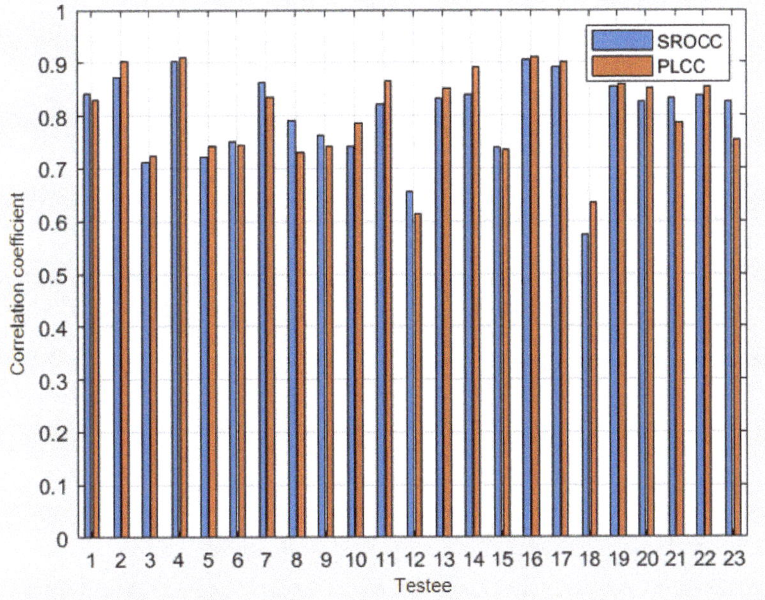

Figure 8. The correlations between each subject and the MOS.

The scores of the remaining 21 subjects were further examined by data saturation validation [32]. Due to the randomness of user scores, insufficient subjects would lead to inaccurate MOS values. To check whether the subjects are enough, data saturation validation was proposed. For a subjective test with K subjects, it randomly selects $k = 1, 2, \ldots, K$ subjects to calculate the correlation between their averaged score and the MOS. If the correlation value converges to one as k increases, the subjects are considered sufficient. In our test, this correlation value is very close to 1 with $k = 13$ subjects, as shown in Figure 9. Therefore, the remaining 21 subjects are sufficient to represent the averaged opinion of human users.

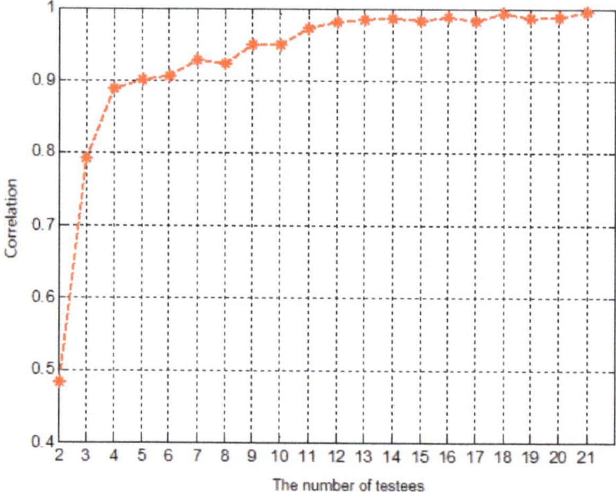

Figure 9. The data saturation validation in our test.

Figure 10 shows the MOS values under different delay settings. Two settings are compared: receiving end with and without our method. In the central part of curves (i.e., −33~66 ms), the delays are unperceivable to human users; thus the two settings achieve very similar MOS values. As the absolute value of delay gets larger, the difference between the two settings becomes more significant. In extreme cases (i.e., ±330 ms), our synchronization method improves the MOS values by around four, which shows the high capability of anti-interference under severe network conditions. On average, the MOS value is increased by 1.6169, with MOS variation decreased by 3.1315. This fact demonstrates the significant improvement of our synchronization method that is agreed by the majority of human users. In conclusion, the proposed method can guarantee the user experience in case of haptic–visual asynchronization.

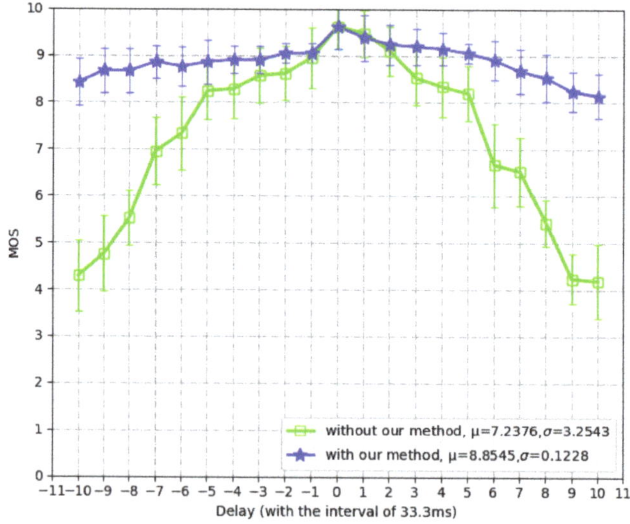

Figure 10. The subjective improvements with our method.

5. Conclusions

In this paper, we explore the haptic–visual correlations in a haptic-aware interaction system. Based on the observations, we propose a timestamp-independent synchronization method for haptic–visual signals, which consists of haptic signal analysis, learning-based vision analysis, perception-based thresholding and an overall method for asynchronization detection and removal. It should be pointed out that the example of virtual hand (i.e., the ball) and target (i.e., the box) can be extended to more types of objects with retrained models. Therefore, our model is still applicable in more general scenarios. To our best knowledge, this is the very first attempt to design a haptic-aware multimedia synchronization model by considering the special characteristics of haptic interaction. It can also be utilized as a reference to design new synchronization models for emerging sensorial media such as olfactory signals. We envision a more widespread use of multiple sensorial media that benefits the immersive user experience in the foreseeable future.

Author Contributions: Conceptualization, Y.X. and T.Z.; methodology, Y.X. and L.H.; software, Y.X. and L.H.; validation, Y.X. and T.Z.; formal analysis, L.L.; investigation, Y.F. and L.L.; resources, Y.X. and T.Z.; data curation, Y.X. and L.H.; writing—original draft preparation, Y.X. and L.H.; writing—review and editing, Y.X. and L.H.; visualization, Y.X. and L.H.; supervision, Y.X. and T.Z.; project administration, Y.X. and T.Z.; funding acquisition, Y.X. and T.Z. All authors have read and agreed to the published version of the manuscript.

Funding: This work were supported by the National Natural Science Foundation of China (No. 62171134) and Foundation for Middle-aged and Young Educational Committee of Fujian Province (No. JAT200024).

Institutional Review Board Statement: Not applicable.

Informed Consent Statement: Not applicable.

Data Availability Statement: Some or all data, models, or code generated or used during the study are available from the corresponding author by request.

Conflicts of Interest: The authors declare no conflict of interest.

References

1. Aijaz, A.; Dohler, M.; Aghvami, A.H.; Friderikos, V.;Frodigh, M. Realizing the Tactile Internet: Haptic Communications over Next Generation 5G Cellular Networks. *IEEE Wirel. Commun.* **2017**, *24*, 82–89. [CrossRef]
2. Antonakoglou, K.; Xu, X.; Steinbach, E.; Mahmoodi, T. Toward Haptic Communications Over the 5G Tactile Internet. *IEEE Commun. Surv. Tutor.* **2018**, *20*, 3034–3059. [CrossRef]
3. Qiao, Y.; Zheng, Q.; Lin, Y.; Fang, Y.; Xu, Y.; Zhao, T. Haptic Communication: Toward 5G Tactile Internet. In Proceedings of the 2020 Cross Strait Radio Science & Wireless Technology Conference (CSRSWTC), Fuzhou, China, 13–16 December 2020; pp. 1–3.
4. Steinbach, E.; Strese, M.; Eid, M.; Liu, X.; Bhardwaj, A.; Liu, Q.; Al-Ja'afreh, M.; Mahmoodi, T.; Hassen, R.; El Saddik, A.; et al. Haptic Codecs for the Tactile Internet. *Proc. IEEE* **2019**, *107*, 447–470. [CrossRef]
5. Xu, Y.; Huang, Y. Chen, W.; Xue, H.; Zhao, T. Error Resilience of Haptic Data in Interactive Systems. In Proceedings of the 2019 11th International Conference on Wireless Communications and Signal Processing (WCSP), Xi'an, China, 23–25 October 2019; pp. 1–6.
6. Bortone, I.; Leonardis, D.; Mastronicola, N.; Crecchi, A.; Bonfiglio, L.; Procopio, C.; Solazzi, M.; Frisoli, A. Wearable Haptics and Immersive Virtual Reality Rehabilitation Training in Children With Neuromotor Impairments. *IEEE Trans. Neural Syst. Rehabil. Eng.* **2018**, *26*, 1469–1478. [CrossRef] [PubMed]
7. Zhou H.; Wei, L.; Cao, R.; Hanoun, S.; Bhatti, A.; Tai, Y.; Nahavandi, S. The Study of Using Eye Movements to Control the Laparoscope Under a Haptically-Enabled Laparoscopic Surgery Simulation Environment. In Proceedings of the 2018 IEEE International Conference on Systems, Man, and Cybernetics (SMC), Miyazaki, Japan, 7–10 October 2018; pp. 3022–3026.
8. Chen, Y.; Zhu, J.; Xu, M.; Zhang, H.; Tang, X.; Dong, E. Application of Haptic Virtual Fixtures on Hot-Line Work Robot-Assisted Manipulation. In *Intelligent Robotics and Applications*; Yu, H., Liu, J., Liu, L., Ju, Z., Liu, Y., Zhou, D., Eds.; Publishing House: Hefei, China, 2019; pp. 221–232.
9. Durai, V.S.I.; Arjunan, R.; Manivannan, M. The Effect of Audio and Visual Modality Based CPR Skill Training with Haptics Feedback in VR. In Proceedings of the 2019 IEEE Conference on Virtual Reality and 3D User Interfaces (VR), Osaka, Japan, 23–27 March 2019; pp. 910–911.
10. Decré, G.B.; Cloonan, C. A touch of gloss: Haptic perception of packaging and consumers' reactions. *J. Prod. Brand Manag.* **2019**, *11743*, 117–132. [CrossRef]

11. Wong, H.; Kuan, W.; Chan, A.; Omamalin, S.; Yap, K.; Ding, A.; Soh, M.; Rahim, A. Deformation and Friction: 3D Haptic Asset Enhancement in e-Commerce for the Visually Impaired. *Haptic Interact. AsiaHaptics* **2018**, *535*, 256–261.

12. Lisini Baldi, T.; Scheggi, S.; Aggravi, M.; Prattichizzo, D. Haptic Guidance in Dynamic Environments Using Optimal Reciprocal Collision Avoidance. *IEEE Robot. Autom. Lett.* **2018**, *3*, 265–272. [CrossRef]

13. Da Fonseca, V.P.; Monteiro Rocha Lima, B.; Alves de Oliveira, T.E.; Zhu, Q.; Groza, V.Z.; Petriu, E.M. In-Hand Telemanipulation Using a Robotic Hand and Biology-Inspired Haptic Sensing. In Proceedings of the 2019 IEEE International Symposium on Medical Measurements and Applications (MeMeA), Istanbul, Turkey, 26–28 June 2019; pp. 1–6.

14. Li, S.; Rameshwar, R.; Votta, A.M.; Onal, C.D. Intuitive Control of a Robotic Arm and Hand System With Pneumatic Haptic Feedback. *IEEE Robot. Autom. Lett.* **2019**, *4*, 4424–4430. [CrossRef]

15. Zeng, Q.; Ishibashi, Y.; Fukushima, N.; Sugawara, S.; Psannis, K.E. Influences of inter-stream synchronization errors among haptic media, sound, and video on quality of experience in networked ensemble. In Proceedings of the 2013 IEEE 2nd Global Conference on Consumer Electronics (GCCE), Tokyo, Japan, 1–4 October 2013; pp. 466–470.

16. Aung, S.T.; Ishibashi, Y.; Mya, K.T.; Watanabe, H.; Huang, P. Influences of Network Delay on Cooperative Work in Networked Virtual Environment with Haptics. In Proceedings of the 2020 IEEE Region 10 Conference (TENCON), Osaka, Japan, 16–19 November 2020; pp. 1266–1271.

17. El-Helaly, M.; Amer, A. Synchronization of Processed Audio-Video Signals using Time-Stamps. In Proceedings of the 2007 IEEE International Conference on Image Processing, San Antonio, TX, USA, 16 September–19 October 2007; pp. VI-193–VI-196.

18. Staelens, N.; Meulenaere, J.D.; Bleumers, L.; Wallendael, G.V.; Cock, J.D.; Geeraert, K.; Vercammen, N.; Broeck, W.; Vermeulen, B.; Walle, R. Assessing the importance of audio/video synchronization for simultaneous translation of video sequences. *Multimed. Syst.* **2012**, *18*, 445–457. [CrossRef]

19. Kikuchi, T.; Ozasa, Y. Watch, Listen Once, and Sync: Audio-Visual Synchronization With Multi-Modal Regression Cnn. In Proceedings of the 2018 IEEE International Conference on Acoustics, Speech and Signal Processing (ICASSP), Calgary, AB, Canada, 15–20 April 2018; pp. 3036–3040.

20. Torfi, A.; Iranmanesh, S.M.; Nasrabadi, N.; Dawson, J. 3D Convolutional Neural Networks for Cross Audio-Visual Matching Recognition. *IEEE Access* **2017**, *5*, 22081–22091. [CrossRef]

21. Yang, M.; Bourbakis, N.; Chen, Z.; Trifas, M. An Efficient Audio-Video Synchronization Methodology. In Proceedings of the 2007 IEEE International Conference on Multimedia and Expo, Beijing, China, 2–5 July 2007; pp. 767–770.

22. Yang, L.; Song, Q.; Wang, Z.; Hu, M.; Liu, C. Hier R-CNN: Instance-Level Human Parts Detection and A New Benchmark. *IEEE Trans. Image Process.* **2021**, *30*, 39–54. [CrossRef] [PubMed]

23. He, K.; Zhang, X.; Ren, S.; Sun, J. Spatial Pyramid Pooling in Deep Convolutional Networks for Visual Recognition. *IEEE Trans. Pattern Anal. Mach. Intell.* **2015**, *37*, 1904–1916. [CrossRef]

24. Ullah, A.; Xie, H.; Farooq, M.O.; Sun, Z. Pedestrian Detection in Infrared Images Using Fast RCNN. In Proceedings of the 2018 Eighth International Conference on Image Processing Theory, Tools and Applications (IPTA), Xi'an, China, 7–10 November 2018; pp. 1–6.

25. Gomzales, R.; Machacuay, J.; Rotta, P.; Chinguel, C. Faster R-CNN with a cross-validation approach to object detection in radar images. In Proceedings of the 2021 IEEE International Conference on Aerospace and Signal Processing(INCAS), Lima, Peru, 28–30 November 2021; pp. 1–4.

26. Ahmad, T.; Chen, X.; Saqlain, A.; Ma, Y. EDF-SSD: An Improved Feature Fused SSD for Object Detection. In Proceedings of the 2021 IEEE 6th International Conference on Cloud Computing and Big Data Analytics (ICCCBDA), Chengdu, China, 24–26 April 2021; pp. 469–473.

27. Wang, Z.; Xie, K.; Zhang, X.; Chen, H.; Wen, C.; He, J. Small-Object Detection Based on YOLO and Dense Block via Image Super-Resolution. *IEEE Access* **2021**, *9*, 56416–56429. [CrossRef]

28. Chen, H.; He, Z.; Shi, B.; Zhong, T. Research on Recognition Method of Electrical Components Based on YOLO V3. *IEEE Access* **2019**, *7*, 157818–157829. [CrossRef]

29. International Telecommunication Union. *Methodology for the Subjective Assessment of the Quality of Television Pictures*; International Telecommunication Union: Geneva, Switzerland, 2002.

30. Huang, P.; Sithu, M.; Ishibashi, Y. Media Synchronization in Networked Multisensory Applications with Haptics. In *MediaSync*; Montagud, M., Cesar, P., Boronat, F., Jansen, J., Eds.; Publishing House: Grao de Gandia, Spain; Amsterdam, The Netherlands, 2018; pp. 295–317.

31. Thirumalai, C.; Chandhini, S.A.; Vaishnavi, M. Analysing the concrete compressive strength using Pearson and Spearman. In Proceedings of the 2017 International Conference of Electronics, Communication and Aerospace Technology (ICECA), Coimbatore, India, 20–22 April 2017; pp. 215–218.

32. Yang, H.; Bao, B.; Guo, H.; Jiang, Y.; Zhang, J. Spearman Correlation Coefficient Abnormal Behavior Monitoring Technology Based on RNN in 5G Network for Smart City. In Proceedings of the 2020 International Wireless Communications and Mobile Computing (IWCMC), Limassol, Cyprus, 15–19 June 2020; pp. 1440–1442.

MDPI AG
Grosspeteranlage 5
4052 Basel
Switzerland
Tel.: +41 61 683 77 34

Sensors Editorial Office
E-mail: sensors@mdpi.com
www.mdpi.com/journal/sensors

Disclaimer/Publisher's Note: The statements, opinions and data contained in all publications are solely those of the individual author(s) and contributor(s) and not of MDPI and/or the editor(s). MDPI and/or the editor(s) disclaim responsibility for any injury to people or property resulting from any ideas, methods, instructions or products referred to in the content.

www.ingramcontent.com/pod-product-compliance
Lightning Source LLC
LaVergne TN
LVHW070115170726
843515LV00007B/1692